MUTATING CONCEPTS, EVOLVING DISCIPLINES: GENETICS, MEDICINE, AND SOCIETY

Philosophy and Medicine

VOLUME 75

The titles published in this series are listed at the end of this volume

MUTATING CONCEPTS, EVOLVING DISCIPLINES: GENETICS, MEDICINE, AND SOCIETY

Edited by

LISA S. PARKER

Associate Professor of Human Genetics and
Director of Graduate Education of the Center for Bioethics and Health Law,
University of Pittsburgh,
Pittsburgh, Pennsylvania, U.S.A.

and

RACHEL A. ANKENY

Lecturer and Director,
Unit for History and Philosophy of Science,
University of Sydney, Australia

KLUWER ACADEMIC PUBLISHERS
DORDRECHT / BOSTON / LONDON

A C.I.P. Catalogue record for this book is available from the Library of Congress.

ISBN 1-4020-1040-0

Published by Kluwer Academic Publishers,
P.O. Box 17, 3300 AA Dordrecht, The Netherlands.

Sold and distributed in North, Central and South America
by Kluwer Academic Publishers,
101 Philip Drive, Norwell, MA 02061, U.S.A.

In all other countries, sold and distributed
by Kluwer Academic Publishers, Distribution Center,
P.O. Box 322, 3300 AH Dordrecht, The Netherlands.

Printed on acid-free paper

TABLE OF CONTENTS

ACKNOWLEDGEMENTS

R.A.A. would like to thank Jason Tong, Research Assistant of the Unit for History and Philosophy of Science at University of Sydney, for his assistance, particularly on the index; Nicolas Rasmussen for informal advice and constant support; and contributors to this volume for their patience and perseverance. L.S.P. would also like to thank our contributors, as well as Ryan Sauder, University of Pittsburgh, for his insightful assistance in editing the volume and index; Karen Ferris, University of Pittsburgh, for her meticulous preparation of individual manuscripts; Lisa Keranen, University of Pittsburgh, for her research assistance; Lisa Rasmussen, for her editorial and moral support, as well as her work formatting the final volume; and David Sogg for his encouragement and his proofreading assistance, in two languages.

CHAPTER 1

LISA S. PARKER AND RACHEL A. ANKENY

INTRODUCTION

The year 2000 marked what is now being called "the century of the gene" (Keller, 2000), and 2003 will mark the fiftieth anniversary of the "discovery" of the double helix. One hundred years after the so-called rediscovery of Gregor Mendel's laws of inheritance, the science of genetics has developed to the point where a working draft of the entire sequence of the human genome is now available. These advances have brought with them numerous challenges and opportunities related to our understanding of genetics and its relation to medicine, ethics, and broader social practices and institutions. Scholarly and popular literature addressing these issues has burgeoned, with numerous important contributions being made to the history of the gene (e.g., Keller, 2000; Morange, 2001) and to our philosophical understanding of the gene concept (e.g., Beurton, Falk, and Rheinberger, 2000), as well as to continued exploration of ethical implications of this new genetic age (e.g., to name just two, Buchanan *et al.*, 2000; Andrews, 2001). These dialogues have occurred somewhat in isolation from each other, however, without much explicit effort to juxtapose historical, conceptual, and ethical analyses of the ever-mutating concept of the gene and related topics, or to examine the influence these analyses exert on the disciplines, fields, and sectors of science and society likely to be affected by these changing concepts.

Our intent in crafting this collection was to reflect the spirit of the series in philosophy and medicine to which it belongs; in particular, we have interpreted in a broad way that which counts as a philosophical contribution to the understanding of medicine with the hope of continuing and reinvigorating productive communication among the many constitutive subdisciplines of philosophy and medicine. With examination of the evolution of concepts related to medical genetics as its centerpiece, this volume also reflects our desire, as teacher-scholars working in interdisciplinary settings, to prompt more dialogue between scholars working in the

L.S. Parker and R.A. Ankeny (eds.), Mutating Concepts, Evolving Disciplines: Genetics, Medicine and Society, 1-7.
© *2002 Kluwer Academic Publishers. Printed in the Netherlands.*

history and philosophy of biomedical sciences and in bioethics and to provide additional cross-disciplinary teaching resource materials.

The collection's first section, which is primarily historical, focuses on key concepts in human genetics and on their origins; hence it engages deep epistemological issues. Garland Allen's contribution provides the backdrop for many of the subsequent chapters by its exploration of the development of the classical gene—or what he argues should be more appropriately considered to be the 'derived Mendelian' gene—from the rediscovery of Mendel's work in 1900 through the mid-1930s. He denies that this concept of the gene has been abandoned, and argues that today's gene concept is still dominated by these classical concerns, largely as promoted in textbooks and popular literature. He traces the roots of the classical gene concept to the historical development of biology as a discipline seeking to emulate physics and chemistry and to the influence of mechanistic materialism on the development of genetics as a separate discipline. Douglas Allchin provides a more explicitly negative assessment of the impact of supposedly Mendelian concepts on current-day genetics, arguing forcefully in favor of eliminating the concept of dominance. The concept, he claims, is inessential to genetics and misleading in terms of heredity, natural selection, and various molecular and cellular processes. Since clearer language is available to us, we should avail ourselves of it. His examination of key disease concepts in molecular genetics reveals that the concept of dominance has come undone, even at the basic level of classical genetics where it might have been thought to still have force. Allen and Allchin's chapters complement each other strongly, and should be of interest not only to historians of biology, but to anyone who is tempted to continue to use the classical gene concept in teaching or scholarship.

Manfred D. Laubichler and Sahotra Sarkar explore the intellectual history and current social context of the genetic concepts of penetrance and expressivity. In an account with many unexpected historical twists, they argue that the introduction of these terms was motivated by the interaction of research programs in human neuroanatomy and in the evolutionary genetics of fruitflies, together with a project to establish the source of the genius embodied in V.I. Lenin's preserved brain. Their examination reveals that the vision of the discipline of neurobiology endorsed by particular scientists at the time made it inevitable that these two fields would intersect, despite their apparent incongruity. In addition, pursuing the theme introduced by Allen and Allchin—the problematic nature of long-standing concepts in genetics—Laubichler and Sarkar discuss 'penetrance' in contemporary human behavioral genetics, in light of the influence of ideological factors as well as conceptual considerations. As they put it, our post-Human Genome Project era is powered by an atmosphere of 'geneticism,' or what elsewhere has been termed 'geneticization' (Hubbard, 1990; Lippman, 1991, esp. pp. 18-19; Wolf, 1995): there is now a gene for everything. 'Penetrance' serves to mask the difficulties in maintaining this claim, much as Allchin argues the concept of dominance serves to obscure the biological details inherent in current-day medical genetics.

Using the history of genetics and the history of bioethics, Diane Paul's contribution examines the transformation in concepts associated with reproduction from the early 1900s. Focusing on 'reproductive responsibility' and its seeming disappearance from public discourse by the mid-1970s in favor of 'reproductive autonomy,' she argues that although public discourse indeed was affected, the explicit change in concepts masks what was actually continuity in the underlying norms. The changes that did occur through various sociopolitical movements in the 1960s and 1970s did not play out in the same way across different classes and different professions. She concludes that the call for responsibility in reproduction was a central focus throughout the twentieth century, even in the recent era apparently dominated by concern for autonomy. We should be attentive to the intricacies in such historical stories when assessing apparent changes in scientific and bioethical worldviews, particularly with regard to fundamental concepts such as freedom, autonomy, and responsibility.

The second section is explicitly devoted to perspectives from the philosophy of science, but its contributors use interdisciplinary techniques to ground their discussions. Taking up the theme of geneticization introduced in the first chapter, and focusing continued attention on the concepts of penetrance and expressivity, Fred Gifford draws attention to the problems inherent in our current understanding of causation in genetics, and particularly in the genotype-phenotype relation. He argues that formulations of questions and arguments in bioethics and health policy can be criticized because of flaws in the implicit understandings of various conceptual matters being invoked, a premise that is shared by the underlying vision motivating this collection. Pointing to difficulties involved in isolating both particular traits and the genetic contributions to them, as well as to issues raised by the need to take environmental contributions to so-called genetic diseases more seriously, Gifford presents a series of challenges for medical genetics, both current and future, in light of the increasing geneticization of medicine. Rachel A. Ankeny develops the theme of the need to contextualize the implications of advances in medical genetics within the philosophy of medicine to argue that our vision of particular diseases, and of disease in general, will be significantly altered with the availability of genetic sequencing information. As a consequence, the traditional debates in the philosophy of biomedical sciences about reduction of classical genetic concepts to molecular ones must be recast, with more attention being given to holistic views of disease processes involving a variety of biological and conceptual levels. Hence her chapter is a call to resist geneticization rather than to assume its inevitability, given the recent advances in genetics.

Joseph L. Graves, Jr., addresses reductionism and adaptationism as ideologies against the backdrop of traditional medical research into the relationships between the supposed biological category of 'race' and disease prevalence. He argues that historically a variety of conceptual errors have occurred in such research, and that effective and ethical genetic analysis of disease and human genetic variation can only be pursued if sound evolutionary and population genetic thinking is integrated into medicine. At least one concrete outcome of this argument is that the focus of these sorts of investigations should be individual and population variation, rather than antiquated descriptions of socially-defined "racial" categories. These conceptual and biological conclusions are particularly trenchant for bioethics, as they will affect populations that have been historically underserved, if not ill-used, by the medical and scientific communities. In a similar vein, Helen E. Longino examines the prevailing conceptual and methodological frameworks in use within behavioral genetics in order to reveal their presuppositions and limits. Her goals are markedly different (and much less prescriptive) than those of many philosophers of science who have examined this domain; thus, especially given its metaphysical emphasis, her chapter makes a novel contribution to this literature. Her analysis identifies and articulates the commonalities against which different views within the debates on aggression and sexual behavior are framed, and in particular, highlights the way in which behavior often comes to be conceptualized as pathology—that is, as affliction. Once again we see how various conceptual frames and methodological approaches have constrained our abilities to conceptualize and understand the objects of biomedical study.

The third section explores how the adoption of particular conceptual frames, research goals, and approaches reflects various social interests and, in turn, influences ethical analyses and social policies and practices. The selections in this section also pursue an underlying theme of this volume—the need to recognize the normative and fundamentally social context within which genetic discoveries are made, interpreted, and accorded their significance. Licia Carlson's contribution engages debate about the ethics of prenatal testing and selective abortion and explores the concerns grounding the 'expressivist objection' to this practice— namely, the view that terminating a pregnancy because a fetus has a particular genotype expresses disregard for, or devaluation of, existing people who have phenotypes associated with that genotype. Against the social backdrop of the apparent tension between developments in genetics that encourage the geneticization of both disease and disability—that is, their construction as biological, medical "fact"—and arguments advanced by disability activists that encourage shifting from a purely medical to a more social understanding of disability, Carlson explores how the prenatal visibility of genotypes, combined with the indeterminability of the severity of the condition tested for, results in the creation of 'prenatal prototypes' and how the use of these prototypes can give rise to the concerns expressed by the so-called expressivist objection. Carlson suggests that even if philosophical refutations of the expressivist objection are strictly speaking valid, the social and ethical concerns captured by the expressivist objection nevertheless should command our attention.

Reflecting her previous analyses of 'normal functioning' and disability as socially relative constructions (e.g., Silvers, 1998), Anita Silvers comments on the legacy surrounding negative and positive eugenics and on present-day genomics as a promising means to improve human life. Like Paul, who casts doubt on the popular understanding of prenatal testing as an historical outgrowth of reproductive autonomy, Silvers calls into question both the popular view of genetic testing and screening as largely acceptable and the notion that any step toward genetic enhancement necessarily involves flirting with the creation of gods and monsters. Contrary to much current public and scholarly opinion that genetic technologies may be used to avoid disease and disability but not to engage in so-called genetic enhancement (Walters and Palmer, 1997, esp. pp. 47-9, 134), Silvers argues that use of genetic technologies to *enhance* individuals' capabilities (at the molecular level) may be supported by both medicine's meliorist goals and the social goals of promoting fairness and ensuring individuals' abilities to engage in the activities of a participatory democracy. It is practices that have already gained widespread acceptance—e.g., negative eugenic programs that reduce the incidence of disease by preventing the birth of individuals with particular genotypes—that instead may be contrary to medicine's goals of enhancing human well-being. Building on themes in the arguments of Carlson and Silvers, David Wasserman examines the ethics of interventions that may prenatally correct or enhance at the molecular level. His focus differs from that of other contributors, however, for he is concerned first with the effect of such interventions on personal identity and next with the moral evaluation of harms that may result from such interventions (or perhaps, by extension, the choice not to intervene). Does prenatal therapy that alters genetic structure change the identity of the zygote, fetus, or person that is subjected to the intervention? What claims might be made on behalf of the fetus either *for* therapeutic intervention or *against* replacement (i.e., intervention that changes personal identity)? Wasserman's chapter thus contributes substantially to philosophical (and legal) debate about 'wrongful life,' while like Carlson, he emphasizes the practical import of such interventions, arguments, and potential harms in people's actual lives and decision making.

The final two contributions explicitly examine changes wrought by genetics for the important social practices of medicine and law. Paul Han discusses the challenges and opportunities that genetic findings offer for the prevention of disease and promotion of health and for the evolution of the practice of medicine. His chapter argues that not only are these preventive possibilities far from unequivocally positive, but medical *genetics* is far from being appropriately the sole focus of a critique exposing the negative sequellae that can accompany a preoccupation with prevention. Han argues against such genetic exceptionalism by questioning the distinct significance attributed to 'genetic risk' and by revealing assumptions made in the course of debates about the downside of genetic medicine—specifically, assumptions about genetic determinism, the proper goals and scope of medical intervention, and benefits or evils of medicalization. As Gifford argues with respect to the notion of genetic causation, Han reveals with respect to these concepts that much of ethical import in health policy and medical practice depends on how such concepts are understood. By attending to the shift in focus in preventive medicine from the health of individuals to the health of populations, Han sheds light on the ethical tensions inherent in public health genetics, an arena of public health research and practice that is garnering political support and financial resources (Khoury, Burke, and Thomson, 2000).

To close the volume, John H. Robinson and Roberta M. Berry draw our attention to another fundamental social practice influenced by developments in genetics: the law. They focus on challenges to traditional concepts in criminal law that are presented by advances in behavioral genetics and by the consequent evolution in our understanding of concepts concerning behavior, mental states, and morality. These challenges, they recognize, come from "below"—the realm of behavioral genetics as allied with neurophysiology—and from "above"—the realm of behavioral genetics as allied with sociobiology. They argue that the appropriate response to these challenges presented to traditional notions of moral and legal responsibility—and indeed to our conception of the moral person as the sort of entity capable of such responsibility—is to defend our social judgments of wrongdoing. The conception of personhood that informs our criminal code, our social practice of punishment, and our concepts of responsibility is a major cultural achievement, one deeply embedded in a web of moral and social norms, so that this view of moral personhood may not be easily defeated, or rendered rationally untenable, by increased understanding of the "genetic person."

This final selection highlights a key theme of the volume: discoveries in genetics do not determine the course of dialogues about such fundamental concepts as health, normality, disability, responsibility, causation, nature and nurture, the environmental, or the genetic. Genetic discoveries inform and constrain these conceptual debates and the disciplines and discourses encompassing them; however, we determine—as acts of individual and collective choice—the meanings and the import that are accorded to genetic findings. We incorporate these findings and the mutating concepts associated with genetics into various disciplines and social practices, including law, medicine, and obviously, science itself. In turn, these disciplines and practices themselves not only evolve to encompass these new findings and different or more nuanced meanings, but also to pose and pursue constantly evolving ethical, social, and intellectual goals.

University of Pittsburgh
Pittsburgh, Pennsylvania, U.S.A.
and
University of Sydney
Sydney, New South Wales, Australia

REFERENCES

Andrews, L.B., *Future Perfect: Confronting Decisions About Genetics*. New York: Columbia University Press, 2001.

Beurton, P.J., Falk, R., Rheinberger, H.-J., eds. *The Concept of the Gene in Development and Evolution*. Cambridge: Cambridge University Press, 2000.

Buchanan, A., Brock, D.W., Daniels, N., Wikler, D., *From Chance to Choice: Genetics and Justice*. Cambridge: Cambridge University Press, 2000.

Hubbard, R., *The Politics of Women's Biology*. New Brunswick, NJ: Rutgers University Press, 1990.

Keller, E.F., *The Century of the Gene*. Cambridge: Harvard University Press, 2000.

Khoury, M.J., Burke, W., Thomson, E.J. "Genetics and Public Health: A Framework for the Integration of Human Genetics into Public Health Practice." In Genetics and Public Health in the 21st Century: Using Genetic Information to Improve Health and Prevent Disease, M.J. Khoury, W. Burke, E.J. Thomson, eds. New York: Oxford University Press, 2000.

Lippman A. Prenatal genetic testing and screening: constructing needs and reinforcing inequities. American Journal of Law and Medicine 1991; 17:15-50

Morange, M., *The Misunderstood Gene*. M. Cobb trans. Cambridge: Harvard University Press, 2001.

Silvers, A. "A Fatal Attraction to Normalizing: Treating Disabilities as Deviations from 'Species-Typical' Functioning." In Enhancing Human Capacities: Conceptual Complexities and Ethical Implications, E. Parens, ed. Washington, D.C.: Georgetown University Press, 1998.

Walters, L., Palmer, J.G., *The Ethics of Human Gene Therapy*. New York: Oxford University Press, 1997.

Wolf S.M. Beyond "genetic discrimination": toward the broader harm of geneticism. Journal of Law, Medicine, & Ethics 1995; 23:345-53

PART ONE:
HISTORICAL REFLECTIONS
ON CORE CONCEPTS

CHAPTER 2

GARLAND E. ALLEN

THE CLASSICAL GENE:
ITS NATURE AND ITS LEGACY

1. INTRODUCTION

Much has been written in recent years by biologists, historians, and philosophers of biology about the gene concept and its evolving meaning over time (Burian, 1985; Darden, 1991; Falk, 1986, 1995; Kitcher, 1982; Portin, 1993; Gifford, 2000; Rheinberger, 2000). These studies have provided a useful review of the literature tracing the term through almost a century of its changing meaning (Portin, 1993) and various philosophical usages. The terms 'classical gene,' 'developmental gene,' 'biochemical gene,' and 'molecular gene' all have appeared in the recent historical and philosophical literature. But it has not always been clear how these terms differ from one another, whether they map onto specific chronological periods, or the extent to which they overlap in meaning (e.g., how similar or different are the 'classical' and the 'developmental' gene?). Few contemporary geneticists or historians/philosophers of science would deny that the concept of the 'gene,' referring to the material basis of heredity, has undergone significant changes in the course of its evolution during the past century. The key question is: what is the nature of that change? Is it quantitative or qualitative? Has there been divergence into competing gene concepts either in the past or in recent times? How can the history of gene concepts help to illuminate some of the current questions and problems scientific and ethical confronting modern researchers and policy makers about issues such as genetic manipulation, privacy rights, patenting of genes, and the medicalization of social behaviors that are claimed to have a "genetic" basis?

In this chapter, I explore one facet of this complex history: the development of the 'classical,' or what I will call the 'derived Mendelian gene,' from the rediscovery

L.S. Parker and R.A. Ankeny (eds.), Mutating Concepts, Evolving Disciplines: Genetics, Medicine and Society, 11-41.
© 2002 Kluwer Academic Publishers. Printed in the Netherlands.

of Gregor Mendel's work through the mid-1930s. My major thesis is that this view of the gene, especially as it became codified in biology and genetics textbooks and in popular literature (both genetic and the associated eugenical writings of the time), contained many connotations and implications that are still associated with the term 'gene' as it is understood today. I argue further that these connotations developed out of two related movements in the early decades of the century:

(1) the attempt to fashion a new image of biology as a hard, experimental science, akin to, and based on, physics and chemistry, put forward by a younger generation of investigators (born after, roughly, 1860); and

(2) the adherence by this same younger generation to a strong mechanistic materialism that became both the foundation and rationale for the new science.

In this paper, I focus particularly on the influence of mechanistic materialism on biology and genetics during the early decades of the twentieth century.

Mechanistic materialism, or the mechanical philosophy, was hardly new to biology at the turn of the twentieth century. In one way or another it had been a cornerstone of much of Western science since the scientific revolution. It had been imported into areas of the life sciences in the seventeenth century by René Descartes, William Harvey, Robert Boyle, and Giovanni Borelli (among others); in the eighteenth century by Stephen Hales, Joseph Priestley, Luigi Galvani, and others; and in the mid-nineteenth century by Hermann von Helmholtz, Emil Du Bois-Reymond, and the Berlin medical materialists. It also formed a focal point for the "new" biology of the early twentieth century. Even among researchers familiar with the more subtle developments in the field of genetics in the last half-century, especially popularizers of genetics and textbook writers, the mechanistic notions surrounding the concept of the gene persist and influence the very way we conceptualize our understanding of the hereditary process. Because of my own interest in the way genetics has evolved in the area of medical and social policy, I will focus the last part of this chapter on the implications of the mechanistically-conceived gene for present-day research into the supposed inheritance of many personality and behavioral traits in human beings.

For the purposes of convenience and clarity I will follow Petter Portin's chronological division of the history of genetics since 1900 into three periods (Portin, 1993, pp. 173-4):

(1) the period of the 'classical gene,' based on Mendel's original work up through the work of the Morgan and Emerson "schools," as well as that of Barbara McClintock, Theodosius Dobzhansky, and H.J. Muller (1900-1930);

(2) the period of the biochemical or developmental gene, including the work of Boris Ephrussi, George Beadle, Edward Tatum, Conrad Waddington, and to some extent, that of Richard Goldschmidt (1930-1955); and

(3) the period of molecular genetics, beginning with discovery of the structure of DNA and continuing through to the Human Genome Project (HGP) and beyond,

and including the work of James Watson and Francis Crick, Matthew Meselsohn and Frank Stahl, Severo Ochoa, Joshua Lederberg, Sydney Brenner, Marshall Nirenberg, Jonathan Beckwith, Scott Gilbert, and many more concerned with the molecular structure of the gene and its functioning in transcription and translation (1955-present).

I am tempted to add a fourth period, one which we may be just entering at the present time, which could be termed the period of 'molecular/developmental genetics,' characterized by what many see as the concrete prospects of returning to the century-old problem of embryonic development in terms of genetic regulatory elements such as signal transduction pathways between the developing embryo and its external environment (e.g., in the work of Shaun Carroll, Christiane Nüsslein-Vollhard, and Eric Wieschaus, among others). It is, however, primarily the first period that I examine in this chapter, largely because I think the paradigm of the gene laid down during that time is the one that has persisted in the conceptualization of genetics by all but those working directly in the laboratory—and even many of those—studying inheritance in complex (eukaryotic) organisms.

2. THE CONCEPTUAL AND PROFESSIONAL TRANSFORMATION OF BIOLOGY, 1880-1930

The field of biology (not including the medical/physiological sciences as they are currently defined) underwent a major transition in the late nineteenth and early twentieth centuries. In 1880 biology was a descriptive enterprise, dominated by areas such as comparative anatomy, the classical embryology of Francis M. Balfour and Carl Gegenbaur, taxonomy, and, with the growing acceptance of an evolutionary framework, areas of natural history such as biogeography and phylogeny. By the 1930s however, a new generation of biologists considered their discipline to be radically different: more experimentally-based, more rigorous, and concerned more with fundamental—even universal—principles and less with merely describing the structural, functional, and taxonomic relationships of organisms. As I have discussed this transformation at length in a number of previous publications (e.g., Allen, 1983, 1991), I will only summarize here the broad features of what I think it involved. In the sense of professionalizing the field for the current century, however, this transitional period may be characterized as "the coming of age of biology."

Long dominated by concerns of descriptive natural history and taxonomy, biology had experienced its first major paradigm shift with the introduction of Darwin's theory of evolution by natural selection during the latter half of the nineteenth century. As it replaced the older theory of the origin of species by special creation, Darwin's paradigm showed that the most profound concerns of natural history, the origin of life's immense diversity, could be accounted for by natural laws derived, in the best sense of the term, by the process of consilience—the convergence of data from a variety of different disciplines onto a single explanatory scheme. Darwin's theory was synthetic, drawing together observations from taxonomy, field natural history, comparative anatomy and physiology, embryology, geology, and paleontology, among other fields. While the Darwinian paradigm had served as a rallying point for many biologists in the late nineteenth century, it had also created a new set of problems. The difficulties faced by the Darwinian theory, even those generated by its biological claims were many, and have been discussed in detail by Peter Bowler (1983), among others (see also Allen, 1978a). Underlying all of the specific problems, however, was a deeper methodological issue. There was a widespread belief that evolutionary theory, and the morphological research program associated with it and especially identified with the avid pursuit of phylogenies, was at heart a soft, qualitative, non-rigorous science. While granting that Darwin's theory was synthetic, critics claimed it was also non-experimental and therefore non-testable, and ultimately would remain consigned to the realm of speculation.

The critics of Darwinian theory, of morphology, and of what was seen as speculative science, were the younger generation of biologists, especially in the United States, who saw "phylogenizing" as old-fashioned. Numerous phylogenies could all account for the same evidence, and with no way to distinguish among them, younger critics sought to move away from what they saw as the overriding concern with evolutionary and morphologically-based questions. The history of life was a one-time sequence of events that could never be reconstructed in detail, and in many cases not even in broad outline. Moreover, many important biological problems in their own right—for example, the causes of embryonic differentiation or the nature of heredity—were being ignored while the senior investigators in the field, from Louis Agassiz to W.K. Brooks, Ernst Haeckel, and August Weismann, were engaged in endless debates about the origin of one or another taxonomic group (asking questions such as "are the vertebrates most closely related to the annelids or arthropods?") or in doubts about the ability of natural selection to actually produce new species. In the eyes of the younger generation, it was the domination of ultimate over proximate causation that formed the focus of the criticism leveled at the "older biology."

Younger critics also had some unkind words to say about the purely descriptive fields of taxonomy, comparative anatomy, and embryology, and what we today would refer to as the nascent science of ecology. They did not view these areas as unimportant, but they saw them as dominating biology at the expense of newer, more exciting, and promising questions. For example, younger biologists were intrigued by the early experimental studies of embryological differentiation initiated by Hans Driesch and Wilhelm Roux (Allen, 1978a, pp. 29-33). By experimentally separating the first two blastomeres (daughter cells formed by division of the fertilized egg) of developing embryos (Roux with frogs and Driesch with sea urchins), these experimentalists were testing the mosaic theory of development, that is, whether cells during embryonic development retained their ability to form a complete organism (i.e., were totipotent, in today's terminology), or whether they became increasingly restricted in their developmental capacities due to loss of all but their own particular tissue determiners. While there were complications that made the results of these experiments subject to various interpretations (e.g., the use of different species and different methods of manipulating the blastomeres), the focus on specific, proximate causation (i.e., how the organism functions *currently,* not millions of years back in evolutionary history) examined using experimental methods appeared to the younger generation of biologists as a fresh and exciting avenue for investigation.

The methodological paradigm against which Darwinian and morphological theory was judged was that of the physical sciences. The hallmark of this methodology was the use of hypothetico-deductive reasoning ("if A and B are two mutually exclusive states, and if A is true, then B *must* be false"—i.e., the conclusion follows *necessarily* from the premises), experimentation, the collection of quantitative rather than qualitative data, and the use of mathematics, including the nascent science of statistics. Hypothetico-deductive reasoning established alternative predictions from competing hypotheses, allowing those alternatives to then be tested by experimentation. The process was rigorous. Experimentation, especially through the use of controls, made it possible to narrow down the number of variables being investigated to one at a time, and thus isolate the causal agents. Quantitative data were not only more precise than qualitative data but allowed more meaningful comparison between experiments and the repetition of experiments by different investigators. Quantitative data also made it possible to discover and express more precise and general relationships in mathematical terms (e.g., Vito Volterra's mathematical formulation of competitive exclusion delineated more rigorous and precise relationships among species in a given environment than Darwin's qualitative analogy of species as "wedges").

It was a rather naive view of the nature of the physical sciences on which many biologists' views of proper scientific methodology were based. In general, it can be said that the younger generation of biologists were modeling their view of science not on the physics being practiced by cutting-edge physicists in the early 1900s but more on the classical positivist view of physics and chemistry presented by textbooks and current philosophical discussions of the nature of science. It should be pointed out that at least some of the younger biologists involved in championing the "new" biology (e.g., Jacques Loeb, Hans Driesch, and William Bateson) were aware of the debates among physicists and philosophers of science regarding monism versus dualism, and idealism versus materialism—that is, issues raised about classical physics by empirio-critics such as Ernst Mach and Pierre Duhem, who questioned any statements about the *real* nature of matter. So far as I can tell, however, between 1890 and 1920, these sometimes esoteric debates did not significantly enter into the formulation of a new version of mechanistic biology between 1890 and 1930, except within the holistic biology movement (to be discussed in more detail later) as has been so well-explored by Anne Harrington (1996). Holistic biologists such as Jakob von Uxeküll and Ludwig von Bertalanffy saw in the doctrine of vitalism—at least in some parts of it—a significant alternative to the strong mechanistic stance that had pervaded much of German physiology in the early decades of the twentieth century. To the extent that holistic arguments became intertwined in many cases with vitalistic claims, the issues became increasingly confused, so that much of the holistic movement was ignored by younger biologists, especially those working in the United States.

For most biologists at the time, the area of the life sciences that most approached the ideal of physics in terms of methodology and rigor was general physiology, and it was often held up as an example of the proper methodology to be extended to *all* areas of biology still tied to old-fashioned descriptive methods (such as evolution, heredity, embryology, cytology, and comparative anatomy). In addition to figures such as Jacques Loeb, who was the most dramatic spokesman for a mechanistic physiology, others such as T.H. Morgan, Gary Calkins, W.J.V. Osterhout, H.J. Muller, and, somewhat surprisingly, plant taxonomist and successor at Harvard to Asa Gray, W.G. Farlow, strongly encouraged biologists to take up physiological approaches as a means of making biology as a whole a rigorous science. J.C. Arthur, in his 1895 vice-presidential address to Section G of the American Association for the Advancement of Science told botanists that "[t]he present great advance in the science [of plant physiology] may, in large measure, be traced to the wonderful advances in the sciences of chemistry and physics, which have supplied facts and methods to assist the physiologist in his [*sic*] study of life processes" (Cittadino, 1980, p. 179). By the turn of the century, for all sorts of biologists the physical sciences, via physiology, were becoming the model for all scientific methodology.

3. MECHANISTIC MATERIALISM
IN EARLY TWENTIETH-CENTURY BIOLOGY

From the seventeenth century to the present, Western science has been based on the philosophical underpinnings of what has become known as the mechanical philosophy or mechanistic materialism. It is the philosophy underlying Robert Boyle's *Sceptical Chymist* (1680), and embodied in full in Isaac Newton's *Opticks* (1704). Materialism as a general philosophical stance came to replace non-materialism or "idealism" in the West during the scientific revolution. The materialist worldview can be embodied in the following four propositions:

(1) Material reality exists outside of human perception and has existed prior to our knowledge of it.

(2) Ideas about the world are derived from our interaction with material reality, and not from some *a priori* source. This is not to say, of course, that once we have developed ideas we do not impress them on reality and try to change aspects of the real world. Nonetheless, the main direction of such activity passes always from material reality to human conceptualization.

(3) All change in the universe is a result of matter in motion—that is, the action of one material entity on another. The classic example of matter-in-motion is the atomic theory as elucidated in the nineteenth century.

(4) Non-physical forces or mystical causes are inadmissible as explanations of any phenomena. For example, Driesch's 'entelechy' and *élan vital* (vital force) or Henry Bergson's 'creative evolution' would be excluded as unknowable and therefore unallowable explanations for biological processes.

Historically, there have been two major schools of materialist thought over the past 150 years: mechanistic and dialectical materialism. It will be useful to define these two views and distinguish between them. Mechanistic materialism is one particular form of materialism, which can be summarized as follows:

(1) The parts of a complex whole are distinct and separate from one another: for example, the atoms in a molecule or the gears and levers in a clock.

(2) It then follows that the proper method for studying the whole is to break it down into its component parts, each of which can be investigated independently of its more complex involvement with other parts. This method of investigation is often referred to as analysis (from the Greek *lysis*, to "break apart or separate."

(3) Behind the method of analysis lies the general assumption that the whole is equal to the sum of its parts and no more. There are no mystical or "emergent" properties coming from the association of parts. Thus, if we know everything about each part it should be possible to reconstruct the whole in its totality; nothing more is needed.

(4) Systems change over time due largely to constant forces impressed on them from the outside. For example, the planets move in definable orbits because of their gravitational interaction with other bodies; organisms die through accident or through the accumulation of waste metabolites or chance mutations; and populations evolve because organisms are constantly presented with challenges from a changing environment. It is important to point out that in this view change is seen as ever-present, but does not arise necessarily out of conditions existing within the system (organism, population) itself, but through changes presented to the system by its external environment.

(5) Finally, the mechanistic worldview is basically atomistic, viewing phenomena in terms of a mosaic of separate, interacting, but ultimately independent parts.

Although it is beyond the scope of this paper to discuss the principles of dialectical materialism in detail, a brief outline will help to emphasize the ways in which it differs from mechanistic materialism, and thereby provide, by contrast, a better understanding of the mechanistic approach in the earlier parts of the century. The point is, of course, that there were alternative ways of viewing the same biological phenomena at the same time, but that mechanistic materialism held sway among Western biologists, especially in the United States (dialectical materialism, as the official Soviet philosophy of science after 1920, held sway in the Soviet Union and later in the Eastern block countries and China).

Dialectical materialism shares all the basic premises of materialism, but differs in several significant ways from mechanistic materialism. Its basic propositions are contained in the five points listed below:

(1) The parts of a complex whole are interconnected, and thus cannot be studied only in isolation from each other. One of the major characteristics of any part is its interactions with other parts within the whole. It is, therefore, not enough to study the part by itself, but the part must be studied in its dynamic interaction with other parts that make up the whole. Thus, in addition to methods of analysis, it is also necessary to devise new methods and techniques for studying component parts in their interactions.

(2) From proposition (1) it follows that the whole is more than the mere sum of its component parts; it is composed of the parts plus their interactions. The emergent properties which result from the interaction of parts are not mystical or abstract. It is simply that the whole of a process, for example, a functioning organism, is more than the additive values of its separate organs and tissues. There is, in a word, hierarchical organization of matter that means that different levels show different properties.

(3) Processes in the world are dynamic and developmental. Change is a fundamental part of any system, built into the interaction of the parts within the whole, and not merely impressed on the system from the outside. The simplest forms of change are wear and tear, and deterioration. But most systems change in other ways as well: for example, the development of an embryo from egg to adult, the evolution of a population from one species into two or more species, or the succession of communities on a sand dune. Dialectical materialism sees change as something more than merely the response of a system to its external environment, recognizing that systems contain within themselves the basis for their own changing states.

(4) The internal process of change within any system can be understood as dialectic, an interaction of opposing forces or tendencies. I consider it largely a moot point whether such opposing forces are actually present (ontologically) within the system, or simply remain as one of our own inventions for understanding and describing certain kinds of dynamic change—though debate on this very question has spilled much ink from the pens of Marxist writers from Friedrich Engels to V.I. Lenin and Louis Althusser. As examples of the dialectical approach, we can describe key metabolic processes such as growth or senescence as an interaction between anabolism (chemical synthesis) and catabolism (chemical breakdown); we can describe evolution as an interaction between the opposing forces of heredity and variation, or specialization and generality; we can describe enzymatic regulation as an interaction between constrained and relaxed allosteric states of protein molecules; and finally, we can describe the process of ecological success as an interaction between stable and unstable associations of communities and their environments.

(5) Quantitative changes lead to qualitative changes. This means that as small, or quantitative, changes within a system accumulate, eventually large-scale, or qualitative, changes or states emerge. A classic example of this principle is the boiling of water. As water is heated, the temperature gradually rises (quantitative changes), but eventually somewhere between 99.999^0C and 100^0C boiling begins (a qualitative change). A qualitative change has taken place, since steam is a different state of matter from very hot water. A more biological example comes from evolutionary theory: the gradual accumulation of small variations such as base-pair substitutions, or mutations (that is, quantitative change) at the DNA level eventually leads to reproductive isolation and thus the formation of separate species (qualitative change). According to this principle, quantitative changes always can eventually lead to qualitative change, thus making the evolution of new states a necessity with time. Conversely, all qualitative changes are seen to result from antecedent quantitative changes, a relationship that only can be uncovered, in each instance, by study of the history of the process. History thus becomes an essential component in uncovering, and thus understanding, the dynamics of any system.

The dialectical approach to processes provides a way of understanding the dynamic nature of change in any system. It explains why change is to be expected, and why change in systems is not largely accidental, random, or due only to external causes. Dialectics provides a way to understand how the materialist view of matter in motion leads to non-random, developmental (but distinctly not teleological) change.

For the mechanist, the history of matter is irrelevant to understanding the proximate causation of any particular phenomenon. For chemists, the combinations in which an atom has existed prior to its present combination are irrelevant; they have no bearing on its fundamental properties. The atom bears no traces of its history. The characteristics of combinations of atoms into molecules can be resolved into the additive effects of the components. Thus, for example, mechanists argue that if we knew enough about the properties of hydrogen and oxygen, we could predict the properties of water without ever having seen a water molecule. That such predictions were often difficult to make in the real world was not seen as a problem of philosophical outlook but rather of a gap in our knowledge.

Concomitant with the mechanistic outlook was advocacy of the analytical or, in some terminologies, reductionist methodology. Analysis, as the opposite of synthesis, involved taking apart complex processes into their component parts, which could then be studied one at a time. It was the idealized method of the physical sciences transferred over into biology. Organisms could be "analyzed" into their component organs or tissues, tissues into their component cells, and cells into their component molecules. The analytical method had produced striking results in chemistry and physics, and was therefore seen as the appropriate method for investigation in all the sciences, including biology. If applied rigorously, the analytical method could convert biology from a largely descriptive, non-explanatory science into an enterprise that rested, as Morgan was fond of putting it, "on the same footing as chemistry and physics" (see Allen, 1978b). (Biologists at the time were sometimes accused of "physics envy.")

It should be pointed out that in the first several decades of the twentieth century, mechanistic materialism gained an even more pronounced following because of a concomitant rise in a new vitalistic philosophy associated with (or often confused in people's minds with) a movement toward holistic thinking in biology, medicine, and psychiatry (Harrington, 1996). An ongoing controversy between mechanism (mechanistic materialism) and vitalism emerged among biologists and non-biologists alike. The issue was not clarified when Driesch, the arch-mechanist of the 1890s with his sea urchin experiments, left experimental biology, took up a post as Professor of Philosophy (first Heidelberg, then Cologne, and finally Leipzig), and became a major champion of vitalism. Vitalists held that some non-physical, non-chemical force energized living systems and made them something more than simply "chemical machines." Mechanists such as Loeb argued that living systems did not involve any processes that could not be understood completely in physico-chemical terms. Organisms only seemed "vitalistic" because they encompassed a highly complex array of chemical processes that, although still largely unknown, were not unknowable. The existence of a serious and ongoing vitalistic camp served only to harden the mechanistic resolve even further, since they saw vitalism as akin to mysticism, a form of old-fashioned *Naturphilosophie* that threatened to undercut the attempt to place biology on the same footing as chemistry and physics.

A particularly important aspect of mechanistic materialism was its atomism and its insistence that matter as a whole was a mosaic of separate and separable component parts. The organism, after all, was a mosaic of cells, which in turn were a mosaic of organized biological molecules. Any quick survey of the various theories of heredity prevalent in the late nineteenth century—for example, those of Darwin (pangenes), Weismann (idants, ids, and biophors), Haeckel (gemmules), Hugo De Vries (pangenes), Gustav Jäger, and dozens of others—indicates the degree to which biologists tended to see the organism as a mosaic of discrete units. This was, after all, compatible with, and indeed an expression of, the atomistic nature of matter itself.

A final factor that contributed to the upsurge of a new and aggressive form of mechanistic thinking was the professionalization of the sciences in general, and the life sciences in particular, during the late nineteenth and early twentieth centuries. This was the period in which the first generation of biologists trained at universities in the United States rather than Europe came to professional maturity, and participated in the "struggle for authority" within biology to establish the field's place among the true, or hard, sciences (Bourdieu, 1975; see also Sapp, 1987). Biologists were increasingly being elected to major organizations such as the National Academy of Sciences in the United States and the Royal Society in England and were taking a more active leadership role in organizations such as the American and British Associations for the Advancement of Science, and the Max-Planck-Gesellschaft. New professional organizations such as the American Society of Zoologists and the American Physiological Society (which were founded in the 1880s and 1890s) attested to the growing number of biologists and the professional identity accorded to them. In the United States, especially in areas related to agriculture (nutrition, soil science, bacteriology, mycology, plant physiology, and breeding), new positions were opening up for biologists both at universities (the land grant colleges were expanding significantly from the 1890s onward) and government agencies such as the United States Department of Agriculture and the various state agricultural experiment stations (Fitzgerald, 1990; Marcus, 1985; Rossiter, 1975; Rosenberg, 1976). In the quest for greater professional legitimacy, biologists saw that their field could become, like physics, *causal-analytical*. Following the lead of Roux, whose program for *Entwicklungsmechanik* (translated variously as 'developmental mechanics' or 'experimental embryology') was grounded solidly and unabashedly in the mechanistic materialist outlook, younger and professionally-oriented biologists emphasized the mechanistic ideology as a way to distinguish their goals from those of the older generation of naturalists. Thus mechanistic materialism served not only as a guide to research, but also as an ideology around which younger supporters could be rallied in the quest to bring biology into the fold of a truly modern, rigorous science.

In summary, then, the new biology embraced a complex of ideas associated with the mechanistic worldview around the turn of the century: a strong commitment to experimental methods, hypothetico-deductive reasoning, the power of prediction, an analytical approach, the importance of quantitative data, the formulation of general relationships (in mathematical terms, where possible), and adherence to an updated form of materialist thinking that saw organisms as complex chemical machines whose most basic processes were ultimately knowable by human beings.

4. ENTER MENDELIAN GENETICS

Given the dynamics of professionalization within biology in the early part of the twentieth century, it is not difficult to understand why Mendelian genetics became such an important area of research shortly after its rediscovery in 1900. The Mendelian paradigm appeared not only to provide the sort of general law of heredity that had been so sorely lacking in the nineteenth century, but also served as a model for how biology could and should be pursued as a true scientific discipline. It was atomistic (factors, later called 'genes,' were the ultimate units of analysis), experimental, predictive, quantitative, and mathematical. In short, it satisfied virtually all of the requirements of a mechanistically-based, hard science that would at last allow biology to be placed on the same footing as physics and chemistry. No wonder younger biologists found in Mendelism what had been lacking in much of the "older biology" of August Weismann, Ernst Haeckel, Carl Nägeli, and others. This is not to say, of course, that all younger biologists immediately accepted the Mendelian paradigm. As I have shown in the case of Morgan and others, there was sometimes a period of skepticism toward the Mendelian scheme, often in part due to associating Mendel's factors (or *Anlagen*) with the old tradition of inventing hypothetical particles to explain hereditary and developmental processes (Allen, 1978a). And there were those, especially among embryologists of the time (Ross G.

Harrison, E.G. Conklin, Hans Spemann, and Albert Dalcq, for example), who never found the Mendelian scheme attractive, or at least deserving of the attention (and financial support) it received. However, by 1912, or certainly by the start of World War I, a significant number of biologists had adopted the Mendelian scheme and converted it into specific research programs aimed at establishing its general applicability to a variety of organisms and seemingly anomalous breeding results. Throughout the early development of Mendelian theory, the young science of genetics was being held up as a model for how modern biology should be carried out.

Since many authors (e.g., Carlson, 1966; Falk, 1986; Portin, 1993) have traced the evolution of the concept of the gene from the rediscovery of Mendel to the era of molecular genetics, I will not repeat that story here. However, some important details along the way illustrate clearly and explicitly how Mendelism found congruence with mechanistic materialism in ways that have persisted to the present day. It is this lasting legacy of an earlier mechanistic materialism that must be faced squarely today as we seek to explore how the new genomics is to be understood by workers in biology, the medical profession, and government, as well as how it is presented to the general public. It is my contention that persistence of the mechanistic materialist view of the gene has serious and detrimental consequences for understanding the meaning and applications of new genetics research.

One of the clearest ways in which Mendelian theory fit into the prevailing mechanistic materialist philosophy was in its particulate, atomistic nature. This becomes especially apparent in the early debates, discussed by both Raphael Falk (1986) and Marga Vicedo (2000), regarding the intervening variable versus the hypothetical construct notion of the Mendelian gene (these terms come from MacCorquodale and Meehl, 1948, via Falk, 1986, pp. 133-4). The term 'intervening variable' refers to the view of the gene primarily as a heuristic device that helped to account for the data from breeding experiments, and posited nothing specific about the "real" existence, let alone molecular nature, of genes or transmitted elements in the germ plasm. This view was particularly apparent in the early writings of Danish plant breeder Wilhelm Johannsen (who coined the term 'gen' or 'gene' in 1909). Johannsen wrote, "No hypothesis about the nature of this 'something' [*Etwas,* in Johannsen's German text] should therefore be constructed or supported," and went on to say:

> No certain ideas about the nature of the gene are at present well enough established. This, however, is of no consequence to the efficiency of research of heredity. It is enough that it can be asserted with certainty that such 'genes' are available. (Johannsen, 1909, p. 124; in Roll-Hansen, 1978, pp. 202-3; quoted by Falk, 1986, p. 140)

Johannsen's gene was purely notational, but it sufficed to account for the basic patterns of Mendelian inheritance. E.M. East made similar claims, although, as explained below, he served as a transitional figure in the adoption of a more realist perspective. East wrote:

> As I understand Mendelism it is a concept pure and simple. One crosses various animals or plants and records the results. With the duplication of the experiments under

> comparatively constant environments these results recur with sufficient definiteness to justify the use of a notation in which theoretical genes located in the germ cells replace actual somatic characters found by experiments. (East, 1912, p. 633)

Even as late as his Nobel acceptance speech in 1933, Morgan emphasized the uncertainty and lack of real concern among practicing geneticists about the nature of the 'unit of heredity':

> Now that we locate [the genes] in the chromosomes are we justified in regarding them as material units; as chemical bodies of a higher order than molecules? Frankly, these are questions with which the working geneticist has not much concerned himself, except now and then to speculate as to the nature of postulated elements. There is no consensus of opinion amongst geneticists as to what the genes are—whether they are real or purely fictitious—because at the level at which the genetic experiments lie it does not make the slightest difference whether the gene is a hypothetical unit or whether the gene is a material particle. In either case the unit is associated with a specific chromosome, and can be localized there by purely genetic analysis. (Morgan, 1935, pp. 7-8)

Morgan himself did not believe that genes were fictitious entries, as I have emphasized elsewhere (Allen, 1978b). Morgan was a mechanistic materialist but a sophisticated enough thinker in such matters to realize that ultra-simplistic mechanical views were biologically misleading, and that in terms of pursuing the experimental work, one's philosophical position on such matters did not make much difference. It is useful to recall that such agnostic views about the ultimate nature of the phenomena being investigated paralleled the approach taken by chemists earlier in the nineteenth century (and revived by Wilhelm Ostwald and Jacobus Henricus Van't Hoff in the early twentieth century) regarding the nature of atoms, and of physicists, such as the Curies, about the physical nature of atomic radiation.

By contrast, the hypothetical construct view saw the gene as a physical entity, a discrete bit of particulate matter that was passed from parent to offspring in the gametes, but about which little or nothing was yet known concerning its composition or nature. Muller was one of the major spokespeople for this position. In 1922 he wrote in no uncertain terms that genes must have a real, chemical nature:

> ...there are present within the cell *thousands* of distinct substances—the 'genes'; these genes exist as ultra-microscopic particles; their influences nevertheless permeate the entire cell...the genes are in the chromosomes. (Muller, 1922, p. 33)

Falk has shown how in the period prior to 1910 the intervening variable view tended to predominate among those adopting the Mendelian scheme, while by the time of World War I, and especially by the 1920s, the hypothetical construct view had gained ascendance. Both Falk and Vicedo see East at Harvard's Bussey Institution as a key figure in that transition. East developed what Falk has called "an instrumental view of the gene" (Falk, 1986, pp. 144-6). While admitting that the gene could be regarded as a purely fictitious unit, East also defined the gene as "...that substance present in the germ cell which represents potentially the 'unit-character' or whatever it may be called that acts as an entity in heredity" (East and Hays, 1911, p. 21; quoted in Vicedo, 2000). This expression hardly suggests an unequivocal position on the purely notational nature of genes.

The force behind the transition, of course, was the chromosome theory of inheritance developed by the Morgan and Emerson groups initially, and expanded significantly by others during the 1930s-1950s. As Morgan's Nobel speech indicated, the mapping experiments localize a genetic effect to a particular locus on a chromosome; the experimental results are identical whether there is something material posited to reside at that spot contributing to the ultimate phenotypic effect or not. But, as T.H. Morgan, A.H. Sturtevant, Calvin Bridges, and H.J. Muller indicated in the introduction to their path-breaking book of 1915, *The Mechanism of Mendelian Inheritance,* to deny the evidence suggesting that Mendelian factors are material parts of chromosomes would be patently foolish:

> Why, then, we are often asked, do you drag in the chromosomes? Our answer is that since the chromosomes furnish exactly the kind of mechanism that the Mendelian laws call for; and since there is an ever-increasing body of information that points clearly to the chromosomes as the bearers of the Mendelian factors, it would be folly to close one's eyes to so patent a relationship. (Morgan *et al.*, 1915, p. viii)

It was a clear preference on the part of Morgan and his group to seek a material basis for the complex processes postulated in the Mendelian paradigm.

Mechanistic materialism is nowhere more clearly evident in the establishment of the Mendelian paradigm than in the strong emphasis on corpuscularity and atomism—and most explicitly the analogy of factors or genes to the chemists' atoms. To many early Mendelians, the analogy of genes to the atoms of physics or chemistry was obvious; indeed, they saw that atomizing the hereditary process made it comprehensible, predictable, and subject to quantitative and mathematical formulation. Whether or not such an explicitly atomistic view was inherent in Mendel's own conception has been debated with great fervor in recent years. Vitezslav Orel, Mendel's most recent biographer, and geneticist Daniel Hartl have argued that Mendel did think of factors as material entities (Hartl and Orel, 1992), while Robert Olby (1979) and Floyd Monaghan and Alain F. Corcos (1985) have argued that Mendel was primarily an empiricist unconcerned about the physical nature or even existence of his *Anlagen*. According to this view, Mendel's major interest was the patterns formed during the specific process of hybridization, not in constructing fundamental theories about the nature of heredity in general. Whatever Mendel himself may have believed on this issue, it is clear from the literature that his followers in the early twentieth century, from Bateson to Muller, Sturtevant, and Dobzhansky, saw Mendel as their predecessor in a mechanistic view of heredity. The literature of genetics between 1901 and 1930 abounds with references to genes as the biologist's atoms, and genetic methodology as analogous to that of the chemist in the combination, dissociation, and recombination of atoms in strict mathematical proportions. Bateson perhaps was first to see the comparison, referring to it as early as 1901:

> In so far as Mendel's law applies, the conclusion is forced upon us that a living organism is a complex of characters of which some, at least, are dissociable and are capable of being replaced by others. We thus reach the conception of *unit characters*, which may be rearranged in the formation of reproductive cells. It is hardly too much to say that the experiments which led to this advance in knowledge are worthy to rank with

those that laid the foundation of the Atomic laws of Chemistry. (Bateson, 1901; quoted in Punnett, 1928, vol. 2, p. 1)

A similar claim was made by W.E. Castle, who wrote that before the Mendelian scheme could be accepted unequivocally, it was necessary to know whether

...all observed inheritance phenomena can be expressed satisfactorily in terms of genes, which are supposed to be to heredity what atoms are to chemistry, the ultimate, indivisible units, which constitute gametes much as atoms in combination constitute compounds. (Castle, 1919, p. 127)

East echoed the same view when, in discussing the notational nature of Mendelism, he noted that "Mendelism is therefore just such a conceptual notation as is used in algebra or in chemistry" (1912, p. 633). Johannsen, originally trained in pharmacology, used the comparison of heredity to chemistry quite overtly, though he continued to think of Mendelian theory as still only a useful intervening variable, even in the face of the chromosomal evidence. H.S. Jennings (1920) and C.B. Davenport (1906) expressed similar views about the analogy of genes to atoms.

Early depictions of genes in the formal Mendelian notation showed discrete, independent units, sorting and resorting in a classically kinetic model. The later representations portray genes as a linear array (beads on a string); even the more refined chromosomal maps of the 1930s retained the element of genes as discrete units, or particles of definite molecular dimensions. Muller's target theory reinforced the same view of the corpuscularity of the gene: the more mutations within a gene created by a given x-ray dosage, the larger the gene (target) in question.

Another important aspect of the analogy of genes to atoms was that of genes' ability to combine and recombine in different ways to produce different outcomes, or phenotypic effects (i.e., analogous to atoms combining in different ways to produce different physical and chemical compounds). Furthermore, like atoms in chemical combination, genes emerged from each association unchanged in their fundamental properties (the doctrine of the 'purity of the gametes,' as it was known in early Mendelian lingo). Recessive and dominant genes were not altered (contaminated) by their association with each other. A recessive allele came out of a heterozygous combination just as recessive as before. Although later work on such cases as epistasis, modifying factors, and position effect added elements to the Mendelian scheme for which there was no exact counterpart in chemistry, these nuances did not undermine the most basic property of the gene as a stable unit that could combine and recombine, always in predictable ratios, generation after generation.

A consequence of the ability of genes to emerge unchanged from various combinations was the view that their history was unimportant. Like atoms, whose history has always been considered irrelevant to their current combining properties, the combinations in which genes have existed in pre-existing organisms have no bearing on their properties in the organism in which they now reside. Johannsen (1911, p. 138) found this aspect of the gene particularly important in countering the (then) still-prevalent Galtonian law of ancestral inheritance (the idea that each organism is composed by some fraction of hereditary elements from all of its previous ancestors: 1/2 from each parent and through them 1/4 from each grandparent, 1/8 from each great-grandparent, and so on). Galton's law was inconvenient for making predictions about future offspring, and had little heuristic value for breeders or for those interested, as Francis Galton was himself, in the inheritance of complex human phenotypes, especially "behavior" and "personality." More importantly, perhaps, Galton's law appeared to invoke a role for the entire history of a family line in influencing the traits of the present generation, a view that remained both impractical and intellectually distasteful to those trying to break out of the historically-dominated phylogenetic paradigm of late nineteenth-century morphology. If all ancestral contributions played a role, no reliable prediction of the traits in future generations was possible. However, if each gene starts life anew in its present combinations, then predictions of the sort that Mendelian theory embodied were reliable and highly useful, especially to breeders. Of course, from our present, post-synthesis perspective, genes do have histories, and though they are not "contaminated" by their particular combinations, they do undergo change of a sort that is cumulative and incorporates an aspect of their history—the additive effect of mutations, crossovers, insertions, etc.—for which there is no exact counterpart in the chemical world (though radioactive decay might be considered an analogous case of historical change within the life history of a radioactive element).

A key feature of the Mendelian scheme that gave it great currency in the promulgation of the new biology was that, like atoms that are discrete and whose effects, at least, are measurable, Mendelian genes could be treated quantitatively and statistically. Thus, two other conditions of the new biology were fulfilled by the Mendelian paradigm: the genetic data were quantitative and through the use of simple statistics could be manipulated to show both regular, underlying processes and to make general predictions. The early Mendelian schemes of mono- and di-hybrid crosses fit easily into this sort of simple scheme. Later developments, especially those involving quantitative inheritance (i.e., traits that are not phenotypically discrete, such as a range of kernel colors in wheat), required additional hypotheses—namely those of modifying factors or multiple alleles acting epistatically (i.e., additively). There were numerous debates about these auxiliary hypotheses, and some geneticists at least—such as Castle and Goldschmidt—found the constant addition of new hypotheses troublesome. Goldschmidt, for example, complained that phenomena such as position effect (the effect on a gene's phenotypic expression as a result of changing its position within the chromosome) entirely undermined the notion of the discrete Mendelian gene (Allen, 1974). He was correct in one respect: the paradigm of corpuscularity was becoming such an integral part of the Mendelian paradigm that highly continuous characters or qualitative differences were being forced into discrete pigeon-holes in a way that many thought obscured the underlying biological processes that led from genotype to phenotype. This same criticism was also the basis of the hostility (or at least skepticism) emanating from various embryologists (such as F.R. Lillie, Conklin, Harrison, Dalcq, and numerous others) about the contribution Mendelian genetics could make toward understanding the basic processes of embryonic growth and differentiation. As we will see below, various problems emerged from the embryologists' perspective about the mechanistic thrust of the Mendelian paradigm, not the least of which was its failure to take into account a variety of interactions— between genes, between genes and environment, and between the various hierarchical levels of organization within the organism (gene to chromosome to nucleus to cell, differentiated tissue, etc.). Suffice it to say, the strong emphasis on discrete and separable units interacting additively, as opposed to synergistically, posed serious problems for understanding developmental processes.

Growing out of the discrete and mathematical precision embodied in Mendelism was the mechanistic view of the whole organism—its phenotype—as a sum of separate parts, the genes (a mosaic, if you will) of discrete entities (genes at one level, phenotypic traits at another). A corollary of any mosaic conception is that the parts are dissociable and can be profitably studied individually by the analytical method. Thus, as Bateson put it in addressing the New York Horticultural Congress in 1902:

> The organism is a collection of traits. We can pull out yellowness and plug in greenness, pull out tallness and plug in dwarfness. (Lewontin and Levins, 1985, p. 180)

What Bateson had in mind was the process of analysis, or structural reductionism, where complex processes or entities are taken apart and studied one part at a time under controlled laboratory conditions. Mendelism, of course, fulfilled this criterion ideally. The complex of characters that made up the organism could not be studied as an entirety. Indeed, that had been one of the major problems encountered in the pre-Mendelian breeding programs: breeders had tried to follow the transmission of too many traits at a time. Mendel's success had been based largely in his willingness to focus on one, two, or at the most three traits at a time. These traits were then assumed (especially in Mendel's own formulation) to act independently (as in the so-called law of independent assortment). But this independent action could never be discovered without analysis of the breeding process into individual traits.

A consequence of this sort of analytical dissection was the dissociation of the unitary process of heredity into two separate processes: transmission and development. That this split occurred at the professional as well as conceptual level only highlights the pervasive way in which the mechanistic materialist approach dominated the very organization of the new biology itself. Both geneticists and embryologists, from the 1920s onward, encountered considerable difficulty in applying Mendelian genetics to problems of embryological differentiation or, conversely, in understanding genes as agents controlling developmental processes. As surely as the organism was analyzed into separate genes, the cell into nucleus and cytoplasm, and development into heredity and environment, so the professional division of labor into genetics and embryology split asunder an integrated view of the reproductive process that had held sway for the better part of the last half of the nineteenth century. It was a conceptual block that has continued to be an issue among biologists from that day until now, though at the molecular level significant strides are now being made (as with homeotic genes) to reintegrate the study of gene transmission, development, and evolution, separated for so long into distinct fields and professional enterprises.

The mechanistic and analytical method also provided an important incentive to treat Darwinian theory in terms of discrete Mendelian genes in the work of R.A. Fisher, R.H. Lock, J.B.S. Haldane, and other pioneers of what came to be known later as the 'evolutionary synthesis.' Fisher was the most explicit, and most overtly mechanistic, in his analysis of populations as collections of discrete Mendelian genes, and his analysis did indeed provide a powerful tool for advancing the Darwinian cause at a time when the efficacy of natural selection was being questioned on all fronts (Bowler, 1983; Allen, 1983, pp. 89-92; Provine, 1971, pp. 140-50). The investigation of natural selection via analysis of populations into discrete genes existing in different frequencies, Fisher boasted in 1922,

> ...may be compared to the analytic treatment of the Theory of Gases, in which it is possible to make the most varied assumptions as to the accidental circumstances, and even the essential nature of the individual molecules, and yet to develop the general laws as to the behavior of gases, leaving but a few fundamental constants to be determined by experiment. (Fisher, 1922, pp. 321-2; quoted in Provine, 1971, p. 149)

The analytic separation of genotype and phenotype introduced by Johannsen in 1911 at the individual level was completed by Fisher at the population level by statistical means. In Fisher's population models the phenotype disappeared completely, leaving only the discrete Mendelian genes (aggregated into gene "pools") as the objects of evolutionary change. A species or population became an aggregate of genes, interacting randomly, much as do atoms or molecules in an idealized gas. The somewhat derogatory appellation of 'beanbag genetics,' applied to Fisher's kinetic theory by Ernst Mayr, clearly reflects the mechanistic basis on which Fisher's approach seemed to be based: whole populations (indeed species) were abstracted into discrete atomized components (not even individual organisms, but genes, encountering one another by chance in a random universe). This was billiard-ball physics at its most extreme applied to a biological process. Although Fisher did discuss the interaction of genes in producing composite traits, such a process did not occupy center stage in his worldview. While he knew full well that many cases of epistasis and complex interactions occurred, his legacy remained that of one gene-one trait. More field-oriented evolutionists such as Mayr and Dobzhansky came to view Fisherian population genetics as oversimplified to the point of ignoring critical factors that affected organisms in nature (population size, structure, breeding patterns, and so on). Nevertheless, historically, Fisher's approach gave evolutionary theory a rigorous basis for the first time. Evolutionary processes such as selection, population size, migration, and adaptation (including the considerable British debates on mimicry and warning coloration) could now be treated quantitatively, mathematically, and most importantly, eventually through experimental analysis (for example, Dobzhansky's use of population cages filled with *Drosophila* and functioning in the laboratory as an experimental tool). Natural selection no longer remained in the pejorative realm of speculation and fantasy.

In short, then, introduction of the mechanistic materialist approach into genetics led to the expansion of many areas of biological work that previously seemed mired in purely descriptive and speculative methodologies. With the work of R.A. Fisher,

Sewell Wright, J.B.S. Haldane, and other mathematical population geneticists, the problem had come full circle: Darwinian theory, once considered the paragon of old-fashioned, non-testable, and non-rigorous biology, had now been placed on as solid an epistemological footing as the kinetic theory of gases. However, what remained outside the pale of the synthesis was embryonic development, the process by which genetic information is transformed into adult phenotypic traits. Developmental biology has been severely hampered by its mechanistic materialist past, not only in terms of the mechanism of this process, but even the way in which it has been conceptualized in much of the biological literature from the 1920s to the present. The extent of this influence and how I think it affects the way we formulate questions about modern genomics form the basis for the next section of this chapter.

5. MECHANISTIC MATERIALISM, CAPITALISM, AND AGRICULTURE: THE ECONOMIC CONTEXT OF MENDELIAN GENETICS

Mechanistic materialism arose not only in the context of seventeenth-century science and technology, but also more generally in the context of the evolution of feudal to capitalist modes of production. Numerous writers have dealt with this economic and social transition, and the associated philosophical transition from idealism to materialism (specifically mechanistic materialism, such as Bernal, 1971 [1954], pp. 373-8; Cornforth, 1971, p. 14ff; Afanasyev, 1987, pp. 1-5). Thus there is no need to repeat that history here. Rather, I would like to relate this rather general background to the present discussion of the gene by focusing on a crisis in capitalist agriculture in the West (particularly the United States) at the turn of the twentieth century—at the very time Mendelian theory was being rediscovered (Lewontin and Berlan, 1986).

Western capitalism and mechanistic materialism co-evolved in such a way as to share a number of basic assumptions about the way the world—social and natural—functions. Capitalism embodies an atomistic view of economics and the social interactions that go with it. The basic unit of capitalism is the individual (as buyer or seller), interacting in a free marketplace unfettered by externally-imposed guiding hands or regulations (in the idealized conception of classical *laissez-faire* capitalism). The market behaves as a whole, following certain general rules (supply and demand, the relation between wages and profits, etc.) not because they are imposed but because they emerge from individual, atomized activities that are based on certain fundamental principles (for example, the right to buy and sell, to enter into contracts, to "maximize profit"). The whole in a capitalist system is reckoned as the sum of its parts: gross national product (GNP), for example, is the sum of the buying and selling activities of all individual units (individual people and/or corporate groups) per unit time. Individual economic activity cannot be predicted, but overall activity (to a certain degree, that is) is theoretically predictable if there is a large enough population to analyze and if enough of the parameters are known. Sound familiar? It is, indeed, nothing less than the principle of the kinetic theory of gases working itself out in the economic marketplace (many would argue, of course,

from a social constructionist point of view, that the kinetic theory of gases reflects
the thinking emerging from an atomized economics rather than the other way
around). It is perhaps no accident that the detailed system of modern bookkeeping
arose with the spread of capitalism, a quantitative expression of mechanistic
philosophy: all items are accounted for and "balanced" one for one; the whole
remains equal to the sum of its parts and no more.

With the primacy of capitalist economics residing in the exchange of material
goods, it was a system hospitable to philosophical materialism (also, of course, to
vulgar, popularized materialism as a spin-off). With its concern for processes
occurring through the independent action of atomized components, capitalism also
was hospitable to the explicitly mechanistic form of materialism. The integral role
that technology—machines—played (and still play) in expanding production and
profit also made capitalism and its social milieu hospitable to thinking about
processes of all sorts, including living processes, in mechanical terms. For example,
analogies of organisms to machines have been prominent in the life sciences from
seventeenth-century writers like William Harvey, Giovanni Borelli, and Pierre
Gassendi, to early twentieth-century writers such as Jacques Loeb (probably the
most extreme case for his era, he once likened phototropic insects to "chemical
machines enslaved to the light"), W.J.V. Osterhout, and bio-engineer J.H. Hammond
(who delighted Loeb and others with his robotic selenium-eyed dog). The
mechanistic materialist philosophy that became so prominent in early twentieth-
century biology was thus not a new creation, but the re-emergence and capturing by
a new generation of an old tradition within the sciences, including biology.

Before moving to the specifics of early twentieth-century Mendelism and
agriculture, let me emphasize that as sociological writers such as Karl Marx,
Friedrich Engels, and later Robert K. Merton have pointed out, in social evolution
various facets of any society tend to be brought (forced, in some cases) into some
sort of harmony with the needs of the dominant class: what Marx explicitly saw as
the parallel between base and superstructure in any society ('base' refers to the
economic structure and social practices that support it; superstructure refers to all
other developments in a society—politics, art, culture, religion—that are shaped by
the base, but in turn support and reinforce its maintenance). The philosophical
principles, moral dictates, and educational practices of a society tend not to be far
out of step with the economic and social practices of that society. This integration,
which is brought about partly by conscious, planned activities, partly by less
conscious processes such as social approbation, characterizes all societies, whatever
their economic or social forms. When the integration becomes disrupted, the society
enters a crisis period: either concomitance is restored or a major economic/social
upheaval takes place. The major point relevant to our story is that the mechanistic
materialism that became so important in late nineteenth- and early twentieth-century
biology re-emerged as part of a much larger process of reintegrating facets of
capitalist society that were in varying degrees of crisis at that time (1890s). At its
core, the crisis arose in the socio-economic sphere over the issue of retaining
classical *laissez-faire* principles, which had led to enormous concentration of capital
(monopolization), wild swings in prices, unpredictable periods of boom and bust

(significant depressions virtually every decade between 1874 and 1903-4), and increasingly confrontational, violent disputes between labor and capital. The crisis had its ramifications in the sciences—especially in biology—in the attempt to industrialize agriculture and make it a more profitable sector for investment.

The solution to these crises (especially in the United States) was seen to be "Progressive Era" reform, the idea of managed capitalism and of social planning and control based on the principles of scientific management (Weibe, 1967; Weinstein, 1968). The role of the various American foundations in catalyzing and implementing this newer view has been told well by several authors, most notably Lily Kay (1993), Robert Kohler (1991), and Pnina Abir-Am (1982). The "scientific" part of scientific management was based on an understanding of the science of the day; that is, that it was based on empirical data—experimental where possible, quantitative, mathematical, and predictable. Especially with its emphasis on molecular biology and biochemistry, it stressed the mechanistic ideology underlying the "new" biology. Application of the principles of scientific management extended from increased support for fundamental research to practical studies in breeding, soil management, and workplace efficiency (Taylorism), and the means of controlling labor unrest. It was the notion of using science to control not only nature, but human social development as well. Scientific management was a global concept, a modification of classical *laissez-faire* principles by the early 1900s that included agriculture as a cornerstone. Despite the shifting ideology, however, the underlying principles of mechanistic thinking remained a constant. Managed capitalism eventually replaced rampant *laissez-faire* capitalism in virtually all Western societies by the time of World War I.

To illustrate how a revived mechanistic materialist philosophy actually played a major part in constructing our current view of the gene, consider the relation of agriculture in the United States in the late nineteenth and early twentieth centuries to theories of heredity. During the period 1880-1920 in the United States, agriculture was undergoing a particularly rapid transformation: a decrease in the number of farm workers (due both to mechanization of farm work and migration to industrial jobs), increase in the use of purchased inputs (fertilizer, pesticides, and gasoline), and the increasing cost of machinery (Rosenberg, 1976). Although production was on the rise throughout the 1890s, the depression of 1894 hit farmers particularly hard, and falling prices forced abandonment of many family farms. Scientific agriculture in the post-Civil War period had improved production, but its effects had reached a plateau: there was only so much fertilizer that could be added to a field or nutrients to the animal feed bin. Moreover, there was no more profit to be gained beyond reaching the plateau, since fertilizers and feed supplements had to be added every week, month, or year. The breeding of new, more vigorous, high-yield varieties provided one of the incentives to move capital into agriculture in a much larger way than in the previous decades (Lewontin and Berlan, 1986, p. 26ff). The advantage of new genetically engineered (for their day) breeds was that once heredity was fixed and the organisms bred true, the yield from that point onward was sheer profit. No new technological inputs (other than feeding and caring for the organisms) were required. The expense of the breeding program would continue forever (Allen, 2000, pp. 1085-86).

This point was recognized by those who were involved in moving capital into the agricultural sphere around 1900—for example, the Secretary of Agriculture James V. Wilson, Andrew Carnegie (whose nascent Carnegie Institution of Washington provided some of the first scientific grants to practical breeders like Luther Burbank), John H. Kellogg (the cereal magnate in Battle Creek, Michigan), and Henry C. Wallace, Secretary of Agriculture (1921-24)—and as a result, considerable energy and funds were directed toward breeding. It is in this context that Mendel's results were "rediscovered;" they provided the exact strategy for reproducing pure lines that agricultural capitalists were seeking. Mendel's works assumed an importance in 1900 that they had lacked in the quite different economic context of both Europe and the United States in 1865. Breeding new strains now became the hope for making the agricultural sector at last a profitable arena for private investment.

For those who thought about economics in the traditional *laissez-faire* mode and were imbued with the spirit of applying modern scientific methods to agricultural breeding, Mendel's work fit the bill perfectly, just as it did for research biologists imbued with the same values of applying science to the solution of contemporary agricultural problems. Particularly important for the agriculturalist, Mendel's paradigm was predictive and provided the basis for designing breeding programs; it clearly showed the importance of selecting for the genotype, rather than the phenotype, and claimed that genes were not contaminated by existing with each other in a hybrid. Most importantly, Mendel's work spoke to the breeder's biggest issue: the nature of hereditary transmission. Breeders were not concerned with embryonic development (except in a certain practical sense, as with nutrition), but rather with linear transmission of traits from parent to offspring. The atomistic nature of Mendel's theory attracted breeders for many of the same reasons it attracted the biologists: it was quantitative, mathematical, experimental, and predictive. It seemed to those breeders who first encountered it to be the Newtonian theory of animal and plant husbandry. That Mendel's work did not prove to be of such immediate practical value as many had initially hoped neither undermines the importance of the initial enthusiasm, nor the ultimate realization of very successful commercial breeding programs based on the later elaboration of Mendelian principles—for example, Wallace's Pioneer High Bred Corn. The agricultural context provided the right environment in which Mendel's work could thrive.

The formation of the American Breeders' Association in 1903 was intended to foster cooperative work between practical breeders and more academically-oriented biologists, and later geneticists (Paul and Kimmelman, 1988). Mendelian genes, like individual buyers or sellers in the marketplace, interacted constantly, forming various new combinations and recombinations. But it was in the context of an agricultural need—to move capital accumulated through industrial development into a new arena, the farm as an industrial complex—that gained Mendel's work more than ordinary credence among scientists and breeders alike. That the same philanthropic agency, The Carnegie Institution of Washington, funded not only

Burbank, but also Morgan's fruitfly work, suggests the close connection that was emerging in the early decades of the century between Mendelian theory and its practical application to agriculture. The fact that breeders became imbued with the mechanistic view of the gene as an element transmitted between parent and offspring, and were little concerned with development, only underscores the degree to which the Mendelian program, from at least 1915 through the 1970s or 1980s, participated in divorcing the study of heredity from that of development.

6. THE MENDELIAN LEGACY AND MODERN GENOMICS

Since the advent of molecular genetics in the 1950s, the field of genetics has undergone a conceptual and institutional growth that dwarfs even the most active period of classical genetics in the early decades of the century. And certainly, the concept of the material structure of the gene has been greatly revised and refined from the notion of beads on a string. Genes are no longer recognized as the discrete entities that early Mendelians saw them to be. With non-transcribed DNA, introns, overlapping codes, promoters, stop-start sequences, and other control elements, genes are recognized to be much more complex at both the structural and functional level, and less discrete entities (both structurally and functionally) than was the case during the heyday of the classical Mendelian gene. Today, we have ample evidence for the actual interaction of genes and gene products, various "genes" or DNA segments coding for several domains of a single protein (Portin, 1993, p. 207), along with a variety of regulatory mechanisms, so that an understanding of how genotype may become transformed into phenotype appears more attainable than at any time in the past.

However, the old mechanistic materialist notions of the gene still persist in a variety of ways in both the professional scientific, and especially the popular, literature about heredity. And nowhere is this more prominent, and perhaps more dangerous, than in research on the genetics of human behavioral and personality traits that has increased in frequency and boldness of assertion on the coattails of the Human Genome Project (HGP). The HGP has provided a public visibility and prestige for genetics that is similar to the excitement surrounding the discoveries and purported applications of Mendelian genetics in the first half of the twentieth century. Now, as then, exaggerated claims about the possibility of genetically remaking the human species have capitalized on real and significant, if not publicly well-understood, advances in genetics.

For the past decade or more, the general public has been inundated with almost daily claims about the discovery of a new gene not only for this or that clinical disease (obesity, multiple sclerosis, or Alzheimer disease), but also for personality and social behaviors (schizophrenia, alcoholism, shyness, homosexuality, criminality, and risk-taking, to name just a few). Behind much of the view of genetics presented in these reports lurks the old specter of mechanistic materialism, in particular the notion of the gene as a discrete, atomistic unit, with single genes producing single traits, coupled with the analytic separation of genotype from phenotype. Recognizing this is important for reasons beyond simply setting the historical record straight. We are confronted daily with questions about what conditions will be covered by our medical insurance—by what is a "preexisting condition" (i.e., a genetic condition), and therefore not reimbursable. Even more potentially alarming, especially in the United States, the conclusions about human behavior and its regulation or treatment via social and medical policy are very much dependent on a more sophisticated understanding of genetics than is being promulgated in the popular press, including biology textbooks, at the present time. The most recurrent phrase in this genre of science reporting is "biologists have just found the gene for ...." Such phraseology relates back to mechanistic notions of genes as atomistic units determining adult traits in a one-to-one fashion.

The mechanistic gene model is problematic on several levels. First, definitions of behavioral phenotypes such as alcoholism, criminality, schizophrenia, and the like are both vague and subjective. These are very complex traits, not easily resolvable into a single, unambiguous description, nor assigned a single cause. Stephen J. Gould and others have termed the process of treating a complex process as though it were a single thing 'reification,' that is, giving something a concrete existence it does not have. While the problem of reification is also a danger in dealing with even much simpler physical traits such as hair or flower color, or height, it is an immensely more difficult problem when traits are as multidimensional as human behaviors. The dual tendency to reify the trait as if it were a discrete entity and to associate it causally with a single gene (or two) leads to the highly simplistic view, so rampant today, that every characteristic of our being is directly determined by one or a few genes. We are, indeed, being portrayed as a mosaic of genetic factors that invariably will play out their deterministic roles in our development into adults. This naive view is sometimes referred to as 'biological' determinism, or more specifically, 'genetic' determinism.

The even more central problem in understanding the modern literature on human behavior genetics is that of the lack of any understanding of the biological processes involved in development, especially related to development of the nervous system and the role of sensory input on the structural and biochemical properties of the nervous system itself. The genetic stability that was a cornerstone of the classical Mendelian theory has led to the general assumption that genes simply produce the adult traits almost irrespective of environmental conditions. It was this implicit notion that caused Morgan, as late as 1909, to refer to Mendelism skeptically as the "new preformationism." The old notion that somehow development is reducible to Mendelian genetics, to learning the sequence by which particular genes are turned on or off, has persisted from the 1910s right down to the present. The fact of the matter is that embryologists have known for virtually a century that development is not a mere unfolding of invariant form. Teratology, the occurrence of phenocopies, and environmental stimulations, all attest to the fact that genes are very responsive to changes in the conditions under which they function. Mammalian geneticists know this as a canon of their work: the homozygous strains of mice established by the Jackson Laboratory in Bar Harbor, Maine, for example, serve as standard genetic "backgrounds" against which to test the phenotypic outcome of any new mutant that is discovered. Genes have a way of producing quite different phenotypes under different genetic backgrounds and environmental conditions.

A prominent feature of the mechanistic approach to genetics was that the existing concept of genetic norm of reaction received relatively little attention during the classical period. 'Norm of reaction,' a phrase ("*Reaktionsnorm*") coined by R. Woltereck in 1909, refers to the range of phenotypes that a given gene or set of genes can give rise to under a variety of environmental conditions (Sarkar, 1999). The concept of a norm of reaction suggests that genes are not rigid, stable units that always produce the same phenotypic effect, but are variable, around a 'norm' that is dependent on overall environmental conditions. For example, when *Drosophila* larvae are raised at a variety of temperatures, the adults develop wing structures of many different shapes and patterns (some of these represent forms very similar to known point mutations—hence they are referred to as 'phenocopies'). Plant biologists have recognized the norm of reaction for a long time—plant phenotypes have always shown themselves to be far more plastic than animal phenotypes under similar ranges of conditions. However, even in animals, different genes have different norms of reaction, that is, some produce roughly the same phenotype however variant the background conditions, while others produce quite different results when even slight changes in environment occur. It is telling that the standard genetic experiment under the Mendelian paradigm involved raising a variety of genotypes under the same, or controlled, environmental conditions but did not consider it necessary to do the converse—that is, to raise the same genotype under a variety of environmental conditions. The latter would have revealed the norm of reaction of the various genes—information that is still not available about most of the genes identified in insect or mammalian models today. This "oversight" would seem to be no accident, but rather the result of a strong commitment to the mechanistic view of the gene as a stable unit (like the chemists' atom) that invariably produces the same effect regardless of conditions.

Mechanistic materialism also led biologists in general and geneticists in particular into a reductionist mode of thinking in which all biological phenomena were thought to be explained by reference to smaller, component parts of the system (e.g., the phenotype could be reduced to the expression of specific genes through linear time). The phenotype was thus predictable from the genotype. In a provocative discussion of this issue, James Griesemer (2000) has pointed out that reductionism has traditionally referred to a structural or hierarchical relationship in which more complex entities are reduced to, and hopefully thereby explained by, the properties of their smaller, fundamental components. Such reductionism, Griesemer points out, led to the attempt to "reduce" development to genetics, an enterprise that ultimately proved futile and unrewarding. More promising, he points out, would have been (or would be today) to think of reduction in terms of processes, in which both hereditary transmission and ontogenetic expression of genes were seen as parts of the general phenomenon of reproduction, so that "reduction" could take place in all directions: transmission and development could be the cause as well as the effect of reproduction, providing a more comprehensive, non-linear way of understanding relationships among interacting processes. Classical mechanistic materialism, however, persists in a bias toward linear, hierarchical, and structural reductionism that misses such interactive and reciprocal chains of causation.

In Griesemer's critique, then, genetic systems are best understood as processes, not the sort of structural relationships that have persisted throughout the periods of classical and molecular genetics. Genes are to be understood not so much as loci on a chromosome or units of recombination, mutation, or a linear array of kilobases, but as components of the cell that interact with each other and with other cell components (including those that serve as specific extracellular signals), altering their function as a result of these interactions. Such alterations of function would be an expression of the norm of reaction of that particular genetic-developmental system. Norms of reaction can be studied experimentally (indeed, as pointed out above, failure to do so is one of the significant shortcomings of classical genetics). However, it was not thought that it was necessary to do so, since the doctrine of the stability of the genotype-phenotype relationship was taken for granted. After all, in the simplistic model of the physical sciences on which mechanistic genetics was based, the fundamental elements of any system are expected to change only quantitatively, not qualitatively, under changed conditions (chemical reactions speed up or slow down with changing temperature, but the products are the same). Only under extreme conditions such as very high or low temperature or pH—the very extremes at which living systems cease to function—do the outcomes of ordinary physical and chemical processes appear radically different. Driesch may have opted for flights of mysticism when he referred to the ability of the embryo to adjust itself to experimental manipulation and return to a normal course of development as a "harmonious equipotential system." But he was responding to a problem that the mechanistic materialist philosophy has never been adequately able to handle with ease: self-regulating, interactive systems that displayed non-linear relationships.

The holistic movement, especially in German biology in the 1920s and 1930s, as well as the dialectical materialist movement in the Soviet Union, each attempted to deal with these issues, with varying degrees of success. The problem was (and is), of course, that we have had almost four hundred years of developing mechanistic methods not only of conceptualizing the issues themselves, but also of devising experimental tools to investigate them. By comparison, we have had relatively little time to develop similar methods for conceptualizing and developing experimental techniques for investigating interactive phenomena (systems analysis in modern engineering and computer science represents one attempt to develop a technology for studying complex systems; chaos theory is a still relatively crude, but similar approach, and both are constrained by the predominant mechanistic bias of most modern science).

Linear genetic reductionism plays havoc with our understanding of what we are wont to call 'genetic diseases.' Fred Gifford (2000), among others, has argued that the usual way of framing questions about genetic conditions conflates several different forms of argument, and in reality obscures important differences among so-called genetic traits. In several intriguing examples, he underscores the mechanistic bias that leads us to call some conditions 'genetic' when they are clearly the result of a gene-environment interaction. In humans, the condition known as phenylketonuria (PKU), caused by the lack of the enzyme phenylalanine hydroxylase (coded for by a given DNA segment) leads to build-up of phenylalanine in the body, which causes serious mental retardation. However, in a phenylalanine-free environment, the "genetic defect" is never observed. Is PKU a genetic or dietary disease? In a similar example, researchers have labeled obesity a "disease of civilization" rather than a genetic disease *per se*. DNA segments that code for enzymes promoting conversion of starch to lipids (fat storage pathways) have had great selective value in a species like our own that has evolved under periods of food shortages (the feast and famine environment). Only in a society where fat-rich foods are present in abundance does obesity ever become a visible phenotype. The label 'genetic' for both PKU and obesity reflects the strong mechanistic bias toward linear causal systems, where in fact interactive systems are the rule in most natural systems, especially living organisms.

It is in areas dealing with human behavior genetics that the old reductionism has the potential for playing its most harmful role. If the development of human behavior is truly the result of a non-linear, developmental process (from fertilization throughout life) involving the interaction of genes with other genes, environmental factors, and the potential modification of gene function by each of these interactions, then the process of development is highly complex and not reducible to the activity or primacy of either genes or environment alone. Behavior is both more plastic and less linearly determined than a mechanistic genetic model would imply. The importance of this point lies in the consequences of genetic determinist paradigms in our present social and political setting. Classical mechanistic genetics has long had associated with it the view that if something is genetic it cannot be changed. It is part of the individual's biological make-up, their medical condition. In earlier decades of the last century, individuals with what were thought to be undesirable, genetically-determined conditions (mental problems, alcoholism, etc.) were sterilized in various countries (especially the United States, Germany, and Sweden) as a way supposedly to protect society from their degenerative effects. At the present time, similarly diagnosed conditions are likely to be treated with drugs (Ritalin, Prozac, and others) as part of "a new therapeutic approach." In either case, medical and social policy is being based on an old and outmoded concept of genes and genetics that is a leftover from the heyday of mechanistic materialism in early twentieth-century biology.

In conclusion, what I wish to emphasize is that the mechanistic materialist bias in modern genetics is not just something that was tacked onto Mendelian theory earlier in the twentieth century, nor was it just imbibed as part of the milieu of early twentieth-century biology in which genetics happened to develop. It was integral to the development of genetics itself. As I have argued, Mendelism became the centerpiece in the movement to make biology more rigorous, experimental, and quantitative. In all respects Mendelian theory, as it was understood and enlarged by early twentieth-century biologists, filled that bill exquisitely. The mechanistic materialist bias that was part of the movement to make biology more like physics and chemistry quickly became central to the worldview of Mendelian theory itself. The wedding of Mendel's factor hypothesis and the chromosome theory of heredity, developed from a cytological perspective by Theodor Boveri, E.B. Wilson, Walter Sutton, and others, gave full berth to the view that Mendel's factors were material entities in the cell. Chromosome studies provided the exact vehicle for Mendel's factors that a materialist theory required. That Mendelian theory became the standard-bearer for the attempt to transform biology from a largely descriptive and speculative enterprise to a quantitative and experimental science meant that the course of mechanistic science and the development of the Mendelian theory itself were inextricably linked. They have remained linked for the better part of a century.

Although many developments in both Mendelian and later molecular genetics have added great sophistication to our understanding of hereditary processes, the theory of the gene has remained clearly tied to its mechanistic materialist heritage. Debates about the units to which the term gene most appropriately applies today are rendered highly complex because of the mechanistic bias that still compels us to find *the unit* of heredity, whether it be a cistron, recon, exon, or nucleotide sequence between two promoter or control regions. It will be intellectually and socially difficult to free genetics from the mechanistic, atomistic conception of the gene. Not only is the discrete gene of classical Mendelian theory still a useful icon for public relations and propaganda purposes (Nelkin and Lindee, 1995), it also has a pedagogical simplicity that would be hard to give up. Having tried in my own teaching of biology majors and pre-medical students to introduce the concept of norm of reaction in place of the standard Mendelian gene, I can attest to the much greater difficulty in conveying a more holistic, interactive concept of the hereditary/reproductive process. Everything about our science is still based on mechanistic materialist foundations, so that the oversimplified Mendelian theory fits more comfortably into what students and the general public already knows as a basic foundation.

Yet, I would argue that the alternative, to continue to propagate an erroneous view of the integrated processes of heredity and development in general, and of the "gene" in particular, will ultimately perpetuate on the one hand a serious conceptual obstacle to understanding the relationship between heredity, development, and evolution, and on the other hand a serious social/political obstacle to the application of our biological knowledge to the solution of important medical problems relating to human health and behavior. The challenge is there, but it will take a serious and persistent struggle to overcome the long tradition of mechanistic materialism in genetics.

As difficult as the challenge may be, there are ways to start rectifying the current dilemma. One is to raise for critical discussion mechanistic concepts whenever and wherever they are employed in modern genetics—in seminar talks, meeting presentations, or published journal articles. Another is to work with journalists and science writers to help them avoid falling into the mechanistic trap for which they are all too well prepared. Good, informed science writing would help a lot, over time, to winnow readers away from simplistic genetic determinism. And science writers, like teachers, are key to this process. Still a third approach, for those of us who teach, is to find other ways of presenting genetic concepts that will avoid strict deterministic implications. One way might be to start genetics courses or units with molecular genetics, to emphasize the interactive process of turning genotype into phenotype, with all the factors influencing gene expression. Mendelian genetics—which is, after all, only a special case of inheritance in a relatively few traits—could then come in toward the end of the course and thus avoid the "gene for..." misconception. The struggle to alter the mechanistic view of the gene will be a long one, but already so many new ways of understanding genetic processes are going beyond the classical view, that I have no doubt someday a far more sophisticated and biologically-informed view of the hereditary process will prevail.

Washington University
St. Louis, Missouri, U.S.A.

REFERENCES

Abir-Am P. The discourse of physical power and biological knowledge in the 1930s: a reappraisal of the Rockefeller Foundation's policy in molecular biology. Social Studies of Science 1982; 12:341-82

Afanasyev, V.G., *Historical Materialism.* New York: International Publishers, 1987.

Allen G.E. An opposition to the Mendelian-chromosome theory: the physiological and developmental genetics of Richard Goldschmidt. Journal of the History of Biology 1974; 7:49-92

Allen, G.E., *Life Science in the Twentieth Century.* New York: Cambridge University Press, 1978a.

Allen, G.E., *Thomas Hunt Morgan, the Man and His Science.* Princeton: Princeton University Press, 1978b.

Allen, G.E. "The Several Faces of Darwin: Materialism in Nineteenth and Twentieth Century Evolutionary Theory." In *Evolution from Molecules to Men,* D.S. Bendall, ed. Cambridge, UK: Cambridge University Press, 1983.

Allen G.E. Essay review: history of agriculture and the study of heredity: a new horizon. Journal of the History of Biology 1991; 24:529-36

Allen G.E. The reception of Mendelism in the United States, 1900-1930. Comptes Rendu de l'Academie des Sciences 2000; 323:1081-8

Bernal, J.D., *Science in History: The Scientific and Industrial Revolutions,* Volume 2. Cambridge: MIT Press, 1971[1954].

Bourdieu P. The specificity of the scientific field and the social conditions of the progress of reason. Social Science Information 1975; 14(6):19-47

Bowler, P., *The Eclipse of Darwinism.* Baltimore: Johns Hopkins University Press, 1983.

Boyle, B., *Sceptical Chymist.* 2nd ed. Oxford: Henry Hall, 1680.

Burian, R.M. "On Conceptual Change in Biology: The Case of the Gene." In *Evolution at a Crossroads: The New Biology and the New Philosophy of Science,* D.J. Depew, B.H. Weber, eds. Cambridge: MIT Press, 1985.

Carlson, E.A., *The Gene: A Critical History.* Philadelphia: W.B. Saunders, 1966.

Castle W.E. Piebald rats and the theory of genes. Proceedings of the National Academy of Sciences 1919; 5:126-30

Cittadino E. Ecology and the professionalization of botany in America, 1890-1905. Studies in History of Biology 1980; 4:171-98

Cornforth, M., *Historical Materialism.* New York: International Publishers, 1971.

Darden, L., *Theory Change in Science.* New York: Oxford University Press, 1991.

Davenport, C.B., *Inheritance in Poultry.* Washington, D.C.: Carnegie Institution of Washington, 1906.

East E.M. The Mendelian notation as a description of physiological facts. American Naturalist 1912; 46:633-95

East E.M., Hays H.K. Inheritance in maize. Connecticut Agricultural Station Bulletin 1911; 167:1-141

Falk R. What is a gene? Studies in History and Philosophy of Science 1986; 17:133-73

Falk R. The struggle of genetics for independence. Journal of the History of Biology 1995; 28:219-46

Fisher R.A. On the dominance ratio. Proceedings of the Royal Society of Edinburgh (Series B) 1922; 42:321-41

Fitzgerald, D., *The Business of Breeding: Hybrid Corn in Illinois, 1890-1940.* Ithaca, NY: Cornell University Press, 1990.

Gifford, F. "Gene Concepts and Genetic Concepts." In *The Concept of the Gene in Development and Evolution: Historical and Epistemological Perspectives*, P.J. Beurton, R. Falk, H-J. Rheinberger, eds. Cambridge, UK: Cambridge University Press, 2000.

Griesemer, J.R. "Reproduction and the Reduction of Genetics." In *The Concept of the Gene in Development and Evolution: Historical and Epistemological Perspectives,* P.J. Beurton, R. Falk, H-J. Rheinberger, eds. Cambridge, UK: Cambridge University Press, 2000.

Harrington, A., *Reenchanted Science.* Princeton: Princeton University Press, 1996.

Hartl D., Orel V. What did Gregor Mendel think he discovered? Genetics 1992; 131:245-53

Jennings, H.S., *Life and Death, Heredity and Evolution in Unicellular Organisms.* Boston: Gorham Press, 1920.

Johannsen, W., *Elemente der exakten Erblichkeitslehre.* Jena: Gustav Fischer, 1909.

Johannsen W. The genotype conception of heredity. American Naturalist 1911; 45:129-59

Kay, L., *The Molecular Vision of Life. Caltech, the Rockefeller Foundation, and the Rise of the New Biology.* New York: Oxford University Press, 1993.

Kitcher P. Genes. British Journal for the Philosophy of Science 1982; 33:337-59

Kohler, R.E., *Partners in Science: Foundations and Natural Scientists, 1900-1945.* Chicago: University of Chicago Press, 1991.

Lewontin R.C., Berlan J-P. Technology, research, and the penetration of capital: the case of U.S. agriculture. Monthly Review 1986; 38:21-34

Lewontin, R.C., Levins, R., *The Dialectical Biologist.* Cambridge: Harvard University Press, 1985.

MacCorquodale K., Meehl P.E. On a distinction between hypothetical constructs and intervening variables. Psychological Review 1948; 55:95-107

Marcus, A.I., *Agricultural Science and the Quest for Legitimacy. Farmers, Agricultural Colleges, and Experiment Stations, 1870-1890.* Ames, IA: Iowa State University Press, 1985.

Monaghan F.V., Corcos A.F., Mendel. The empiricist. Journal of Heredity 1985; 76:49-54

Morgan T.H. The relation of genetics to physiology and medicine (Nobel lecture). Scientific Monthly 1935; 41:5-18

Morgan, T.H., Sturtevant, A.H., Muller, H.J., Bridges, C.B., *The Mechanism of Mendelian Inheritance.* New York: Henry Holt and Co., 1915.

Muller H.J. Variation due to change in the individual gene. American Naturalist 1922; 56:32-50

Nelkin, D., Lindee, M.S., *The DNA Mystique.* New York: W.H. Freeman Co., 1995.

Newton, I., *Opticks.* London: Sam Smith and Benjamin Walford, 1704.

Olby R.C. Mendel no Mendelian? History of Science 1979; 17:53-72

Paul, D.B., Kimmelman, B.A. "Mendel in America: Theory and Practice, 1900-1919." In *The American Development of Biology*, R. Rainger, K.R. Benson, J. Maienschein, eds. Philadelphia: University of Pennsylvania Press, 1988.

Portin P. The concept of the gene: short history and present status. Quarterly Review of Biology 1993; 68:173-223

Provine, W.B., *The Origins of Theoretical Population Genetics.* Chicago: University of Chicago Press, 1971.

Punnett, R.C., ed., *Scientific Papers of William Bateson*, Volumes 1-2. Cambridge: Cambridge University Press, 1928.

Rheinberger, H-J. "Gene Concepts: Fragments from the Perspective of Molecular Biology." In *The Concept of the Gene in Development and Evolution: Historical and Epistemological Perspectives,* P.J. Beurton, R. Falk, H-J. Rheinberger, eds. Cambridge, UK: Cambridge University Press, 2000.

Roll-Hansen N. The genotype theory of Wilhelm Johannsen and its relation to plant breeding and the study of evolution. Centaurus 1978; 22:201-35

Rosenberg, C. "Science, Technology and Economic Growth: The Case of the Agricultural Experiment Station Scientist." In *No Other Gods: On Science and American Social Thought,* C. Rosenberg, Baltimore: Johns Hopkins University Press, 1976.

Rossiter, M.W., *The Emergence of Agricultural Science: Justus Liebig and the Americans, 1840- 1880*. New Haven: Yale University Press, 1975.

Sapp, J., *Beyond the Gene: Cytoplasmic Inheritance and the Struggle for Authority in Genetics*. New York: Oxford University Press, 1987.

Sarkar S. From the *Reaktionsnorm* to the adaptive norm: the norm of reaction, 1909-1960. Biology and Philosophy 1999; 14:235-52

Vicedo M. Unit characters, factors, and genes: E.M. East's views on the nature of hereditary units. Unpublished Paper 2000

Weibe, R., *The Search for Order, 1988-1920*. New York: Hill and Wang, 1967.

Weinstein, J., *The Corporate Ideal in the Liberal State, 1900-1918*. Boston: Beacon Press, 1968.

CHAPTER 3

DOUGLAS ALLCHIN

DISSOLVING DOMINANCE

1. INTRODUCTION

The time has come to dissolve the concept of dominance in genetics. The concept is a vestige of history, a frozen accident that may have aided Mendel's important discovery but is hardly essential as a basic principle of genetics (section 2). Moreover, the concept of dominance is ill-framed and often misleading in terms of heredity, natural selection, and molecular and cellular processes (section 3). More direct language is available to refer to the key relevant principles in inheritance and the phenotypic expression of genetic states (section 4).

At first, the concept of dominance seems simple enough: when two different traits are inherited, only one will be expressed—that trait is *dominant*, the other is recessive. But even this simple formulation hides a wealth of implicit assumptions about genotype-phenotype interaction, numbers of available alleles, the typical effects of combining two alleles in diploid organisms, interaction of allelic pairs, definitions of 'similar' alleles, and more. This paper aims to tease apart these conceptual issues and clarify genetics by discussing how to proceed without the concept of dominance and the confusions it frequently generates.

2. WHENCE DOMINANCE?

Today's term 'dominance' originated, of course, in Gregor Mendel's now classic 1865 paper on "Experiments Concerning Plant Hybrids" (1966a). Those who read the original paper over a century later are often impressed with its clarity and modern style, accessible even to high school students. But the conceptual context has changed dramatically since Mendel's time, and contemporary readers often miss

L.S. Parker and R.A. Ankeny (eds.), Mutating Concepts, Evolving Disciplines: Genetics, Medicine and Society, 43-61.

differences in meaning obscured by the use of familiar terms. These differences offer important clues, however, to understanding how the modern concept of dominance emerged, evolved, and has continued to shape our thinking about genetics.

2.1 Mendel's "Discovery"

Mendel introduced the term *dominirende* (translated variously as 'dominating' or 'dominant') to refer to characters "which are transmitted entire, or almost unchanged in the hybridization" of two contrasting parental types (1966a, §4; see also §11). The other traits, of course, he termed *recessive*. While today's popular accounts tend to portray this as a significant and novel claim, Mendel and his contemporaries who conducted breeding experiments would have readily acknowledged that some parental forms are more likely to be found in offspring—a phenomenon they called *prepotency*. Theories at the time, however, often attributed the prepotency to the sex of the parent (i.e., whether the trait was transmitted by the male or female gamete). By doing reciprocal crosses, Mendel was able to underscore the "interesting fact" that "it is immaterial whether the dominant character belongs to the seed plant or to the pollen plant" (§4). He was not wholly novel in this claim or approach. Mendel himself cites work by Carl Friedrich von Gärtner, and there were others earlier in the century (Orel, 1996). In this respect, Mendel's concept of dominant traits would have been important, but hardly revolutionary (and hence not especially noteworthy to his contemporaries). Dominance embodied a familiar notion—familiar even to non-scientists then as much as now—that some specific traits resemble one parent and not the other. By itself, the concept of dominance explained nothing new.

In discussing dominant traits as he did, Mendel thus addressed an existing misconception about parental influence in inheritance. At the same time, however, he provided a foundation for another misconception. That is, he primed a tradition of attributing the appearance of certain traits to the traits themselves. For later interpreters of Mendel, certain traits appeared *because* they were dominant, rather than because of, say, some feature of inheritance, development, or the coupling of traits. The term 'dominant,' originally introduced as a mere descriptive label, became widely regarded as a causal property (precipitating some unexpected consequences and confusions, discussed more fully in section 3 below).

While noting that some traits are dominant, Mendel also noted that other, complementary traits—which he called 'recessive'—"withdraw or entirely disappear in the hybrids, but nevertheless reappear unchanged in their progeny" (§4). The non-dominant traits were not lost by crossbreeding. Rather, they were 'latent.' Later, they reappeared wholly intact. In this case, too, Mendel's results merely illustrated another familiar hereditary phenomenon: the reappearance of ancestral forms, known at the time as *reversion*. By calling such traits 'recessive,' Mendel hardly did more than redescribe a widely-known feature of inheritance in new terms.

Mendel's work was indeed exceptional—though not always in the ways or for reasons most frequently attributed to him. For example, among most biologists now,

Mendel's legacy falls squarely in the abstract principles of inheritance, or genetics. Yet as much as people cite Mendel's original paper, they often overlook the title that reveals Mendel's primary focus: "Experiments Concerning Plant Hybrids." Hybridization was an important field at the time, both for practical breeding purposes and for addressing questions about evolution and the origin of new species. Could hybrids ever breed true, for example? If so, under what conditions? Could they create new species or stable domestic varieties? For those studying hybridization, reversion had been relatively unpredictable and puzzling. Not so for Mendel.

Mendel highlighted the fact that recessive traits not only reappeared (or reverted) in a hybrid's offspring, but reappeared "unchanged," "fully developed," "without any essential alteration," and thus "remain constant in their offspring" (§§4, 5). As if pure, they could once again breed true, even if their hybrid parents did not. Indeed, a dominant trait could also emerge from a hybrid in true-breeding form. The dominant character could have a *"double signification"* (§5), some plants being mixed (hybrids again) and others breeding true (like the original parents). For Mendel, this reappearance from hybrids of true-breeding forms—sometimes recessive, sometimes dominant—was as important as any 'reversion' of the recessive trait. Something allowed both types of traits to be transmitted "unchanged," and for them to reunite on occasions.

But only some offspring were true-breeding. Others behaved like the hybrid parents. Mendel quantified this pattern in the now familiar 1:2:1 (or 2:1:1) ratio, and showed that the pattern repeated itself in successive generations of hybrids (§§5-7). He thereby revealed an unexpected regularity to or 'law' in the development of hybrids (also see Olby, 1997, §III). Several times during his original paper, Mendel repeated his thematic claim in virtually identical phrasing (see Hartl and Orel, 1992):

> ...*it is now clear that the hybrids form seeds having one or the other of the two differentiating characters, and of these one-half develop again the hybrid form, while the other half yield plants which remain constant and receive the dominant or the recessive characters in equal numbers.* (Mendel, 1966a, §6, original italicized; also see §§7, 8, 9)

That is, hybrids produced equal numbers of hybrid and true-breeding offspring; of the true-breeding forms, half showed the dominant trait and half the recessive. Mendel further elaborated the ratios mathematically in terms of a "developmental series" (based on the binomial expansion). Most important, the "foundation and explanation" of this pattern was the formation of *different* gametes, each representing one of the two "pure" characters originally brought together in the hybrid (hence, Aa x Aa ==> A + 2Aa + a). Genes, we say now, segregate and recombine without losing their integrity. That was Mendel's significant insight, not dominance.

Mendel's conception of the 'laws' or mathematical rules of the development of hybrids relied very much on thinking about pairs of different gametes (egg and pollen) and pairs of distinct traits. He thought "in twos" and in combinations of twos. Mendel's insight was thus intimately linked to his choice of dichotomous traits—those that can be designated as either dominant or recessive. Mendel was aware of another common conception of the era: *blending inheritance*. According to this notion, traits mixed (or 'blended'), producing intermediate forms while becoming inseparable in later generations. For Mendel to explain his results, traits could not blend, become impure, or lose their discrete integrity in hybrids. After all, recessive as well as dominant traits were able to reappear in true-breeding forms. Hence, Mendel emphasized that no intermediate forms (which might indicate blending) occurred. The dominant characters "in themselves constitute the characters of the hybrids," he said, with no ostensible contribution from the recessive characters which, though present, are "latent" or "withdraw." The recessive characters do not just partially disappear; they "entirely disappear" (§4). For Mendel, as for others to follow, dominant characters wholly eclipse the corresponding recessive characters. *"Transitional forms were not observed in any experiment,"* he stressed (§5). For Mendel, the discrete distinction between dominant and recessive traits corresponded to the purity of each trait through the various processes of hybridization, gamete formation, fertilization, and development.[1]

In retrospect, we can easily see that Mendel confused genotype and phenotype (a distinction that emerged only much later). Here, he seems to have assumed that any phenotypic combination of traits in intermediates also reflected an irreversible mixture of 'traits' genotypically. Mendel's conclusions about the segregation and recombination of genetic material or genes (in today's terms) still hold, however, even if there is no sharp dichotomy of dominant and recessive traits phenotypically (for example, in cases of 'incomplete dominance' and 'codominance,' reviewed below). Still, one can appreciate how Mendel's own reasoning and original conclusions were likely facilitated by (if not wholly dependent upon) the concept of dominance, with strictly dichotomous traits. In this case, a false model may have been integral to, or even essential for, Mendel's discovery (see Wimsatt, 1987). For us, over a century later, the principles of dominance and segregation are clearly independent. Understanding how they were once closely coupled historically, however, allows us to perceive more clearly how we might abandon the former without disturbing the latter. We need not embrace a mere contingency of history.

Mendel himself certainly recognized that not all traits are expressed in dominant and recessive pairs. Indeed, Mendel essentially admitted that dominance was not the exclusive norm. Even before introducing dominant traits he noted, for example, that "with some of the more striking characters, those, for instance, which relate to the form and size of the leaves, the pubescence of the several parts, etc., the intermediate, indeed, is nearly always to be seen" (§4). Later he commented: "as regards flowering time of the hybrids...the time stands almost exactly between those of the seed and pollen parents" (§8). Mendel certainly saw in the years immediately following his work on peas that his results on dominance in *Pisum* did not generalize to *Hieracium*, or hawkweed (Mendel, 1966b). For Mendel this merely meant that his law of hybrid development applied only to "those differentiating characters, which admit of easy and certain recognition" (§8). Other characters followed another, different rule or law. Dominance, even for Mendel, had a limited domain.

2.2 Mendel's Legacy

Mendel's work became a guide, of course—almost a touchstone—for the pioneers of the new science of genetics at the turn of the century. However, the particular concept of dominance was not uniformly endorsed. Indeed, the scientific reception of this element of Mendel's work in the early 1900s illustrates that its status was never secure. William Bateson, for example, was perhaps the strongest advocate of the new Mendelism among English-speaking researchers. He found Mendel's quantitative style consonant with his own, hailed the recombination of pure Mendelian units as an explanation for both heredity and the source of variation in evolution, and thus boasted that genetics had discovered the fundamental biological units and rules of combination akin to chemical stoichiometry (see Olby, 1997, §VI). At the same time, Bateson demurred, even at the outset, from accepting any principle or law of dominance:

> In the *Pisum* cases the heterozygote normally exhibits only one of the allelomorphs [alternative phenotypic forms] clearly, which is therefore called the dominant. It is, however, clear from what we know of cross-breeding that such exclusive exhibition of one allelomorph in its totality is by no means a universal phenomenon. Even in the pea it is not the case that the heterozygote always shows the dominant allelomorph as clearly and in the same intensity as the pure dominant... (Bateson, 1902, p. 129)

Bateson's own work on inheritance in poultry showed that traits 'mixed' in hybrids, though the traits still segregated neatly in offspring according to Mendel's model. "The degree of blending in the heterozygotes," Bateson declared, "has nothing to do with the purity of the gametes" (1902, p. 152). (Again, intermediate forms were possible, denoted here by Bateson—erroneously—as a form of 'blending inheritance.') Bateson's example of Andalusian fowl—blue-grey hybrids of black and white parents that formed a 1:2:1 ratio in the F_2 generation—soon became a classic case, cited in textbooks throughout the century (e.g., see Russell, 1992, p. 98, and below).

Others objected to dominance as a universal feature of inheritance. Case after case of intermediate form was cited. In contrast to Mendel's work on color in seed pods, seed endosperm, and unripe pods, for instance, hybrids of red and white four o'clock flowers were neither red nor white, as predicted by dominance, but pink. For most informed breeders and geneticists, characters that differentiated into only two forms, such as Mendel's tall/dwarf plant height or green/yellow seed color, were relatively rare. Thus they did *not* form a secure model for interpreting heredity generally. Indeed, the lack of the universality of dominance was perhaps the single most cited reason for rejecting Mendelism outright (see also note 1). Thomas Hunt Morgan and his students summarized the prevailing view by 1915:

> Whether a character is completely dominant or not appears to be a matter of no special significance. In fact, the failure of many characters to show complete dominance raises

a doubt as to whether there is such a condition as complete dominance. (Morgan *et al.*, 1915, p. 31)

By 1926 Morgan had abandoned any special reference to dominance. In his landmark and synoptic *Theory of the Gene*, which summarized over two decades of findings in classical genetics, 'dominance' failed to appear in the table of contents, in the index, or even as part of Morgan's formal statement of the theory of the gene (Darden, 1991, p. 72).

In carrying forward the legacy of Mendelism, textbooks in the ensuing decades and throughout the century have continued to reflect the ambivalence toward dominance as a universal 'law' or basic model. For example, an early text by R.H. Lock (1906) states that dominance is not universal (Darden, 1991, p. 72). Likewise, a 1921 text lauds Mendel's landmark discovery of dominance, then adds ironically, "of course breeding is not so simple as this, and some characteristics do blend or average in the hybrids" (Moon, 1921, p. 543). A 1933 zoology text, too, follows its description of dominance with a cautionary note: "dominance and recessiveness do not, however, characterize all cases of inheritance" (Curtis and Gurthrie, 1933, p. 184), and then introduces the examples of Andalusian fowl and pink four o'clock flowers. In 1969, we find another text carefully detailing "Mendel's law of dominance," then citing the very same two examples, noting that:

Since Mendel's time, we have found that the law of dominance does not always hold...It is clear that we cannot speak of a 'law' of dominance even though dominance occurs frequently. (Kroeber, Wolff, and Weaver, 1969, p. 412)

Could more equivocation be found? Dominance is both a law and not a law. By the 1990s dominance and recessiveness had retreated to the status of a 'feature' in one standard genetics text (Russell, 1992, p. 41).

Despite the ambivalence, dominance continues to be preserved as an essential or core feature of genetics, consistently introduced before it is dismissed or qualified by any exceptions. Why? Why has dominance persisted as a standard or model, even if in disrepute? Whereas Mendel associated dominance with segregation, we now associate dominance with Mendel himself, as a scientist of mythic proportion (e.g., see Brannigan, 1981; Sapp, 1990). Nearly every introductory biology textbook introduces Gregor Mendel with a picture and supplemental comments. They implicitly portray him as an exemplary scientist. He worked alone in an Austrian monastery: scientists modestly seek the truth; they do not ambitiously pursue fame or wealth. Mendel used peas: scientists choose "the right organism for the job" (see Clarke and Fujimura, 1992; Burian, 1993). He counted his peas: scientists are quantitative. He counted his peas for many generations over many years: scientists are patient. He counted thousands and thousands of peas: scientists are hard-working. After all this, Mendel was unfairly neglected by his peers, who failed to appreciate the significance of his work, but was later and justly 'rediscovered': ultimately, scientific truth triumphs over social prejudice. Above all, Mendel was right. By all these measures, Mendel is a model scientist, a biological hero to parade before students. How could we admit that Mendel erred (good scientists don't make mistakes)? Because dominance was part of Mendel's original scheme and, at the same time, we honor Mendel almost religiously, we do not exclude dominance from basic genetics. Dominance has become entrenched in the romantic lore of Mendel.

The acceptance of dominance as a model has not been without consequences, however. Most notably, the many 'exceptions' that emerge by regarding dominance as a norm have led to a proliferation of otherwise needless concepts. That is, textbooks typically begin genetics with the eponymous 'Mendelian' genetics. But then they proceed to note several 'exceptions' or qualifications. For example, Bateson's Andalusian fowl and pink four o'clocks exemplify *incomplete dominance*. As noted above, the basic Mendelian pattern of segregation and recombination still occurs, but with no eclipsing 'dominance'; rather, the hybrid phenotype is intermediate. Texts also commonly distinguish *codominance*, exemplified by blood type, where both alleles contribute concretely to a 'compound' phenotype. The dominance model further implies antagonistic or complementary pairs, so *multiple alleles* must also be mentioned (also illustrated by blood type). And because dominance is presented as absolute (and as a sufficient cause), any occasion when the 'dominant' trait does not appear in all individuals with the allele, or does not appear to the full extent, also requires special note—hence, *penetrance* and *expressivity* (e.g., Russell, 1992, pp. 54, 112-4). When one refrains from recognizing dominance as a prior model, however, all these concepts—incomplete dominance, codominance, multiple alleles, expressivity, and penetrance—become superfluous. Because these concepts have populated the standard repertoire and vocabulary for so long, though, we can easily fail to notice the alternatives. Yet the 'exceptions' dissolve conveniently when one removes dominance as a faulty standard.

It seems that dominance must be 'basic' by some standard. But is it? It is certainly not foundational, in the sense of being simplest or making the fewest assumptions. The most basic assumption would be that each allele is expressed; hence, if two alleles are present, both are expressed. Dominance requires an *additional* assumption about the relationship between two alleles. By contrast, one can describe all the 'exceptions' more simply and uniformly by (a) knowing that diploid organisms have *two* alleles, and (b) noting the characteristic expression of each allele, even if it is a 'truncated' version of another trait (see section 4 below). Dominance is not a model by virtue of simplicity.

Is dominance 'basic,' then, in the sense of being most prevalent? Or might the 'exceptions' even outnumber the 'model?' As early as 1907 Morgan quoted C.C. Hurst as saying that incomplete dominance is twice as frequent·as complete dominance (Darden, 1991, p. 68). A more recent estimate also suggests that fewer than one-third of human clinical genetic conditions follow the dominant-recessive rule (Rodgers, 1991, p. 3). Has there ever been a systematic study documenting this other 'Mendelian' ratio? An indirect measure might be the scarcity of good 'textbook examples' of complete dominance in humans. To illustrate Mendelian traits, we often appeal to attached earlobe, hitchhiker's thumb, short little finger, widow's peak, woolly hair, crumbly earwax, tongue-curling, or PTC-tasting. These are trivial. They hardly reflect important dimensions of human genetics. Nearly all interesting or significant cases have more complex stories (see section 4 for example). Dominance is not a model by virtue of frequency, either.

Viewed retrospectively, then, the concept of dominance is not essential. It was first coupled with other basic patterns of inheritance by Mendel, who likely found it integral to inferring segregation. Since Mendel, authors seem to have been unwilling to challenge Mendel's conceptual precedent, though all cite problems or exceptions. Winterton C. Curtis and Mary C. Guthrie summarized a prospective view well in their 1933 text: "The course of inheritance for characteristics that do not exhibit dominance, therefore, is in no way different from that for characteristics in which dominance occurs" (p. 185). Dominance can be abandoned without loss. Though historically dominance has been intimately linked with Mendelism and the rules of heredity, and perpetuated for this reason, it is not essential to understanding basic genetics (see also Falk 2001).

3. MISCONCEIVING AND MISFRAMING DOMINANCE

What, indeed, does dominance mean? Is it fundamentally a noun, an adjective, or a verb? Is it descriptive or explanatory? Does it refer to the phenotype or the genotype, inheritance patterns or mechanisms of genetic expression? Is it a 'law,' a 'principle,' a 'feature,' or something else?

On such questions the tendency is to refer to Mendel himself as an authority although, as noted above, others have shaped and reshaped our concepts of 'Mendelian' genetics. Unfortunately, Mendel never clearly characterized 'dominant' and 'recessive' as concepts. Rather, he used them as *labels*, identifying certain sets of heritable characters in contrast to one another. Even modern textbooks find themselves in similar situations and typically introduce the terms, not in clear statements, but by ostension or exemplification (using such conditional phrases as "when traits combine in hybrids..."; e.g., see section 1 above). Moreover, Mendel defined these terms from observable behavior. Since then, we have inverted the meaning, such that we now attribute the observed phenotype to dominance: a trait is dominant when we observe it in the hybrid, and the hybrid exhibits this trait because it is dominant. The characterization of dominance is circumspect in its current circularity.

Curiously, perhaps, Mendel only used the adjectival form, *dominirende*. He never used a noun or verb equivalent. That is, he never described a general principle or relationship between two characters as 'dominance,' nor referred to one trait as 'dominating' another. Rather, he merely sorted characters into two categories based on the visible traits of hybrid offspring. This descriptive modesty contrasts with later usage, which commonly characterizes dominance as a principle or even a law (a 'Mendelian' law, no less!), that is, as something more than a convention of nomenclature. The linguistic change marks an important conceptual shift. Dominance has been subtly *reified* into a concrete property that can be causal and explanatory, not merely descriptive.

In addition, it is rarely clear whether dominance refers to the phenotypic trait or the genetic allele associated with it, or both. Mendel labeled only the trait, or visible character. Whether he intended dominance to refer to any abstract underlying gene or 'element' is unclear; his language and notation are certainly ambiguous (and inspire contentious debate among historians!). Nowadays, over a century after Mendel, the referents for dominance are slippery. Sometimes, one *trait* is dominant. At other times, it is the dominant *allele*, or *gene*. The second type of reference reinforces the notion that dominance is a property of the individual allele, not of a larger context (see below). No one considers a phenotype to be causally important genetically, for example; it is the product or effect, not the cause. By contrast, we view genes as causal. Hence, referring to an allele rather than a trait as dominant carries substantially more content semantically.

Mendel also set a precedent with his choice of words. The term *dominirende* is now largely obsolete in German, but in Mendel's time it carried the meaning or connotation (in a modern translation) of 'coming to the fore,' though it was based on the Latin root for 'master' (Charles and Barbara Elerick, personal communication). Similarly, after Mendel first described recessive traits as 'latent,' he noted that he chose the expression 'recessive' because "the characters thereby designated withdraw or entirely disappear" (1966a, §4). Between the paired elements that guide development in their "enforced union" in hybrids, "some sort of compromise is effected" (§11). Whether deliberately or not, Mendel cast the relationship between dominant and recessive characters in terms that could easily be interpreted in terms of power and forces.

All the initial translations of Mendel early in the next century preserved Mendel's original root: *dominant* in William Bateson's English; *dominirt* in Carl Correns' German; although there was pointed disagreement between Hugo de Vries, who preferred the French *dominant*, and Lucien Cuénot, who considered *dominé* more appropriate. They carried forward Mendel's pregnant images, while at the same time changes in the vernacular meaning of 'dominant' only amplified the connotations of power. The interpreted meaning became explicit as Mendelism entered textbooks. A popular 1921 text, for example, calls the dominant traits "stronger," though the recessive traits eventually "overcome" this (Moon, 1921, p. 543). A 1933 text likewise calls the recessive trait "obscured or suppressed" (Curtis and Guthrie, 1933, p. 183). Still, in 1969, the dominant trait "dominates or hides" the recessive (Kroeber, Wolff, and Weaver, 1969, p. 412). Even a late twentieth-century genetics text follows the pattern, describing the "missing" recessive trait as "masked by the visible trait" (Russell, 1992, p. 41). Throughout its history, the meaning of the term 'dominant' (or 'dominance') in genetics has resonated with the term's vernacular meaning.

As a result of the continued use of the term 'dominance,' misconceptions about inheritance abound. They are more typically found among students and non-professionals. Still, the confusions can confound efforts in genetic counseling and social policy decisions, now made more urgent by the Human Genome Initiative. Some conceptions that teachers have confronted regularly include (these appeared widely, for example, on the national college-level Advanced Placement Biology essay exam recently; also see Donovan, 1997):

- Dominant traits are 'stronger' and 'overpower' the recessive trait.
- Dominant traits are more likely to be inherited.
- Dominant traits are more 'fit,' or more adaptive in terms of natural selection. (Also, any recessive adaptive mutant trait will eventually evolve to become dominant.)
- Dominant traits are more prevalent in the population.
- Dominant traits are 'better.'
- 'Wild-type' or 'natural' traits are dominant, whereas mutants are recessive.
- Male or masculine traits are 'dominant.'

None of these claims is necessarily true. Some are false. All are misleading. One might wonder, therefore, why students of genetics (and, in some cases, prominent biologists historically) so readily and commonly assume their validity. Moreover, these preconceptions and images are notoriously resilient—difficult for instructors to rectify even when they note and address them explicitly. Notice, though, how the vernacular meaning of dominance, where one thing 'dominates' another, percolates through every misconception.

Clarifying the meaning of dominance in genetics is further frustrated by other uses of the same term within biology. For example, in ethology, 'dominance' describes political hierarchies among social organisms, just as the vernacular sense of the term suggests. The untutored person imagines 'dominant' genes to 'behave' the same way: they 'dominate' over other alleles or traits through competitive interactions, etc. In ecology, 'dominance' refers instead to the relative biomass of one species in a particular ecosystem: the 'dominant' species is typically viewed as the most influential or important because of its sheer bulk (a combination of size and frequency). Biologically, dominance seems to be a technical measure of influence or power, as one might expect from the term's common usage.

In *Metaphors We Live By* (1980), George Lakoff and Mark Johnson describe how our thinking is shaped by the words we choose and their meanings in other contexts. 'Dominance,' though it may be well *defined* as a term strictly within the field of genetics, unavoidably carries with it the meanings or connotations from other biological as well as non-biological contexts. Hence, a dominant trait is expressed, not 'in lieu of,' but '*over*' the recessive. Likewise, the concept of *predominant* can easily shape expectations about the frequency of *dominant* alleles (or traits) in a population or subsequent generation, for instance. No wonder people can mislead themselves. They reason about the prevalence of alleles, the interaction of genes, heritability, reproductive fitness, normality, and gender using the language available. In this case, using the single term 'dominance' primes the multiple misconceptions enumerated above.

The resonant meanings of dominance are significant in part because the conceptual space is open for them to fill. What *explains* dominance? No clear single molecular or cellular mechanism describes why or how one trait is dominant while another is recessive (Wilkie 1994). No proper explanatory concept can eclipse misconceptions before they develop or replace them afterward.

Many conceive or explain dominance as a form of gene regulation. They suppose that the dominant allele somehow *inhibits* the expression of the recessive allele (e.g., see the popular textbook by Benjamin Lewin, 1997, p. 62). It must produce or induce a repressor protein, say, that *actively prevents* the transcription of the recessive allele's DNA on the homologous chromosome. At the same time, the recessive trait 'withdraws' or is 'latent,' virtually powerless to express itself. Here, the dominance metaphor appears with a vengeance, though filtered through standard biological concepts. One presumes that the dominant allele has some ability to 'shut off' or 'dominate' the recessive allele. While such genetic regulation is conceivable, however, no such direct interaction or suppression is yet known to occur. The common conception of dominance as gene regulation is false, though obviously fueled by the term 'dominance.'

Others conceive or explain dominance as the presence or absence of a trait. That is, the recessive trait is not due to the *presence* of a specific recessive allele or protein but is due instead to the *absence* of a functional dominant allele. This interpretation has a rich history, with roots extending back to a popular 1905 proposal by William Bateson and R.C. Punnett (Darden, 1991, pp. 69-71). Variations persist today (e.g., Lewin, 1997, p. 62). At the level of alleles, this concept is patently absurd. Both alleles are inherited; both are present. The deficit must therefore appear in gene expression: recessives fail to produce a functional protein, or to produce any polypeptide whatsoever (e.g., as illustrated in Lewin's text). In severe cases, the absence might even be lethal (e.g., Russell, 1992, pp. 110-2). But how the absence occurs is typically unexplained. While a change in the function of protein is important, the presence-absence interpretation implies, again, that the recessive allele or gene produces *no* protein and hence contributes nothing to the character of the individual—and thus can be safely dismissed (again, echoing Mendel's notion of 'eclipsing' traits). Again, this is overstated. For example, in cases of Tay-Sachs, cystic fibrosis, and other 'recessive' disorders, geneticists can detect heterozygote carriers specifically because their 'recessive' allele produces a detectable alternative protein. A recessive protein *is* present. In an evolutionary context, the presence-absence conception is flawed because it implies that all changes in genes are losses of function; hence, adaptation and evolution itself would seem impossible. The presence-absence assumption is further challenged by contrary cases where the 'presence' of a specific protein can be dysfunctional, making the normal function oddly due to an 'absence.' Sickle cell anemia, for example, is not merely the absence of hemoglobin. Rather, there is a variant hemoglobin with its own distinctive phenotype, including the transport of oxygen but also the characteristic sickling of cells that blocks capillary circulation. In other cases, such as 'dominant negative mutants,' an ostensibly functional protein is 'present' in the heterozygote, but the variant allele subverts its normal function (e.g., see discussion of osteogenesis imperfecta, section 4 below; many similar cases were debated early in the century as well; see Darden, 1991, p. 70). In these instances, the *absence* of a particular trait is functional. Ultimately, the presence-absence hypothesis cannot explain dominance fully because it does not address the very details of why or how function is lost or changes. A robust interpretation of dominance must explain generally (1) why traits (or alleles) differ in expression, and (2) how those traits overlap when coupled in diploid organisms.

Ideally (as suggested by these two prevalent misconceptions), we might want to explain dominance at the molecular level through some single, well-defined mechanism. We want all dominant traits to be dominant for the same reason. However, genotype is linked to phenotype at several stages or layers of expression, from the levels of biochemistry and the cell to the levels of physiology and social behavior. The familiar textbook notion of 'one gene, one protein' and the processes of transcription and translation only begin to characterize the multi-layered process of gene expression. In a sense, we must rethink our notion of phenotype to include 'traits' from all levels, from biochemistry and the cell to the organism.

Consider, for example, two renowned traits from classical genetics: Mendel's wrinkled peas and Morgan's white-eyed fruitflies (Guilfoile, 1997). The wrinkled (versus smooth) trait in pea seeds has now been isolated to a transposon in the exon of a gene for a starch-branching enzyme (SBE1). In one form ('smooth'), the protein acts enzymatically to convert amylose to amylopectin. As a result, starch accumulates in the developing seed. In another genetic form ('wrinkled'), the gene is presumably transcribed, but it is either not translated (due to the sizable DNA insertion) or the resultant protein does not fold into a similar shape. Consequently, the same reaction is not catalyzed. Instead, unpolymerized amylose and sucrose molecules accumulate in the developing seed, which osmotically imbibes considerably more water, producing a temporarily larger seed. When the seed matures and dries, however, the endosperm contracts and the now-enlarged seed coat wrinkles. The appearance of wrinkled seed is thus due to the cascading downstream developmental effects of a protein without specific catalytic activity. It is a trait with many components, including relative sugar and starch composition as much as visual appearance. But, of course, this is only half of the story. What happens when the two allelic variants appear together in a hybrid? Apparently, one gene of the first ('smooth') type alone can produce enough of the enzyme to convert the sugars fully into starch. Hence, 'smooth-seed' appears dominant, while 'wrinkled-seed' appears recessive. However, there is no 'dominating' influence between alleles, which are each expressed independently. We can expect hybrids to express the wrinkled-seed allele and to contain the wrinkled-seed protein even though they appear smooth. The key information is that one allele alone in this pair is sufficient physiologically for promoting the reactions associated with smooth, starchy seeds. The dominant/recessive label adds nothing to this simplified explanation. 'Dominance' as a distinct property is an artifact.

The case of eye color pigments in fruitflies reveals similar complexities and redundancies (Guilfoile, 1997, pp. 93-4). Pigment development relies on precursors that are transported into the eye cells by various membrane proteins. The gene variant associated with white-eye, located on the fruitfly's X chromosome, seems central to all such transport systems. The specific variation in DNA sequence has not yet been isolated. Still, one can see that even if the protein is synthesized and becomes embedded in the membrane, it need not be shaped like its counterpart in red-eyed individuals. It does not transport pigment precursors into the cell. (Moreover, in an apparent pleiotropic effect, it may also affect courtship behavior in males when expressed in all cells: we might suspect that the transported elements have adopted more than one role in the cell.) In this case, by following inheritance patterns, one can infer that a single copy of the gene can generate sufficient protein for membrane transport. Again, in hybrids, both alleles are expressed. What matters is what each allele leads to independently. How does labeling 'red-eye' as dominant and 'white-eye' as recessive contribute to this explanation?

The two cases of 'wrinkled-seed' in peas and 'red-eye' in fruitflies illustrate that dominance is not a property of the 'dominant' allele. Nor is it a special interaction between two alleles. Rather, what we call 'dominance' is a contingent, emergent feature based on the particular pair of alleles and the expression of each. To understand inheritance patterns fully, then, one must appreciate how the expression of genetic variants can diverge. The two classic cases above begin to exemplify how to conceive genetic expression at many levels. Additional cases from human genetics can help further articulate, at least broadly, the many levels (Table 1).

'Level' of Genetic Expression (Phenotype)	Examples
DNA Transcription	elongated mRNA/blood type A2 [110300.0003] multiple repeats/ Huntington disease
mRNA Editing & Translation	new exon arrangements of mRNA/ alcaptonuria #4 premature translational stop/ phenylketonuria #1
Protein Structure	keratin / skin conditions collagen / osteogenesis imperfecta dystrophin / Duchenne muscular dystrophy
Protein Function	hemoglobin/sickle cell anemia, thalassemia blood clotting Factor VIII / hemophilia insulin or insulin receptor / diabetes
Enzymatic Reactions	lactase / lactose intolerance fructose-1,6-diphosphatase/ hypoglycemia

TABLE 1. LEVELS OF FUNCTION IN THE GENETIC EXPRESSION OF A TRAIT.

'Dominance' appears due to the differential expression of coupled alleles, often with observable differences traceable to certain levels of expression. Here, various human genetic traits are identified with the level, or type, of expression where function diverges.

A phenotypic 'trait' will depend on the nature of the DNA sequence and the role of the protein physiologically or developmentally. Thus, genetic variants may and may not be transcribed, may or may not both produce proteins. They may or may not have the same three-dimensional configuration, and may or may not fit with other protein units in multimers. They may or may not have similar catalytic activity (enzymes), signaling properties (hormones), or activation potential (neurotransmitters, membrane receptors, etc.). Even subtle variations in enzymatic activity may lead to different reaction rates or differing amounts of an enzymatic product. These variations, in turn, may result in further physiological effects, developmental responses, or behavioral differences (e.g., if they are affecting certain types of cells within the nervous system). The expression of variant alleles may potentially vary or diverge at any of these levels. In a sense, every allele is expressed phenotypically. The question is how this expression occurs, and what are the various downstream consequences? Dominance implies that one of the two chains of expression is completely suppressed, with no consequences for the organism. But there is no single property that leads to one gene being expressed in contrast to its homolog. All hybrid phenotypes are compound traits. The key is interpreting the dual contributions simultaneously at the many levels of genetic expression. Hence, it should surprise no one that cases coded as 'complete' dominance are neither in the majority, nor representative, nor fundamental.

The second major aspect of interpreting traits labeled as dominant and recessive is to understand the *coupled* expression of *pairs* of alleles in *diploid* organisms. Genes are not just expressed. They are expressed in pairs. According to the conventional view of dominance, the number of copies of a gene should not really matter: 'dominant' homozygotes and heterozygotes should be identical phenotypically. However, some 'dominant' traits express themselves differently when they occur in single versus double copies (work by Bruce Cattantach and M. Kirk, cited in Rodgers, 1991, p. 5). More familiar may be the dramatic phenotypic effects of a third copy of a chromosome, most notably trisomy 21, or Down syndrome. Less well known may be the fact that the sex of the parent contributing the third chromosome in these cases can affect the phenotype (Rodgers, 1991, p. 5). There are other cases of aneuploidy, as well, including, of course, the sex chromosomes (XO, XXY, XYY, etc.). All autosomal monosomics are apparently lethal (Russell, 1992, p. 598), although, again, according to Mendelian frameworks, the number of alleles should not matter. In fact, a dominant allele is not uniformly expressed; it is not predictably the same in single, double, or triple copies. Dominance does not inhere in an individual allele. And phenotypes are not either-or, based on simple dichotomous features. Rather, traits are based more fundamentally on the alleles that are present (whether one, two, or more) and how each is expressed.

The phenotype of hybrids is compound. Two alleles are expressed. The challenge is to interpret the many possible patterns of double expression, as illustrated below (section 4). 'Dominance' is one possible pattern, seen at a large (organismal) scale. But it is still not caused by any relationship *between* the alleles. Each allele is independent. Once again, there is a tendency to *reify* dominance as a causal property or process, when it is nothing more than an observable pattern. 'Dominance'—or anti-recessiveness—means, modestly, almost just as Mendel stated, that one trait is manifest in a hybrid and, simultaneously, its allele is coupled with another allele whose phenotypic expression is minimal or not visibly significant at the macroscopic level. There are no corresponding molecular overtones. When one properly narrows the definition in this way, the concept loses its current scope, and much of its intended significance.

Misconceptions about dominance persist largely because it is so rarely explained and its domain is not explicitly limited. Explaining the molecular or developmental basis for each trait would contribute to alleviating current misconceptions, yet dissolving the concept of dominance in genetics would help prevent such misconceptions at the outset. Indeed, many professional geneticists are already purging the term from their discourse or simply abandoning it as uninformative. Online Mendelian Inheritance in Man (OMIM), the primary reference for human genetics, for example, discontinued classifying traits as dominant and recessive in 1994. The concept of dominance is obsolete.

4. RECASTING DOMINANCE

Ultimately, dissolving dominance can clarify discussion and interpretations of genetics. For example, our concepts and language should describe both the levels of intergenerational transmission and molecular expression, while facilitating understanding of the relationship between them. In place of the confusions and complexities of dominance and all its exceptions and qualifications, along with the now separate and often inconsistent discussions of its molecular mechanism(s), only two basic elements are needed (as discussed above in section 3). First, what is the expression or developmental meaning of each trait? Second, how does any pair of traits establish a 'joint' phenotype in concert? The remainder is explained by the segregation and recombination of alleles and by the background knowledge of the layers of genetic expression (see Table 1).

Indeed, one can find the prospective model already deployed in the standard conception and discussion of ABO blood types (OMIM #110300). First, many possible alleles are acknowledged: A, B, and O, commonly, and others (including a remarkable form, cis-AB, that shows dual enzymatic function). However, due to the sexual nature of human reproduction, each individual carries just two alleles, one from each parent. The relevant blood-type phenotypes originally became evident from agglutination of blood mixtures, indicating that red blood cells can have specific antigenic properties. The now common and safe use of transfusions between blood types indicates that the O allele does not generate specific antigens, while A and B produce distinct antigens. Hence, an O allele in combination with A or B becomes functionally 'invisible' when considering transfusions, by virtue of producing no antigen. (Note, here, that no one sees the need to call A or B 'dominant' to O.) Nor is there any confusion when, by contrast, A and B alleles combine: each contributes an antigen to the red blood cells. The hybrid is a hybrid.

Furthermore, one can predict possible allelic make-ups from observed blood type, as well as possible phenotypes of the prospective offspring from two parental genotypes. How simple. How clear. (And how free of dominance.)

The discourse of blood types illustrates a model that can be applied uniformly and consistently across all cases. One set of concepts and language can embrace both the limited domain of dominance and its 'exceptions.' The relevant information is always knowing which of many alleles are present and what each means for the individual's physiology or development. Thus, though blood types are typically labeled as a case of 'co-dominance,' the concept of dominance itself is wholly peripheral. Co-dominance dissolves. All phenotypes are dual phenotypes, with each of two genes contributing. Hence, pink flowers and coloration patterns associated with Andalusian fowl, for example, are neither heritable traits on their own (distinct alleles), nor 'half-traits' based on one allele being 'incompletely' expressed. Rather, they result when full saturation of a particular color requires a 'double dose' of a pigment. A single dose (in a hybrid) appears 'diluted' by contrast, but only in relation to the double, more fully saturated color. Incomplete dominance likewise dissolves.

Consider, too, the cases that resonate most closely with the presence-absence hypothesis proposed early in the twentieth century. These are perhaps the best candidates for imagining any residual relevance for dominance (or 'complete dominance'). Alcaptonuria (OMIM #203500) is a dramatic example and noteworthy as one of the earliest recognized inborn errors of metabolism (described as Mendelian by Archibald Garrod in 1902). Alcaptonuria patients pass black urine. They lack a functional enzyme (homogentisate 1,2 dioxygenase, or HGD), one of a series that breaks down phenylalanine into excreted waste. As a result, the compound 'upstream' of the enzyme accumulates, turning black when alkalinized in the urine. Here, one might imagine calling the black-urine alleles 'recessive' and their yellow-urine counterparts 'dominant.' But it is far simpler and more direct to state that any individual has two alleles: when at least one of them produces the enzyme, the urine is yellow; when neither does, the urine is black. Indeed, one need not even refer to the cellular processes at all: one copy of the 'yellow-urine' gene is sufficient to ensure yellow urine. This adequately describes the particular pattern of expression when two alleles combine—and dominance dissolves. All genetic expression—dominant or not—is thereby unified conceptually.

The same simple concepts and direct discourse can apply to cases that otherwise appear awkward or convoluted on the dominance model. Take, for example, osteogenesis imperfecta (OMIM #120160), a condition of extremely fragile bones. Collagen is a triple helix protein, assembled from two procα1(I) chains and one procα2(I) chain. One allele forms an altered procα2(I) chain that does not self-assemble into the helix. In this case, the altered protein chain can bind partially to procα1(I) but subverts the overall structure, even if some functional procα2(I) is present. With insufficient collagen fibers, then, individuals are especially susceptible to bone injury. Here, one copy of the allele can precipitate the condition. In the language of dominance, osteogenesis imperfecta is 'dominant,' by virtue of the trait's expression in hybrids. At the same time, it seems awkward to say that the 'recessive' trait—in this case, the functional trait—has 'withdrawn,' is 'latent,' or is not expressed. After all, the allele *is* expressed and the protein appears in precisely the same form as when fully functional. Here, it is clearer to state that one copy of the variant gene is sufficient to *interfere* with the otherwise standard function (in contrast to other cases where one copy is sufficient to maintain the function). One does not have to invert the meanings implied by first referring to dominance.

Finally, consider the classic case of sickle cell anemia (OMIM #141900.0243). The disease appears in severe form when individuals have no regular hemoglobin (i.e., are homozygous for the 'sickle cell allele'). Thus the disease has conventionally been construed as recessive. However, individuals with one of each allele have both forms of hemoglobin and do exhibit distinct physiological symptoms. The hybrid has a third phenotype, mild hemolytic anemia. The condition is also noted for a pleiotropic effect, resistance to malaria, associated with the 'sickle cell allele.' Even one copy of this gene can provide benefit. These complexities wreak havoc with assignments of dominance. Hence, if the trait is a life-threatening disease, the 'non-sickle hemoglobin' allele is dominant. If the trait is physiology, then perhaps it is incompletely dominant. If the trait is malaria resistance, then the same gene is recessive. The same allele can exhibit three forms of dominance, depending on how one delineates the character. How paradoxical. But, of course, nothing important hangs on the label. All of the essential information can be conveyed by noting (a) the properties of expressing each allele, and (b) which alleles are paired. In this case, there are three distinct phenotypes based on combining two specific hemoglobins, with each allele contributing something significant. Once again, the current discourse on blood types provides a fully functional and unifying model.

Ultimately, once one understands molecular genetics, the whole concept and language of dominance unravels, even at the level of transmission (or classical) genetics. The dominant-recessive concept depends on being able to sort traits neatly as expressed or unexpressed. Phenotypes must be essentially dichotomous, at least at the macroscopic, or observable, level. Fuller familiarity with molecular biology, however, reveals that phenotype exists at all levels simultaneously (see Table 1). Traits that seem unexpressed at one level may certainly be expressed at another. Indeed, as noted above, focusing on 'traits' only at the organismal level can hide or obscure important phenotypic differences at the molecular or cellular level. Any distinction that privileges one level of expression as *the* phenotype is arbitrary and potentially misleading. The Mendelian who hopes to characterize a trait as 'dominant' because of observed differences in the organism thereby risks mischaracterizing the trait. Hence, even without detailing the molecular story of each gene, one needs to ensure that the multiple levels of interpretation will at least be commensurable. Dominance fails, even for classical genetics, because it draws an artificial or overstated boundary between the expressed and the unexpressed. Inferences about suppression or latency of traits (dominance or recessiveness), in view of molecular understanding, are false. Dominance is an artifact of interpreting phenotype only macroscopically. Coupled alleles are each expressed independently. Our language can—and should—reflect that.

Dominance does not mark any important property beyond the trait itself and how it is expressed. We can abandon the concept without loss, while preserving the other basic principles of inheritance that Mendel noted. Again, one can echo Curtis and Guthrie in their 1933 text: "The course of inheritance for characteristics that do not exhibit dominance, therefore, is in no way different from that for characteristics in which dominance occurs" (p. 185). One merely needs to keep in mind the coupled alleles and the dual phenotype. Dominance is superfluous. It fosters misconceptions. Simple alternative language is available. We are ready to dissolve dominance.

University of Minnesota
Minneapolis, Minnesota, U.S.A.

ACKNOWLEDGMENTS

My appreciation to John Beatty, Louise Paquin, and Barbara and Charles Elerick for sharing their expertise. Special thanks to Bob Olby for encouragement and for guiding me to several valuable historical references. Work was supported in part by NSF Grant #DUE-9453612.

NOTES

1. Distinguished English geneticist William Bateson echoed Mendel's misconceptions in his own conceptual development in 1902. At first Bateson praised Mendelism as an explanation only for discontinuous, or non-blending, variation. He, too, interpreted the purity of the genes as requiring the dominant-recessive relationship. Later, he was able to conceive of intermediate forms (expressed in heterozygotes) as contributing to continuous variation, especially where there might be many alleles that could form a series of phenotypic forms (Olby, 1987, pp. 414-5, 417).

REFERENCES

Bateson, W. "The Facts of Heredity in the Light of Mendel's Discovery." In *Reports to the Evolution Committee of the Royal Society*, London (I), W. Bateson, E.R.Saunders, eds. London: Harrison and Sons, 1902.
Brannigan, A. "The Law Valid for *Pisum* and the Reification of Mendel." In *The Social Basis of Scientific Discoveries*. Cambridge: Cambridge University Press, 1981.
Burian R.M. How the choice of experimental organism matters: epistemological reflections on an aspect of biological practice. Journal of the History of Biology 1993; 26:351-67
Clarke, A.E., Fujimura, J., eds., *The Right Tools for the Job: At Work in Twentieth-Century Life Sciences*. Princeton: Princeton University Press, 1992.
Curtis, W.C., Guthrie, M.J., *Textbook of General Zoology*, 2nd ed. New York: John Wiley & Sons, 1933.
Darden, L., *Theory Change in Science: Strategies from Mendelian Genetics*. Oxford: Oxford University Press, 1991.
Donovan M.P. The vocabulary of biology and the problem of semantics. Journal of College Science Teaching 1997; 26:381-2
Falk R. The rise and fall of dominance. Biology and Philosophy 2001; 16:285-323
Guilfoile P. Wrinkled peas and white-eyed fruit flies: the molecular basis of two classical genetic traits. American Biology Teacher 1997; 59:92-5
Hartl D.L., Orel V. What did Gregor Mendel think he discovered? Genetics 1992; 131:245-53

Kroeber, E., Wolff, W.H., Weaver, R.L., *Biology*, 2nd ed. Lexington, MA: D.C. Heath and Co., 1969.

Lakoff, G., Johnson, M., *Metaphors We Live By*. Chicago: University of Chicago Press, 1980.

Lewin, B., *Genes VI*. Oxford: Oxford University Press, 1997.

Lock, R.H., *Recent Progress in the Study of Variation, Heredity, and Evolution*. New York: Dutton, 1906.

Mendel, G. "Versuche über Pflanzenhybriden [Experiments Concerning Plant Hybrids]," 1866. Reprinted in Fundamanta Genetica, J. Kíeneck, ed. Prague: Anthropological Publications, Oosterhout; Moravian Museum, Brno; and Czech Academy of Sciences, 1966a. Translation reprinted in *The Origin of Genetics: A Mendel Source Book*, C. Stern, E. Sherwood, eds. San Francisco: W.H. Freeman, 1966. Translation by C.T. Druery and W. Bateson. Available at: MendelWeb, www.netspace.org/MendelWeb.

Mendel, G. "On Hieracium-Hybrids Obtained by Artificial Fertilization," 1869. Translation reprinted in *The Origin of Genetics: A Mendel Source Book*, C. Stern, E. Sherwood, eds. San Francisco: W.H. Freeman, 1966.

Moon, T.J., *Biology for Beginners*. New York: Henry Holt, 1921.

Morgan, T.H., Sturtevant, A.H., Muller, H.J., Bridges, C.B., *The Mechanism of Mendelian Heredity*. New York: Henry Holt, 1915.

Olby R.C. William Bateson's introduction of Mendelism to England: a reassessment. British Journal for the History of Science 1987; 20:399-420

Olby R.C. Mendel, Mendelism and genetics, 1997. Available at: MendelWeb, www.netspace.org/MendelWeb/.

OMIM™. Center for Medical Genetics, Johns Hopkins University, Baltimore, MD, and National Center for Biotechnology Information. Bethesda, MD: National Library of Medicine, 1997. Available at: www.ncbi.nlm.nih.gov/omim/.

Orel, V. "Heredity before Mendel." In *Gregor Mendel: The First Geneticist*. Translation by S. Finn. Oxford: Oxford University Press, 1996.

Rodgers J. *Mechanisms Mendel never knew*. Mosaic 1991; 22:2-11

Russell, P.J., *Genetics*, 3rd ed. New York: Harper Collins, 1992.

Sapp, J. "The Nine Lives of Gregor Mendel." In *Experimental Inquiries*, H.E. Le Grand, ed. Dordrecht: Kluwer Academic Press, 1990.

Wilkie A.O.M. The molecular basis of genetic dominance. Journal of Medical Genetics 1994; 31:89-98

Wimsatt, W. "False Models as a Means to Truer Theories." In *Neutral Models in Biology*, M. Nitecki, A. Hoffman, eds. Oxford: Oxford University Press, 1987.

CHAPTER 4

MANFRED D. LAUBICHLER AND SAHOTRA SARKAR

FLIES, GENES, AND BRAINS: OSKAR VOGT, NIKOLAI TIMOFÉEFF-RESSOVSKY, AND THE ORIGIN OF THE CONCEPTS OF PENETRANCE AND EXPRESSIVITY

1. INTRODUCTION

The concepts of 'penetrance' and 'expressivity' are widely used in medical genetics today. Even a brief search of Medline will attest to that. Yet despite their popularity in certain types of studies—most commonly ones that try to establish a link between one (or a few) gene(s) and a disease—the origin of the concepts is all but unknown. In this chapter we attempt to trace the intellectual history and some attendant sociological factors surrounding the introduction of those concepts. There are two reasons that motivate this endeavor. First, as historians and philosophers, we find this story fascinating not only because of the peculiar circumstances that marked the introduction of these concepts but also because of the philosophical and historiographical issues it raises. The introduction of 'penetrance' and 'expressivity' was marked by an unlikely conjuncture of research programs (human neuroanatomy and the evolutionary genetics of *Drosophila*) conjoined to a wishful program of establishing the genius embodied in V.I. Lenin's preserved brain. At this level of analysis, the story underscores the historical contingency that underlies the development of scientific practices and theories. However, at a finer level of resolution the story appears almost unsurprising, given the nature of the individuals and research programs involved. The German neuroanatomist Oskar Vogt introduced the terms 'penetrance' and 'expressivity' in 1926 to interpret certain results obtained by Soviet evolutionary geneticists, in particular Nikolai Timoféeff-

L.S. Parker and R.A. Ankeny (eds.), Mutating Concepts, Evolving Disciplines: Genetics, Medicine and Society, 63-85.
© *2002 Kluwer Academic Publishers. Printed in the Netherlands.*

Ressovsky, working on miscellaneous *Drosophila* species. That Vogt was at all aware of the Soviet school of genetics was due to his involvement with the project to section and analyze Lenin's brain; hence the talk of historical contingency above. But once we appreciate the catholic scope of Vogt's vision of *neurobiology*, that is, the integration of functional neuroanatomy and neurophysiology with neuropathology, it almost appears inevitable that, working in the 1920s, he would sooner or later invoke the power of the new and internationally fashionable science of genetics embodied in research on *Drosophila*.

Second, the history we reconstruct here also shows exactly why the two concepts, especially penetrance, remain problematic even today. We believe that there is widespread (though usually only implicit) disquiet about the use of these terms among geneticists. This unease usually remains implicit because (1) the definitions of these terms are still murky and different definitions contradict each other; and (2) the conceptual, rather than ideological (see below), role of these concepts in genetics is far from clear. Both concepts are invoked when any simple relation between a specific genotype and phenotype breaks down. A standard glossary of genetic terminology defines expressivity as "the phenotypic expression (kind or degree) of a penetrant gene or genotype which may be slight, intermediate or severe and may be described in either qualitative or quantitative terms," and penetrance as "the frequency (in percent) with which a (dominant or homozygous recessive) gene or gene combination manifests itself in the phenotype of carriers" (Rieger, Michaelis, and Green, 1991, pp. 178, 372). According to these definitions, expressivity and penetrance are both properties of genes or groups of genes.

These definitions are deceptively simple but, as stated, either specify almost entirely useless concepts (as in the case of expressivity) or ones that can be fully captured using dominance and without recourse to such terminological excess. It is hard to see how expressivity is to be quantified except by fiat, and even harder to see of what use a qualitative ordering of alleles according to the degree of their expressivity could be. Furthermore, the degree of expressivity of an allele seems hard to distinguish from its degree of dominance. Turning to penetrance, for homozygous recessive alleles, the penetrance of an allele is identical, according to the definition given above, to the frequency of the recessive homozygote genotype. The penetrance of a dominant allele is similarly the sum of the frequencies of the heterozygote and dominant homozygotes.

However, a textbook of methods for human genetic linkage studies provides a quite different and potentially useful definition for penetrance: "Penetrance…is defined as the conditional probability $R(xg)$ that an individual with a given genotype g expresses the phenotype x" (Ott, 1991, p. 146).[1] Now, penetrance is no longer a property of a gene, but only of an entire genotype, and this definition is not easily reconciled with the one given before. As we shall show below, while the frequencies by which this probability can be measured can indeed be estimated in the field, even this definition is adequate only if the genetic background is entirely fixed, for instance, in a pure line. In some experimental contexts this can be achieved, and we are willing to entertain a possible role for penetrance in those contexts. However, when the genetic background cannot be fixed—as, for instance, in humans—what penetrance means is far from clear. Yet the main use of expressivity and penetrance today is not in experimental contexts but in human behavioral genetics, and that subdiscipline, within and outside medical contexts, is the target of our attention here.

We argue that the basic role that 'expressivity' and 'penetrance' play in contemporary human behavioral genetics is ideological: they help maintain a genetic etiology for traits in the face of recalcitrant detail. If we want to establish such a genetic etiology, then we have available to us the traditional techniques of segregation and linkage analysis, and some more modern ones such as allelic association studies, allele-sharing methods, and quantitative trait locus mapping (Sarkar, 1998). When we turn to complex behavioral traits of organisms, in particular human cognitive or mental traits (alcoholism, attention deficit disorder, autism, bipolar affective disorder, schizophrenia, sexual orientation, and so on), these methods do not yield unequivocal results. Yet, in the wake of the Human Genome Project, we live in an atmosphere infused with geneticism ("one adjective-one gene"): there is a gene for everything. Expressivity and penetrance rescue us from the problem of equivocal, even absent, data. If there is a gene for a trait, but the presence of the trait is nevertheless capricious in the presence of the gene, there is still a gene *for* that trait but the gene has *variable expressivity*. If the trait does not deign to manifest itself at all, there is still a gene for the trait: it is just that the gene is *incompletely penetrant*.

Our story will show how we managed to get to this state. Furthermore, we will argue that the concepts of 'penetrance' and 'expressivity' were first developed in the context of a more complex theory of gene action, one that placed greater emphasis on the interaction between different genes and between the gene and its cellular and developmental environment than is generally the case in the context of modern behavioral genetics. However, in addition to representing the intricate complexities of developmental systems, Vogt, the idiosyncratic originator of these concepts, already intended penetrance and expressivity to rescue a straightforward pattern of inheritance in the face of ambiguous pedigrees of phenotypic characters (such as brain pathologies). The confusion that these concepts later caused in medical genetics was thus present from their inception (see Laubichler, 2000).

2. BERLIN PRELUDE:
THE RESEARCH PROGRAM OF OSKAR AND CÉCILE VOGT

At first sight it might seem somewhat surprising that a neuroanatomist, Oskar Vogt, defined two concepts that became central to classical genetics and remain ubiquitous in medical genetic contexts. However, when we consider the details of Oskar and Cécile Vogt's rather inclusive research program, most of this mystery disappears. Furthermore, Oskar Vogt's intensive contacts with Soviet scientists during the 1920s placed him at a critical conjuncture between the (largely) Western tradition of attempts to localize specific functions in the brain (e.g., see Breidbach, 1996;

Hagner, 1994, 1997), and the specifically Soviet tradition of evolutionary genetics that produced Sergei Chetverikov, Theodosius Dobzhansky, and Nikolai Timoféeff Ressovsky, among others (Adams, 1980).

Oskar Vogt was born in 1870 in Husum, Schleswig-Holstein, then part of Prussia, into a family of Lutheran ministers. Because of an early interest in natural history and philosophy, he studied philosophy and medicine first at Kiel (1888-90) and then at Jena. His teachers included eminent scientists such as Ernst Haeckel (zoology), Max Fürbringer (anatomy), and Otto Binswanger (psychiatry). After finishing his studies in Jena, Vogt did postdoctoral work with August Forel in Zürich and Paul Flechsig in Leipzig. It was Flechsig who first suggested that the phenomenon of myelogenesis—the post-embryological acquisition of a myelin cover by nerve fibers in the brain—could be used to localize distinct regions in the cortex. Flechsig proposed grouping clusters of nerve cells together in functionally equivalent classes, such as the projection fibers and association fibers of the respective sensory and associative centers in the cortex that showed the same pattern of successive myelogenesis. This idea was a major new development in the century-old quest to map the brain. Shortly after Vogt left for Paris in 1895 to study with the prominent French neuroscientists Jules and Augusta Dejerine in order to gain more insight into the clinical side of neurology, he challenged Flechsig's myelogenetic criterion for the identification of distinct regions of the brain. Contrary to Flechsig, Vogt (1897) demonstrated that the associative centers in the cortex also contained projection fibers. Even this early in his professional career, Vogt was deeply involved in methodological discussions regarding the appropriate criteria for the localization of distinct regions in the brain, a question that would remain central to his research throughout his entire career and eventually trigger his interest in genetics.

In Paris Vogt also met his future wife, Cécile Mugnier, a medical student working with the neuroanatomist Pierre Marie. In 1897 he left Paris for Berlin, where he set up his own research facilities.[2] Cécile Mugnier joined him there the next year, and they were married in 1899. Their ensuing private and professional partnership would last for almost sixty years. Oskar Vogt had exceptional organizational skills that helped him pursue his scientific vision. Over the years he was the editor of three scientific journals and a series of monographs;[3] he also founded or helped to found five research institutes that were all more or less devoted to his own scientific agenda.[4] In 1902, Oskar Vogt and August Forel started the second series of their joint journal under the new title *Journal für Psychologie und Neurologie, Mitteilungen aus dem Gesamtgebiet der Anatomie, Physiologie und Pathologie des Zentralnervensystems, sowie der medizinischen Psychologie* [*Journal for Psychology and Neurology: Communications from all Areas of the Anatomy, Physiology and Pathology of the Central Nervous System and of Medical Psychology*]. In the accompanying editorial on psychology, neurophysiology, and neuroanatomy, Vogt (1902) gave the first clear outline of his future research program (see also Vogt, 1897). He argued that every psychological phenomenon has a material correlate in form of a vital process within the brain and that the anatomical structures in the brain are closely correlated with their respective functions. Accordingly, neuroanatomy and neurophysiology would have to be integrated to form what he referred to as neurobiology (the anatomy and physiology of the nervous system). His boldest claim was that in certain cases it should even be possible, as knowledge of the brain increased, to infer the function of a specific part of the brain from its structure and *vice versa*. The last statement, in its recognition of a close connection between form and function, is an early indication of how Vogt would later conceptualize the problem of biological characters and organize his research program, including the eventual turn to genetics that led to the introduction of the concepts of 'penetrance' and 'expressivity.'

Attempts to identify the function of any given biological object have always depended on biological or pathological variation (the so-called 'gene knock-outs' are the latest example of such a technique). Oskar and Cécile Vogt's program was no different. The study of pathologies was a logical extension of their search for the material basis of brain functions. From the proposed connection between the structure of a biological character and its function, it follows that any variation in structure will lead to a correlated variation in function and, correspondingly, that any observed variation in function will be accompanied by a variation in structure. In their view, the difference between so-called 'normal variations' and 'pathologies' is just a quantitative one. However, a prerequisite for any comprehensive study that would correlate variations in structure with variations in function was the identification of those individual regions in the brain that could be correlated with specific functions.

The project of mapping the brain was not new (e.g., see Breidbach, 1996; Hagner, 1994, 1997; Harrington, 1987). Its success depended on the availability of operational criteria for the identification of specific regions. Oskar Vogt chose to focus on brain architectonics, the analysis of the number, size, and form of the nerve cells within the laminae of the cortex, with special attention to the specific arrangement of these elements in various regions. Individual areas were then identified by local differences in composition, number, and kind of the structural elements identifiable in specific preparations (see also Hagner, 1994). Using this method, which required the analysis of whole series of microscopic preparations and their three-dimensional reconstruction, the Vogts and their collaborator, Korbinian Brodmann, were able to identify up to 200 separate regions in the cortex.

While the research program in brain architectonics gave the Vogts a method to identify structurally defined regions, it did not, by itself, provide an answer to the question of whether the regions so identified could be linked with corresponding functions. To establish these connections, the Vogts used a variety of experimental approaches. On the one hand they employed the technique of electrical stimulation of specific regions of the cortex. This work was done primarily with monkeys (macaques) (Vogt and Vogt, 1907, 1919a, 1926; Brodmann, 1909). But they also analyzed post-mortem the brains of human patients who had suffered from known neurological and psychiatric conditions and attempted to correlate these conditions with the pedigrees and medical records of their patients. Eventually they brought the results of both lines of investigation together and produced maps that were characterized by both structural and functional criteria (for a synthesis of the results, see especially Vogt and Vogt, 1937).

Neurological and neuropathological studies were common at the beginning of the twentieth century partly due to the high number of brain injuries during World War I (for further detail, see Harrington, 1996). Many neuroscientists, presented with these often highly localized lesions, began to map the human brain. Prominent among them was Otfried Foerster, a neuropathologist from Breslau (Foerster, 1923, 1925, 1926), who also was to become one of Lenin's most trusted physicians. After completing his electrophysiological mapping of the human cortex, Foerster proposed a functional map of the cortex that agreed to an amazing degree with the one produced by the Vogts on the basis of known cytoarchitectonic fields combined with the functional assignments that were derived from their studies with monkeys. For the Vogts, this was a triumph not only for their method, but also for their idea of combining animal models and human pathologies to study the human brain.

For Oskar Vogt, the study of brain pathologies was linked to the question of the classification of psychiatric diseases which, in turn, he believed to be related to zoological and botanical systematics. In both domains, the problem of biological variation figured prominently. Geographical variation had been put forward as an important argument in favor of a Darwinian picture of gradual evolution by means of natural selection, though the extent of gradual and continuous variation in nature was contested (Bateson, 1894). Similarly, there was widespread disagreement about the causes of variation even within a population, resulting in the highly charged debate between biometricians and Mendelians in the United Kingdom (e.g., see Provine, 1971). Oskar Vogt was keenly interested in all of these issues. He had been collecting bumblebees on his travels throughout Europe since early youth, and by the time he finished publishing his ideas on the species question in 1911, his collection contained more than 75,000 specimens. In these papers (Vogt, 1909, 1911) he argued for the importance of the inner organization in constraining the possibilities of variation, a conclusion he drew by arranging all known variations of a character, such as coloration, in a systematic way (e.g., see Vogt, 1909, p. 57). He then interpreted this arrangement as a sequence of transitions from one character state to the next.

In the context of the discussions of evolutionary mechanisms during this period, Vogt interpreted such constrained linear transitions between character states as evidence for orthogenesis, which was still widely accepted at that time (see Bowler, 1983). But his observations inevitably broached the question of inheritance. Variation, constrained or not, is obviously connected to genetics and development. Genetics—the study of patterns of inheritance—and embryology—the study of the physiology and mechanics of ontogenesis—both deal with the causes of variation. In turn, the existence of natural and artificially induced variations proved essential to progress in both disciplines. Vogt was keenly aware of the importance of genetics and embryology to his own research agenda. Indeed, Oskar and Cécile Vogt's theory of biological organization was based on a synthetic integration of both embryology and genetics in an attempt to derive criteria for the identification and individuation of biological characters and their variations (Vogt and Vogt, 1937, 1938; for an analysis of their views, see Laubichler, 2000).

During this period, studies of the inheritance of pathological conditions, especially of mental and psychological deficiencies, were a major part of the eugenics movement (e.g., see Kevles, 1985; Adams, 1990; Paul, 1995). The Vogts were aware of this trend; however, their own research was only marginally relevant to eugenics. Nevertheless, they often phrased their results in the language of 'improvement,' for instance, when they argued that "the practical goals of brain research are the care and selective enhancement [Höherzüchtung] of the human brain" (Vogt and Vogt, 1929, p. 438). Even though their involvement with the ideology of 'improvement' was largely theoretical, it was based on specific ideas about the mechanisms of inheritance.

As Jonathan Harwood (1993) points out in his study of the German genetics community, there were at least two different schools of genetics within Germany during this period: the 'pragmatics,' who were eager to adopt a Morgan-style genetics with its emphasis on transmission, and the 'comprehensives,' who preferred to study genetics in the context of evolution and development. Harwood argues that these scientific preferences in genetics can, to some extent, also be correlated with more general "styles of thought" that include the breadth of biological knowledge, ideas about what counts as science (*Wissenschaft*), and the socio-economic and political position of the respective researchers. Within this classification of geneticists, Oskar Vogt is somewhat of an oddity. While his ideas on the mechanism of inheritance are most certainly 'comprehensive'—his framework is thoroughly developmental and emphasized the role of extra-nuclear and environmental factors even to the point where he considered the possibility of an inheritance of acquired characters including the existence of so-called *Dauermodifikationen* [permanent modifications]—his insistence on a strict correlation of structural and functional variations and his materialistic belief in a localized and material basis for higher brain functions puts him closer to the 'pragmatists,' in Harwood's sense of the term. In any case, what was most central to Vogt's ideas on inheritance was his synthetic outlook. He combined a Morgan-type Mendelism with a developmental view of gene action. Consequently, he was critically involved in attempts to adapt the framework of classical genetics in order to accommodate more complicated biological characters and more ambiguous patterns of inheritance. It is, therefore, not surprising that Oskar Vogt would expand his study of model organisms in the 1920s to include genetic studies with *Drosophila*.

3. LENIN'S BRAIN

The story now shifts from Berlin to Moscow, and to Lenin's brain. In the early 1920s, thanks to the Western boycott of German scientists and the West's visceral hostility to the Soviet Union, scientific contacts and collaboration between German and Soviet scientists expanded rapidly. V.I. Lenin, who had been complaining of fatigue, intense headaches, nausea, and insomnia, had already placed himself under the care of German doctors starting in March 1922, with Foerster becoming his most trusted physician (Richter, 1991, 1996; Weindling, 1992). Lenin's first stroke, on March 25, 1922, came shortly after the initial examination by Foerster; a second stroke followed in December. In January 1923, while Vogt was attending a neurological congress in Moscow, he was called in as a consultant to Lenin's physicians. A third paralytic stroke followed in March 1923, and several German physicians were immediately called to Moscow. Nevertheless, Lenin died on January 21, 1924.

In Moscow, Vogt had begun to develop extensive scientific and medical contacts with Soviet scientists. These were probably highly influential since, in December 1924, he was asked to undertake the dissection of Lenin's brain. In April 1925, with the full approval of the German Foreign Office, Vogt accepted the Soviet offer. The Soviet government promised to create and place at his disposal a new and well-equipped brain research institute in Moscow. This became the Lenin Institute for Brain Research and, although Vogt was its nominal director, executive responsibility largely lay with his Soviet colleagues. Drawing on prevalent views of the extraordinary anatomical attributes of so-called 'elite brains,' the Soviet government's intention was to demonstrate the exceptional nature of Lenin's genius from his brain's neuroanatomical features. Between 1925 and 1927, Lenin's brain was sliced into a series of sections. In 1927, Vogt delivered a preliminary report of the histoanatomical examinations of the stratigraphy of the cortex of Lenin's brain (Richter, 1991). From the presence of numerous, large pyramidal cells in the third layer, he concluded that Lenin had outstanding mental capacities.

Though Vogt apparently did not know it, this report was of considerable political significance, as Richter (1991) has pointed out. In 1922, shortly after his second stroke, Lenin had dictated a testament that was particularly critical of Josef Stalin. Contrary to Lenin's explicit instructions, this document was not published in the Soviet Union until 1956, though it was circulated to the Central Committee. Meanwhile, Vogt's report argued for Lenin's continued mental capacity up to the time of his death, which could not have pleased Stalin who kept a close watch on the situation through the Lenin Institute's Deputy Director, I.P. Tovstuhka who, secretly, was one of his three personal secretaries. Diane Paul and Costas Krimbas note that

> in 1929, Vogt issued a...report of their findings, noting the hypertrophy of pyramidal cells in the association cortex. By analogy to the muscular hypertrophy that results from strenuous physical exercise, he lauded Lenin as an 'association athlete.' In a later article and lecture Vogt also compared Lenin's brain with that of a murderer and concluded that the differences must be inherited (since nerve cells do not divide after birth), and were thus unaffected by social environment. (Paul and Krimbas, 1991, unpublished)

Clearly, Vogt was not actively concerned with the political reception of his findings.

Between 1900 and 1920, Vogt had become convinced that various psychiatric conditions, which he believed to be inherited, were mediated by genes and that the same rules that governed the ordered expression of variations in individuals of insect morphological pattern formation would also apply to the temporal sequence of nervous degeneration in brains within individuals showing psychopathologies.[5] This was, ultimately, a modification of Haeckel's doctrine of the connection of morphogenesis and evolution, with degeneration replacing morphogenesis, and genotypic variation replacing evolution. Vogt wanted to push this putative insight further but could not find a trained geneticist anywhere in Germany with whom to do collaborative work on this problem.

In sharp contrast to Germany, where he remained largely an outsider to the experimental genetics community, Vogt established close connections with the Institute of Experimental Biology in Moscow founded and run by the zoologist and eugenicist, Nikolai Kol'tsov (Adams, 1990). A very active genetics group had formed there around the pioneering population geneticist, Sergei Chetverikov. It included Chetverikov's wife, Anna Ivanovna, Nikolai W. and H.A. Timoféeff-Ressovsky, D.D. Romashov, S.R. Tsarapkin, and A.N. Promptov. Two sets of results produced by this group particularly intrigued Vogt. Romashov had discovered the mutation *Abdomen abnormalis* in *Drosophila funebris* in the autumn of 1922, though systematic work on it did not begin until 1924 (Romaschoff, 1925).[6] This mutation affected the pigmentation of the abdomen of the flies (see Figure 1).

Abb. 1.   Abb. 2.

Drosophila funebris, Genenvariation „*Abdomen abnormale*".
Abb. 1 in Trockenheit gezüchtet und daher äußerlich
normal. Abb. 2 in Feuchtigkeit gezüchtet und daher
typisch verändert.

Figure 1.

It resulted in the degeneration of the normally uniform striped pigmentation. However, in mutant flies, this pattern was not replaced with some other uniform pattern. There was considerable variability, which Romashov interpreted as a difference in the strength of the effect of the mutation. Moreover, occasionally, there were mutant flies that did not show the effect at all, although they were known to be carrying the mutation (because Romashov had created pure lines). The manifestation of the mutation thus depended on environmental conditions—in particular dryness as well as the liquid content of food eaten—though Romashov also noted the possibility that it could depend on the variability of other genetic factors. The environmental dependence would have been particularly resonant with Vogt's one-time faith in the importance of the inheritance of environmentally modified acquired characters though, by this time, Vogt had begun to join mainstream Mendelism at least by doubting the importance of such inheritance.[7]

Meanwhile, Timoféeff's results emphasized the importance of additional genes in the manifestation of a phenotype associated with one particular gene (1925). Timoféeff worked with the recessive *Radius incompletus* mutation of *Drosophila funebris*; in mutant flies which were homozygous for the mutation, the second longitudinal vein did not reach the end of the wing (see Figure 2). From homozygotes carrying the mutation, Timoféeff created several pure lines for additional study. However, when some of these pure lines were bred further, their

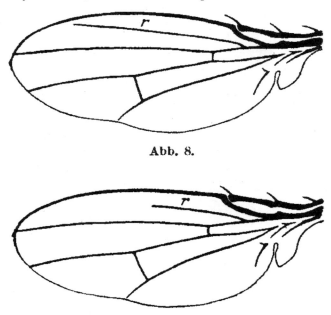

Abb. 8.

Abb. 9.

Drosophila funebris, Genenvariation ,,*Radius incompletus"*.   Abb. 8 schwache, Abb. 9 starke Expressivität.

Figure 2

descendants included phenotypically normal flies. The percentage of phenotypically normal flies was fixed for any particular line, and varied between zero and forty percent; external factors seemed to have very little influence on the percentage. Thus the differences in the percentage of normal flies in different lines appeared to be due to additional genotypic factors. More importantly, some lines that had a high percentage of affected flies nevertheless manifested the mutation weakly, and the converse was also true. Timoféeff concluded "it shows us that the phenotypic manifestation of a hereditary trait by itself and the strength and form of this manifestation can be independent of one another. However, this question requires further specific studies" (1925, p. 309).[8]

4. 'EXPRESSIVITY' AND 'PENETRANCE'

Vogt arranged for the publication of both sets of results in the *Journal für Psychologie und Neurologie* in 1925, which represented one of the first glimpses available to Western scientists of this important Soviet work. He offered positions to the Timoféeffs and to Tsarapkin at the Kaiser-Wilhelm-Institut für Hirnforschung in Berlin.[9] The Timoféeffs moved to Berlin in 1925 or 1926 (Paul and Krimbas, 1991, 1992). Timoféeff was only twenty-five years old, and did not even possess an undergraduate degree. Nevertheless, he soon became one of the most prominent geneticists of his generation.

In 1926 Vogt published a long and tendentious, but nevertheless critically important, paper entitled "Psychiatrically Relevant Facts of Zoological and Botanical Systematics [*Psychiatrisch wichtige Tatsachen der zoologisch-botanische Systematik*]." He began this paper by explicitly stating that he was going to take a "personal standpoint [*persönlichen Standpunkt*]" (Vogt, 1926, p. 805). That standpoint was little short of idiosyncratic. Zoological and botanical systematics, he claimed, were based on a classification of "reaction-types [*Reaktionstypen*]." These types were determined by evolutionary history as mediated through the germplasm. To explain his position, Vogt turned to two tenets of his evolutionary faith: that the differences between species, sub-species, or lower classificatory groups occurred in jumps [*Sprünge*], even when they appeared to be continuously graded, and that all such variation could be put in a definite linear order which he called "eunomies."[10] Both the gradual variation and the linear order were generalizations that Vogt had arrived at in his earlier work on the geographical variation of morphological patterns in insects (see above and Figure 3). In his 1926 paper, Vogt provided many examples of such eunomies. Furthermore, his reinterpretation of the work of Romashov and Timoféeff reconciled these generalizations with Mendelism as the source of the 'jumps.'

Romashov's report that a mutation was sometimes not manifested phenotypically, and Timoféeff's measurements of the fraction of individuals not showing a mutation in a pure line inspired Vogt to terminological innovation: "There are mutations that prevail under the most different circumstances. One must sharply distinguish this tendency from dominance. I propose that this tendency to prevail be called '*penetrance*'"[11] (1926, p. 809). 'Penetrance' thus first slipped into genetics as a property of a mutation; Vogt's discussion of examples, which followed this introduction, shows that he conceived of it as a property of an allele (which in itself is not problematic since, for example, the Morgan school used 'mutation' in the same way). The crucial move here was to ignore that Timoféeff's data showed that the manifestation of the mutation not only varied from pure line to pure line, but that the mutation was characterized by a different definite percentage of manifestation in each of these pure lines. Instead, Vogt simplified the situation: "Our collaborator, N.W. Timoféeff-Ressovsky has shown for the weakly penetrant mutation 'radius incompletus' of *Drosophila funebris* that the 'rest' of the germ-plasm hinders the manifestation of the mutation in up to 33.10% of the bred animals"[12] (1926, p. 810). What was at best a relation between an allele and a pure line (and even this is not clear, since Timoféeff does not report data showing that the same pure line consistently gave the same percentage) becomes a property of a gene itself, with no admission of the range of complexity that had initially been reported. A number (Timoféeff's percentage) representing (at best) the relation between an allele and a specific genotypic background (a pure line) now became a property of a single allele itself.

Next, Vogt seized upon Romashov's and Timoféeff's rather tentative remarks that the variation in the manifestation of a mutation (its 'strength') could be linked to the percentage of its manifestation in a pure line and reconstituted it as a regularity of nomological import. The basic distinction is attributed to Timoféeff:

> In addition to penetrance and specificity,[13] a third different feature of a mutation is distinguished by Timoféeff, for which one can introduce the term 'expressivity.' In the case of stronger penetrance, that is, in the case of a proportionally more frequent manifestation of a trait, this trait can appear in a weaker form (weak expressivity). Conversely, in the case of low penetrance, that is, in the case of a proportionally rare manifestation of the trait, the manifestation of this trait can be rather strong (strong expressivity).[14] (Vogt, 1926, p. 811)

Earlier in the paper, still in a discussion of penetrance, Vogt had presented only two of Romashov's eleven diagrams, claiming:

> In Diagrams 1 and 2 [see Figure 1], taken from the work of Romashov, Diagram 2 shows a completely disfigured abdomen. Such animals, which have been modified to the point of inviability, regularly arise from the mutation 'abdomen abnormale' through a medium degree of humidity. In contrast, should one breed the animals in great dryness or at very great humidity, there appear externally normal animals such as the one depicted in Diagram 1. Intermediate ranges of environment create intermediate ranges of forms between those depicted in Diagrams 1 and 2.[15] (Vogt, 1926, pp. 809-10)

Neither Romashov nor Timoféeff had quite suggested this fixed linear order, although they both had spoken of the strength of the modification. Even according to Vogt, Diagram 1 (see Figure 1) represents the results of two different directions in which variation of humidity could go (extreme dryness and extreme humidity). Therefore, Diagram 2 (also see Figure 1) represents an intermediate form with respect to environmental parameters. What Vogt had achieved, intentionally or not, was to reconstruct diffuse phenotypic variation into a single parameter-governed phenomenon, the parameter being expressivity.

Compared to the consequences of these two new definitions, Vogt's ultimate, explicit goal in this paper appears innocuous: "All these factors are more or less capable of concealing utterly discrete mutation in the phenotypes of a population"[16] (1926, p. 812). Throughout Vogt's discussion, expressivity and penetrance are occasionally properties of genotypes, sometimes of traits, but usually of mutations and, by implication, of alleles. But what is most interesting is that penetrance became a property of the gene itself. Other genes may prevent the manifestation of the trait for which there is a gene, but in contrast to Timoféeff's original report, Vogt suggests no difference between different lineages for the same gene and ignores the fact that these lineages were pure lines. It is not surprising that these critical details were lost in the subsequent history of the concept in behavioral genetics. Meanwhile, expressivity as a singular property capturing the variability of a trait's manifestation was an almost pure invention arrived at by means of a reconstruction of the original reports as a linear series of images.

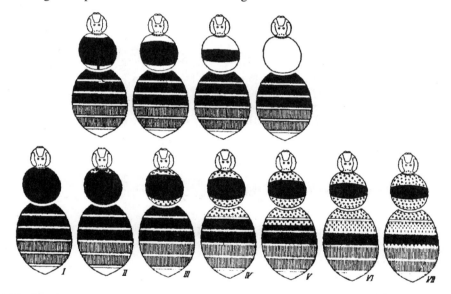

Abb. 17. Untere Reihe: übliche Eunomie des Hellwerdens des *Bombus lapidarius*. Obere Reihe: abweichende Eunomie der kaukasischen Tiere.

Figure 3.

After introducing the concepts 'penetrance' and 'expressivity,' Vogt went on to present examples for eunomical series of geographical variation in morphological patterns in many insects (Figure 3). He then turned to neurological degeneration in macaques and humans and interpreted the sequential decrease of cells in specific regions of the cortex as a eunomie (Figure 4). By means of this analogy, the connection between genetics and psychiatric illness was complete. But this was no mere analogy, Vogt claimed; rather, "we restrict ourselves to deductions on the basis of a true identity between psychoses on the one hand, and the somatic variations and all the elementary mutations of zoological-botanical lineages within species or subspecies"[17] (1926, p. 830). This pointed to the final methodological conclusion: "In all complicated hereditary pathways, research into inheritance is no longer restricted to the mixing of genes and their dominance and recessivity. All [such research] must take those and also the newly recognized features of genes, penetrance, specificity and expressivity into account"[18] (Vogt, 1926, p. 831).

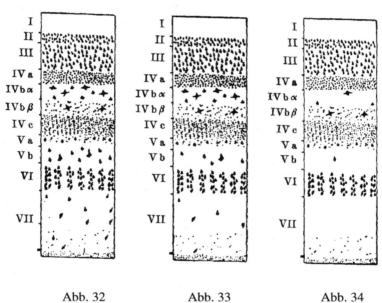

Abb. 32   Abb. 33   Abb. 34

Abb. 32-34. Eunomie einer encephalitischen Rindernerkrankung der *Area striata* des Menschen

Figure 4

The proposed identity between genetic variations and psychiatric illnesses and the role of the concepts of penetrance and expressivity in establishing this link can only fully be understood in the context of Vogt's more inclusive theory of biological characters. This theory rests on two crucial assumptions. The first is the postulate that all functional differences are based on structural differences and that there is no fundamental difference between pathological and natural variation. Correspondingly, all psychiatric conditions are based on a material change of some relevant regions of the brain. But this postulate also justifies the methodological principle that natural variation can serve as a model system for pathological variation and *vice versa*. It is also the main reason for Vogt's interest in questions of biological systematics. The second assumption is about the nature of biological characters: for Vogt, all characters are essentially dynamic and not static entities. Each individualized biological character follows a specific ontogenetic trajectory throughout the life history of an organism, a process he termed *"Bioklise."* Biological characters are defined as "topistic" units (derived from the Greek word 'topos' [place]) that originate and are maintained and modified by specific formative processes. A variety of factors, including genetic ones, contributes to these formative processes. Variations in any one of these factors then lead to correlated variations in the expression of the character and in its function. However, according to Vogt, these changes in the expression of characters do not occur at random, but rather show characteristic sequences of transitions, the so-called eunomical series. The explanation for these limitations on the expression of variations can be found in the rules of interactions among the many formative factors that influence a given character. They are the reason why fundamentally discontinuous mutations are so often masked in their expression, a phenomenon that Vogt tried to capture with the introduction of the concepts of penetrance and expressivity.

The introduction of these two genetic concepts in the context of his overarching theory of biological characters allowed Vogt to utilize the increasingly popular study of inheritance patterns in *Drosophila* as a model system for the study of psychiatric diseases. But brain pathologies were typically complex diseases that rarely yielded straightforward patterns of inheritance. Endowing presumed genetic factors for these diseases with those additional characteristics (penetrance and expressivity) enabled Vogt to search for patterns in the expression of complex psychiatric conditions within the pedigrees of his patients. But this resulted in an emphasis on genetic factors as the causes of mental diseases, something that was in conflict with the complex and multi-causal theory of biological characters that was the basis of Vogt's overall research program. He never fully reconciled these two views. But while his theory of biological characters is now mostly forgotten, the idea of genetic etiologies of mental disease and the two concepts that helped to rescue it in the face of contradicting evidence are still with us.

5. THE RECEPTION

The Timoféeffs' response to Vogt's terminological innovations of 1926 was enthusiastic. In an important paper published in 1926 they discussed the mutation *Alae plexus* in *Drosophila funebris*, which has an asymmetrical effect on the cross-veins of the wing (Timoféeff-Ressovsky and Timoféeff-Ressovsky, 1926; see also Vogt, 1926, on unpublished work by the Timoféeffs). In this paper, they presented a schematic diagram that depicted the increasing divergence of the pattern in a linear series. This, they thought, justified the invocation of expressivity. The new terminology was explained in a footnote:

> The ability of the factor [gene] to manifest itself phenotypically generally is described by the term 'penetrance' (O. Vogt 1926). There are factors [genes] (strongly penetrant) which always manifest themselves, while others (weakly penetrant) are expressed only in a certain percentage of cases. The degree to which a factor manifests itself in the phenotype of the trait is described by the term 'expressivity.'[19] (Timoféeff-Ressovsky and Timoféeff-Ressovsky, 1926, p. 150n)

In this paper they attributed variations in penetrance and expressivity to the presence of other genes [*Faktoren*]. The Timoféeffs (1927) continued to discuss penetrance, and N.W. Timoféeff later invoked both terms in two additional papers (1931a, 1931b). One of these two papers (1931a) is almost certainly the source of the terms' eventual incorporation into Anglophone genetics.

By 1930, Timoféeff's work had become well known to the German genetics community, and his reputation began to rise in the Anglophone genetics communities as well, especially after he turned to mutation research. While his advocacy led to the gradual acceptance of the concepts of expressivity and penetrance in the German genetics community in the early 1930s (e.g., see Patzig, 1933), these concepts initially did not become part of Anglophone genetics. A 1932 paper by Vogt, published in the *Eugenics Review* (which included only a very modest implicit endorsement of eugenics but is still troubling because of the political context of the period) contained the first account of these two concepts in English. While Vogt did not translate the terms, the Editor suggested that 'potency' would be the appropriate translation of '*Expressivität*.' Introducing 'expressivity' and 'penetrance,' Vogt states clearly that the concepts had been necessary in order to retrieve a genetic etiology in the face of recalcitrant pedigrees. But as late as 1937, a "Glossary of Genetic Terms" published in the *Journal of Heredity* makes no mention of the terms. A.H. Sturtevant and George W. Beadle's 1939 textbook, *An Introduction to Genetics*, also makes no mention of 'expressivity' or 'penetrance.'

The first uses of 'expressivity' and 'penetrance' in English, in 1938, seem to have been by Conrad H. Waddington, who incorrectly attributed all three concepts (expressivity, penetrance, and specificity) to Timoféeff's 1931 paper (1938, p. 188). As someone working at the intersection of genetics and embryology, Waddington was concerned with the control of processes by genes as early as the 1930s. Like Vogt, he was particularly interested in morphological pattern formation. For Waddington, expressivity and penetrance were both properties of genes that helped to account for the "[v]ariability of [g]ene [e]ffects": "expressivity is a measure of the amount of effect shown by the gene"; "penetrance is the frequency, measured as a percentage, with which the gene shows any effect at all" (1938, pp. 188, 190). In Waddington's view, both depended on the "rest of the genotype" as well as the environment.

The phenomenon of dominance is generally downplayed throughout Waddington's book, and no attempt is made to explain how expressivity is to be distinguished from dominance. With respect to penetrance, Waddington's elaborations are even more puzzling: "Most of the genes usually worked with in genetical experiments are chosen for having a high or complete penetrance; every organism homozygous for a recessive factor shows it" (1938, p. 190). If this is taken literally (and this would be uncharitable), penetrance becomes a property of an organism. What is more pertinent is that, though he is referring to Timoféeff's original experiments, Waddington ignores the point that the phenomenon captured by penetrance, if it is to refer to a description of Timoféeff's original experiment—that is, a stable frequency of manifested phenotypes—was observed only relative to a pure line. This was Waddington at the height of his enthusiasm for the possibility that single loci might provide key insights into phenogenesis (see Waddington, 1940). In Waddington's later work, where genetic *systems* move to the center, the importance of both expressivity and penetrance is implicitly played down (e.g., see 1957, p. 91).

6. DISCUSSION

In Britain, after Waddington's adoption of the terms, expressivity and penetrance gained some popularity. Comparatively little was published in genetics during World War II. After the war, in the first major textbook of genetics to be published, Darlington and Mather (1949) introduced both concepts and, presumably following Waddington, attributed them to Timoféeff's 1931 paper. The terms were included in Darlington and Mather's "Glossary of Important Genetic Concepts." In the United States, Dobzhansky seems to have been responsible for the introduction and popularization of the concepts.[20] Dobzhansky had moved from the Soviet Union to the United States in 1927 where, over the next decade, he established himself as one of the most influential figures in evolutionary genetics. Neither the 1932 nor the 1939 edition of E.W. Sinnott and L.C. Dunn's *Principles of Genetics* contains any mention of these concepts. However, in the 1950 edition, systematically rewritten, now with Dobzhansky as a co-author, the concepts are introduced and attributed, as usual, to Timoféeff's 1931 paper.

Thus, in the 1950s, both concepts found their way into the two dominant textbooks of genetics on both sides of the Atlantic. Darlington and Mather's textbook (*Elements of Genetics*) was influential enough to be reprinted, with a new introduction by Darlington, in 1969. Sinnott and Dunn's (and Dobzhansky's) text went through five editions, the last one published in 1958. Dobzhansky's growing prominence also spread the use of penetrance and expressivity throughout the Anglophone genetics community even though published uses are few in the 1950s (Richard Lewontin, personal communication). Neither of the two textbooks that introduced these concepts to the Anglophone world mentions that penetrance should be relativized to a fixed genetic background.

By the end of the 1950s expressivity and penetrance became staple concepts within human genetics, presumably because of the textbooks mentioned above. Though we have not reconstructed this story in all its details, this assessment of the role of these two concepts in human genetics during this period is based on a letter that Curt Stern (1960) wrote to the editor of the *American Journal of Human Genetics* shortly after Vogt's death in 1959. The intention of this letter was to establish priority for Vogt as the originator of these concepts, but what is important in our context is that it indicated the prevalence of the concepts in human genetics by 1960. It is worth quoting the relevant passage:

> The main field of Vogt's research was neuroanatomy but he also contributed to facts and interpretations in what has become to be known as 'The New Systematics'...Vogt coined two words which have become part of the terminology of genetics, *and particularly human genetics*: penetrance and expressivity. [Author's note: An extensive quotation from the relevant passage from Vogt (1926) occurs here.]...It is of historical interest that the terms penetrance and expressivity were introduced by an investigator whose main work was related to medicine. (Stern, 1960, p. 141; emphasis added)

Stern's prominence in, and familiarity with, the human genetics community lead us to agree with his assessment of the role the two concepts had begun to play in human genetics by 1960.

In human genetics, penetrance came to be estimated from pedigrees, using all the pedigrees that were available (see Ott, 1991, section 7.3). That the value of penetrance, a single number, could capture the problem of phenotypic heterogeneity even in the presence of the gene allegedly responsible for the genesis of a trait, was taken for granted. There was, of course, no question of creating a pure line with humans (as Timoféeff had done with *Drosophila funebris*), but this problem was not even perceived since the original reasoning behind the introduction of such a numerical measure had been lost in a history that was unknown to human geneticists. Most of all, penetrance came to be invoked to save a genetic etiology (the way in which it was invoked in the case of male sexual orientation is just a characteristic example). Expressivity, meanwhile, was yet another property of a gene, supposed to have a definite applicability, which helped to rescue genetic etiologies by explaining away variation in phenotypic expressions.

The real question for today is whether values for penetrance obtained in this fashion have any theoretical significance. Estimating penetrance from individual pedigrees and then combining them, with or without subsequent treatment of the data (for instance by establishing weighting based on some assumption about the prevalence of a pedigree) certainly results in an empirical frequency. Our concern is that these pedigrees neither reflect a random sample of the population, nor involve individuals with a fixed genetic background (pure lines), and that there is no reason to believe that, as the number of pedigrees increase, the estimated frequencies would converge to a definite probability. Consequently, it is simply illegitimate to think that the penetrance of an allele, so estimated, can ever be interpreted as the probability that an individual with that allele will manifest the phenotype in question. This point, which becomes obvious once the origin and subsequent history of the concept of penetrance has been reconstructed, is exactly what has been lost because that history, like the history of many other concepts and practices of human genetics, has often been neglected for a variety of political and ideological reasons. Meanwhile, 'penetrance,' as well as 'expressivity,' have been happily co-opted by those who wish to find a gene for every human characteristic, the enthusiasts for the new geneticism which, for both intellectual and sociopolitical reasons, we find increasingly troubling.

Arizona State University
Tempe, Arizona, U.S.A.
and
University of Texas
Austin, Texas, U.S.A.

ACKNOWLEDGMENTS

S.S. would like to acknowledge the support of a Fellowship from the Wissenschaftskolleg zu Berlin for the period when the research on which this paper is based was performed (1996-97). Both S.S. and M.D.L. would like to acknowledge support from the Max-Planck-Institut für Wissenschaftsgeschichte in Berlin, and especially its co-director, Hans-Jörg Rheinberger, for facilitating this collaboration.

NOTES

1. This is penetrance in the "general sense" or "wide sense." Ott also introduces a "narrow sense" in an effort to clarify some of the definitional problems: "Penetrance is sometimes used in a *narrow sense* as the probability of being affected (with a disease) given a genotype" (1991, p. 147). This distinction is not particularly relevant for the discussions in this chapter. The narrower sense of penetrance is as problematic as the customary sense (see the end of section 2).
2. Initially Vogt funded his research with the income from his own private medical practice. However, he was soon supported by grants from the Krupp family, Germany's most powerful steel manufacturer and defense contractor. The exact details of Vogt's relationship with the Krupp family are still not completely known. Vogt served as a medical consultant to the Krupps, but his research also captured the attention of Friedrich Alfred Krupp, an amateur biologist. Vogt's relationship with the Krupp family is also the subject of Tilman Spengler's novel *Lenins Hirn* (1991).
3. The journals were *Zeitschrift für Hypnotismus, Psychotherapie sowie anderer psychophysiologischer und psychopathologischer Forschungen* (1895-1902, with August Forel), *Journal für Psychologie und Neurologie* (1902-1942, with Forel and after 1920 also with Cécile Vogt), and *Journal für Hirnforschung* (1954-1959, with Cécile Vogt); the series of monographs is the *Neurobiologische Arbeiten*.
4. The research institutes were the *Neurologische Zentralstation* in Berlin (1899-1902), the *Neurobiologisches Universitäts Laboratorium* in Berlin (1902-1914/1919), the *Kaiser-Wilhelm Institut für Hirnforschung in Berlin* (1914/1919-1937, when Vogt was forced to resign as director), the *Institut für Hirnforschung und Allgemeine Biologie in Neustadt* (1937-1959), and the *Lenin Institute for Brain Research* in Moscow (see discussion below).
5. Vogt coined the term '*Eunomie*' for the ordered expression of variations, a concept that bears some resemblance to various theories of orthogenesis popular among evolutionists at that time. The idea behind a eunomic series of variations is that different stimuli will lead to predictable sequential changes in the expression of a character either between individuals or within the life history of a single individual. For a somewhat different interpretation of the relations between Vogt's genetics and psychiatry, see Satzinger, 1996, 1998; Mayr, 1980, p. 281.
6. The terms 'Genovariation,' 'Genenvariation,' and so on, are all being rendered as 'mutation' in this chapter. The former family of terms, derived from the work of the Chetverikov school, refers to the same process/entity as 'mutation' as used in U.S. and British genetics. See N.W. Timoféeff-Ressovsky and H.A. Timoféeff-Ressovsky for the adoption of 'mutation' as a synonym of the other terms (1927, pp. 70n, 106n).
7. See Vogt: "Insofar as extreme environmental relationships have not simultaneously altered the germplasm, modifications will sooner or later disappear again [Soweit die extremen Milieuverhältnisse

nicht gleichzeitig das Keimplasma verändert haben, werden Modifikationen naturgemäß früher oder später wieder verschwinden]" (1926, p. 809).

8.   "Das zeigt uns, daß die phänotypische Manifestierung an sich und die Stärke und Form dieser Manifestierung eines Erbmerkmals unabhängig voneinander sein können. Diese Frage erfordert aber weitere spezielle Studien."

9.   According to Ernst Mayr, Vogt advertised in the Soviet Union for a "young geneticist." Mayr states that "Dobzhansky felt he had a good chance for this position, except that Chetverikov was slow in writing the necessary letter of recommendation" (1980, p. 280). However, there is no evidence for this, and it is unlikely because of Vogt's obvious fascination with N.W. Timoféeff's 1925 paper.

10. See Vogt: "We name a determinately directed series of variation a *eunomie* [Eine bestimmt gerichtete Variationsreihe bezeichnen wir as eine *Eunomie*]," (1926, p. 812), and note 5.

11. "Es gibt Genenvariationen, welche sich unter der verschiedensten Bedigungen durchsetzen. Man muß diese Tendenz zum Sichdurchsetzen scharf von der Dominanz trennen. Ich schlage vor, diese Tendenz zum Sichdurchsetzen als '*Penetranz*' zu bezeichnen" (Vogt, 1926, p. 809).

12. "Unser Mitarbeiter *N. W. Timoféeff-Ressowsky* hat von der schwach penetranten Genenvaration "Radius incompletus" gezeigt, daß der "Rest" bestimmter Keimplasmen die Manifestierung der Genenvariation bis in 33,10% der gezüchteten Tiere hemmt" (Vogt, 1926, p. 810).

13. 'Spezifizität,' standardly translated as 'specificity,' is attributed to Timoféeff and explained as follows: "Then, besides penetrance, Timoféeff also correctly distinguishes different grades of specificity. The author then speaks of strong specificity when the mutation mainly always expresses itself through a strongly identical change of the same trait [Neben der Penetranz unterscheidet Timoféeff dann noch mit Recht verschiedene Grade der Spezifizität. Von starker Spezifizität spricht der Autor dann, wenn sich die Genevariation in der Hauptsache stets durch starke identische Veränderung desselben Merkmals äußert]" (Vogt, 1926, pp. 810-1). Timoféeff only published his definition of specificity later (Timoféeff-Ressovsky and Timoféeff-Ressowsky, 1926). The term 'specificity' is being ignored in this chapter largely because it has disappeared from modern genetics and is not invoked in the debates about genetic etiology that motivate this chapter. Moreover, 'specificity' cannot be clearly distinguished from 'expressivity' (see Rieger, Michaelis, and Green, 1991, p. 459).

14. "Von Penetranz und Spezifität ist dann noch eine dritte, von *Timoféeff* unterschiedene Eigenschaft der Genenvariation zu trennen, für welche man die Bezeichnung '*Expressivität*' einführen kann. Bei starker Penetranz, d. h. bei prozentual häufiger Manifestierung eines Merkmals kann dieses in schwacher Form auftreten (schwache Expressivität). Umgekehrt kann bei schwacher Penetranz, d. h. bei prozentual seltener Manifestierung eines Merkmales dieses bei seinem Auftreten in sehr ausgesprochener Form in Erscheinung treten (starke Expressivität)" (Vogt, 1926, p. 811).

15. "Von den einer Arbeit Romaschoffs entnommenen Abb. 1 und 2 zeigt Abb. 2 ein ganz verunstaltetes Abdomen. Derartige bis zur Lebensunfähigkeit veränderte Tiere entstehen bei der Genenvariation 'Abdomen abnormale [sic]' regelmäßig bei einem mittleren Grad von Feuchtigkeit. Züchtet man die Tiere dagegen in großer Trockenheit oder bei sehr starker Feuchtigkeit, so entstehen äußerlich normale Tiere, wie ein solches in der Abb. 1 abgebildet ist. Zwischenstufen des Milieus schaffen Zwischenstufen zwischen den in Abb. 1 und 2 abgebildeten Formen" (Vogt, 1926, pp. 809 -10).

16. "Alle diese Momente vermögen an sich ausgesprochen sprunghafte Genenvariationen in den Phänotypen einer Population mehr oder weniger zu verdecken" (Vogt, 1926, p. 812).

17. "Wir beschränken uns auf Deduktionen auf Grund einer wirklichen Identität zwischen den Psychosen einerseits und den Somavariationen und ganz elementaren Genenvariationen einer zoologisch-botanischen Sippe innerhalb der Spezies oder Subspezies" (Vogt, 1926, p. 830).

18. "Bei allen komplizierteren Erbgängen wird die erbbiologische Forschung nicht länger sich auf das Mischen der Gene und ihre Dominanz und Recessivität beschränken. Sie muß allen durch die experimentelle Genetik neu erkannten Eigenheiten der Gene Rechnung tragen, so auch der Penetranz, der Spezifizität und der Expressivität" (Vogt, 1926, p. 831).

19. "Mit dem Terminus 'Penetranz' (O. Vogt 1926) wird die Fähigkeit des Faktors, sich phänotypisch überhaupt zu manifestieren, bezeichnet. Es gibt Faktoren (stark penetrante), die sich stets phänotypisch manifestieren, während andere (schwach penetrante) nur in einem gewissen Prozentsatz der Fälle zum Ausdruck kommen. Der Grad der phänotypischen Manifestierung des Faktors im Merkmal wird durch den Terminus 'Expressivität' bezeichnet" (1926, p. 150n). The footnote goes on to introduce the term 'Spezifizität'; although Vogt (1926) had used that term earlier (1926), attributing it to N.W. Timoféeff-Ressowsky, this was the first time that the latter had explicitly used the term.

20. Dobzhansky was similarly responsible for the repatriation of another concept of German origin, the 'norm of reaction,' to the West. See Sarkar, 1999, for a history.

FIGURES

Figure 1: The mutation *"Abdomen abnormale"* in *Drosophila funebris*. Abb. 1 was raised under dry conditions and appears normal; Abb.2 was raised under wet conditions and shows the typical modifications (from Vogt, 1926, after Romaschoff, 1925).
Figure 2: The mutation *"Radius incompletus"* in *Drosophila funebris*. Abb. 8 shows weak expressivity; Abb. 9 shows strong expressivity of the radius *r* on the wing (from Vogt 1926, after Timoféeff, 1925).
Figure 3: Examples for eunomic series in the coloration of *Bombus lapidarius*. The lower series shows a typical sequence of character states demonstrating increasingly lighter coloration; the upper series shows a somewhat altered series of specimens from the Caucasus (from Vogt, 1926).
Figure 4: Eunomic series of a cortical degeneration in the *Area striata*, showing increasing degeneration of nerve cells (from Vogt, 1926).

REFERENCES

Adams, M.B. "Sergei Chetverikov, The Kol'tsov Institute, and The Evolutionary Synthesis." In *The Evolutionary Synthesis: Perspectives on the Unification of Biology*, E. Mayr, W.B. Provine, eds. Cambridge: Harvard University Press, 1980.
Adams, M., ed., *The Wellborn Science*. Oxford: Oxford University Press, 1990.
Bateson, W.B., *Materials for the Study of Variation*. London: Macmillan, 1984.
Bowler, P., *The Eclipse of Darwinism*. Baltimore: Johns Hopkins University Press, 1983.
Breidbach, O., *Die Materialisierung des Ichs, Suhrkamp*. Frankfurt: am Main, 1996.
Brodmann, K., *Vergleichende Lokalisationslehre der Großhirnrinde in ihren Prinzipien dargestellt aufgrund des Zellenbaues*. Barth: Leipzig, 1909.
Darlington, C.D., Mather, K., *The Elements of Genetics*. London: Macmillan, 1949.
Foerster O. Die Topik der Hirnrinde und ihre Bedeutung für die Motilität. Deutsche Zeitschrift für Nervenheilkunde 1923; 77:124-39
Foerster O. Zur Pathogenese und chirurgischen Behandlung der Epilepsie. Zentralblatt für Chirurgie 1925; 25:531-56
Foerster O. Zur Pathogenese des epileptischen Krampfanfalls. Deutsche Zeitschrift für Nervenheilkunde 1926; 94:15-56
Hagner, M. "Lokalization, Funktion, Cytoarchitektonik." In *Objekte, Differenzen und Konjunkturen. Experimentalsysteme im historischen Kontext*, M. Hagner, H.-J. Rheinberger, B. Wahrig Schmidt, eds. Berlin: Akademie Verlag, 1994.
Hagner, M., *Homo cerebralis: Der Wandel vom Seelenorgan zum Gehirn*. Berlin: Berlin Verlag, 1997.
Harrington, A., *Medicine, Mind, and the Double Brain*. Princeton: Princeton University Press, 1987.
Harrington, A., *Reenchanted Science. Holism in German Culture from Wilhelm II to Hitler*. Princeton: Princeton University Press, 1996.
Harwood, J., *Styles of Scientific Thought*. Chicago: University of Chicago Press, 1993.
Kevles, D., *In the Name of Eugenics*. New York: Knopf, 1985.
Laubichler, M. "Oskar and Cécile Vogt: From the Neo-Cortex to Bumble Bees, An Episode in the History of the Biological Character Concept." In *The Character Concept in Evolutionary Biology*, G.P.Wagner, ed. San Diego: Academic Press, 2000.
Mayr, E. "Germany." In *The Evolutionary Synthesis: Perspectives on the Unification of Biology*, E. Mayr, W.B. Provine, eds. Cambridge: Harvard University Press, 1980.
Ott, J., *Analysis of Human Genetic Linkage*, 2nd ed. Baltimore: Johns Hopkins Press, 1991.
Patzig B. Die Bedeutung der schwachen Gene in der menschlichen Pathologie, insbesondere bei der Vererbung striärer Erkrankungen. Naturwissenschaften 1933; 21:410-3
Paul D.B., Krimbas C.B. "The strange career of N.W.Timoféeff-Ressovsky." Unpublished manuscript, 1991.

Paul D. B., Krimbas C. B., Nikolai V., Timoféeff-Ressovsky. Scientific American 1992; 266(2):86-92
Paul, D., *Controlling Human Heredity*. Atlantic Highlands, NJ: Humanities Press, 1995.
Provine, W., *The Origins of Theoretical Population Genetics*. Chicago: University of Chicago Press, 1971.
Richter J. Oskar Vogt, der Begründer des Moskauer Staatsinstituts für Hirnforschung. Psychiatrie, Neurologie und medizinische Psychologie 1976a; 28:385-95
Richter J. Oskar Vogt und die Gründung des Berliner Kaiser-Wilhelm Institutes für Hirnforschung unter den Bedingungen imperialistischer Wissenschaftspolitik. Psychiatrie, Neurologie und medizinische Psychologie 1976b; 28:449-57
Richter, J. "Medicine and Politics in the Soviet-German Relations in the 1920s: A Contribution to Lenin's Pathobiography." In *Proceedings of the XXXIInd International Congress on the History of Medicine Antwerp, 3 -7 September 1990* E. Fierens, J. Tricot, T. Appelboom, M. Their, eds. Bruxelles: Societas Belgica Historiae Medicinae, 1991.
Richter, J. "Das Kaiser-Wilhelm-Institut für Hirnforschung und die Topographie der Großhirnhemisphären. Ein Beitrag zur Institutsgeschichte der Kaiser-Wilhelm-Gesellschaft und zur Geschichte der Architektonischen Hirnforschung." In *Die Kaiser-Wilhelm-Gesellschaft und ihre Institute, Berlin*, B. vom Brocke, H. Laitko, eds. Berlin: Walter de Gruyter, 1996.
Rieger, R., Michaelis, A., Green, M.M., *Glossary of Genetics*, 5th ed. Berlin: Springer Verlag, 1991.
Romaschoff D.D. Über die Variabilität in der Manifestierung eines erblichen Merkmales (Abdomen abnormalis) bei Drosophila funebris F. Journal für Psychologie und Neurologie 1925; 31:323-5
Sarkar, S., *Genetics and Reductionism*. New York: Cambridge University Press, 1998.
Sarkar S. From the *Reaktionsnorm* to the Adaptive Norm: The Norm of Reaction, 1909-1960. Biology and Philosophy 1999; 14:235-52
Satzinger H. Zur Neurobiologie und Genetik im Zeitraum von 1902 -1922 in den Forschungen von CÉCILE und OSKAR VOGT (1875 -1962; 1870 -1959). Biologisches Zentralblatt 1994; 113:183-95
Satzinger H. Vom "Artproblem" zur "Genoplastik": Zur Entwicklung genetischer Fragestellungen im Werk von Cécile und Oskar Vogt 1909-1925. Biologisches Zentralblatt 1996; 115:104-11
Satzinger, H., *Die Geschichte der genetisch orientierten Hirnforschung von Cécile und Oskar Vogt*. Stuttgart: Deutscher Apotheker Verlag, 1998.
Sinnott, E.W., Dunn, L.C., *Principles of Genetics*, 2nd ed. New York: McGraw-Hill, 1932.
Sinnott, E.W., Dunn, L.C., *Principles of Genetics*, 3rd ed. New York: McGraw-Hill, 1939.
Sinnott, E.W., Dunn, L.C., Dobzhansky, T., *Principles of Genetics*, 4th ed. New York: McGraw-Hill, 1950.
Spengler ,T., *Lenins Hirn*. Reinbeck: Rowohlt Verlag, 1991.
Stern C.O. Vogt and the terms "penetrance" and "expressivity." American Journal of Human Genetics 1960; 12:141
Sturtevant, A.H., Beadle, G.W., *An Introduction to Genetics*. Philadelphia: W.B. Saunders, 1939.
Timoféeff-Ressovsky H.A., Timoféeff-Ressovsky N.W. Über das phänotypische Manifestieren des Genotyps. II. Über idio-somatische Variationsgruppen bei Drosophila funebris. Wilhelm Roux' Archiv für Entwicklungsmechanik der Organismen 1926; 108:148-70
Timoféeff-Ressovsky H.A., Timoféeff-Ressovsky N.W. Genetische Analyse einer freilebenden Drosophila melanogaster Population. Wilhelm Roux' Archiv für Entwicklungsmechanik der Organismen 1927; 109:70-109
Timoféeff-Ressovsky N.W. Über den Einfluss des Genotypus auf das phänotypische Auftreten eines einzelnen Gens. Journal für Psychologie und Neurologie 1925; 31:305-10
Timoféeff-Ressovsky N.W. Gerichtetes Variieren in der phänotypischen Manifestierung einiger Genovariationen von Drosophila funebris. Naturwissenschaften 1931a; 19:493-7
Timoféeff-Ressovsky N.W. Über phänotypische Manifestierung polytopen (pleiotropen) Genovariation Polyphaen von Drosophila funebris. Naturwissenschaften 1931b; 19:765-8
Vogt C., Vogt O. Zur Kenntnis der elektrisch erregbaren Hirnrindengebiete bei den Säugetieren. Journal für Psychologie und Neurologie 1907; 8:277-456
Vogt C., Vogt O. Allgemeinere Ergebnisse unserer Hirnforschung. Journal für Psychologie und Neurologie 1919a; 25:277-461
Vogt C., Vogt O. Zur Kenntnis der pathologischen Veränderungen des Striatum und des Palladium und zur Pathophysiologie der dabei auftretenden Krankheitserscheinungen. Sitzungsberichte der Heidelberger Akademie der Wissenschaften. Math. Natwissen. Klasse. Abt B 1919b; 14:1-56

Vogt C., Vogt O. Zur Lehre der Erkrankungen des striären Systems. Journal für Psychologie und Neurologie 1920; 25:631-846

Vogt C., Vogt O. Die vergleichend-architektonische und die vergleichend-reizphysiologische Felderung der Großhirnrinde unter besonderer Berücksichtigung der menschlichen. Naturwissenschaften 1926; 14:1190-4

Vogt C., Vogt O. Hirnforschung und Genetik. Journal für Psychologie und Neurologie 1929; 39:438-46

Vogt C., Vogt O. Sitz und Wesen der Krankheiten im Lichte der topistischen Hirnforschung und des Variieres der Tiere. Erster Teil. Befunde der topistischen Hirnforschung als Beitrag zur Lehre vom Krankheitssitz. Journal für Psychologie und Neurologie 1937; 47:237-457

Vogt C., Vogt O. Sitz und Wesen der Krankheiten im Lichte der topistischen Hirnforschung und des Variieres der Tiere. Zweiter Teil, 1. Hälfte. Zur Einführung in das Variieren der Tiere. Die Erscheinungsseiten der Variation. Journal für Psychologie und Neurologie 1938; 48:169-324

Vogt O. Flechsig's Associationscentrenlehre, ihre Anhänger und Gegner. Zeitschrift für Hypnotismus 1897; 5:347-61

Vogt O. Psychologie, Neurophysiologie und Neuroanatomie. Journal für Psychologie und Neurologie 1902; 1:1-3

Vogt O. Studien über das Artproblem. 1. Mitteilung. Über das Variiren der Hummeln. 1. Teil. Sitzungsberichte der Gesellschaft Naturforschender Freunde zu Berlin 1909; 1:28-84

Vogt O. Studien über das Artproblem. 2. Mitteilung. Über das Variiren der Hummeln. 2. Teil. Sitzungsberichte der Gesellschaft Naturforschender Freunde zu Berlin 1911; 3:31-74

Vogt O. Psychiatrische Krankheitseinheiten im Lichte der Genetik. Zeitschrift für die gesamte Neurologie und Psychiatrie 1925; 100:26-34

Vogt O. Psychiatrisch wichtige Tatsachen der zoologisch-botanischen Systematik. Zeitschrift für die gesamte Neurologie und Psychiatrie 1926; 101:805-32

Vogt O. Neurology and eugenics: the role of experimental genetics in their development. Eugenics Review 1932; 24:15-8

Waddington, C.H., An Introduction to Modern Genetics. London: George Allen & Unwin, 1938.

Waddington, C. H., Organisers and Genes. Cambridge: Cambridge University Press, 1940.

Waddington, C.H., The Strategy of the Genes. New York: Macmillan, 1957.

Weindling P. German-Soviet medical co-operation and the Institute for Racial Research 1927-c. 1935. German History 1992; 10:177-206

CHAPTER 5

DIANE PAUL

FROM REPRODUCTIVE RESPONSIBILITY TO REPRODUCTIVE AUTONOMY

1. INTRODUCTION

That procreation is a basic human right, with which the state has no business meddling, is today the dominant view among Western genetics professionals, bioethicists, and journalists. In their perspective, reproductive genetic services should aim at increasing the choices available to women. Since no reproductive choice is right or wrong, clinicians should be scrupulously neutral in their dealings with clients. Any other approach constitutes "eugenics."

This perspective is encapsulated in a passage from an Institute of Medicine report: "The goal of reducing the incidence of genetic conditions is not acceptable, since this aim is explicitly eugenic; professionals should not present any reproductive decisions as 'correct' or advantageous for a person or society" (Andrews, 1994, p. 15). The authors take for granted that eugenics is bad, an assumption nearly universally shared by geneticists in North America and much of Europe (which is not to say that they agree on what eugenics is). Assessing geneticists' responses to questions in two large, international surveys, Dorothy Wertz writes: "The word *eugenics* almost never appeared in North American or European reasoning, except as an example of unspeakable evil, having been totally discredited by the Nazi's actions" (Wertz, 1998, p. 156).

Yet, earlier in the century (as well as in many countries today), the view that reproduction is entirely a private matter was anything but obvious. Indeed, in the 1910s, 1920s, and 1930s, the opposite view was generally taken for granted even by critics of the eugenics movement of their day (see Paul and Spencer, 1995). Although there was opposition to eugenic sterilization, it did not necessarily imply

L.S. Parker and R.A. Ankeny (eds.), Mutating Concepts, Evolving Disciplines: Genetics, Medicine and Society, 87-105.
© *2002 Kluwer Academic Publishers. Printed in the Netherlands.*

opposition to eugenics *tout court*, much less to the view that reproduction was a matter of social concern. It was generally assumed that mentally deficient individuals should be prevented from breeding. The issue was how best to achieve this end. Sterilization was sometimes criticized as ineffective, or based on unfounded assumptions about the heredity of mental defect, or inevitably biased in its application (e.g., see Davies, 1930, pp. 94-120). A paramount concern was that it would provide a cheap substitute for the more desirable policies of segregation, training, and community supervision. But such criticism of means should not be confused with the claim that procreative liberty is a right.

The Catholic Church was the most influential opponent of sterilization. In his famous 1930 papal encyclical *Casti Connubi* ("On Christian Marriage"), Pope Pius XI criticized those who, "over-solicitous for the cause of eugenics" propose to interfere with "the natural right of man to enter matrimony" (Pius XI, 1931, p. 31). He particularly condemned sterilization as an illegitimate arrogation by the state of power over the bodies of its subjects. In the absence of a crime, he argued, public authorities have no business physically harming their subjects. But the Catholic position on reproduction did not reflect a pro-autonomy perspective. Indeed, immediately after criticizing this use of state power, the Pope noted that private individuals are also barred from tampering with their bodies in order to "render themselves unfit for their natural functions" (p. 33). That the Church hardly considered procreation a private matter is indicated by the condemnation, in the same encyclical, of contraception and abortion.

Nor was the Church unremittingly hostile to eugenics. In the 1931 encyclical, Pius XI conceded that eugenics must "be accepted, provided lawful and upright methods are employed within the proper limits" (p. 31; see also Lepicard, 1998). According to his successor, these included efforts to dissuade (though not forbid) carriers of grave genetic disorders from marrying each other, and use of the rhythm method to prevent conception where married couples risk transmitting a severe disease. While reiterating the ban on sterilization and other means of birth-control, he wrote: "Better warned of the problems posed by genetics and of the gravity of certain hereditary diseases, men of today have, more than in the past, the duty to take account of this increased knowledge so that they might forestall countless physical and moral difficulties for themselves and others" (Pius XII, 1959, p. 13). In the 1950s and 1960s, distinguished Protestant theologians, as well as many progressively-minded geneticists, also argued that individuals at high risk of transmitting a serious genetic disease should refrain from reproducing. Some of them maintained that if individuals failed to discharge their reproductive responsibilities, the state should intervene.

Views common in the United States through the 1960s now horrify when they are expressed by officials in China, where a law explicitly titled "On Eugenics and Health Protection" was proposed in 1993. The government did not anticipate the furor that the law prompted in the West. The 18th International Congress of Genetics, held in Beijing in August 1998, was officially boycotted by the national genetics societies of England, Holland, and Argentina, and shunned as well by numerous individual scientists (Rosenthal, 1998, p. 14). Moreover, many geneticists who attended the Congress did so on the condition that there would be opportunities to criticize the by then renamed Maternal and Infant Health Care Law.

Because it included provisions permitting doctors to sterilize individuals with serious genetic conditions without their consent, the law was said to contravene article 16 of the United Nations Universal Declaration of Human Rights, which states that "men and women of full age, without any limitation due to race, nationality or religion, have the right to marry and to found a family," and to violate basic norms governing the provision of genetic services. In response, the Chinese government eliminated the references to "inferior births," and, following the 1998 Congress, it officially suspended the offensive provisions. According to the new policy, sterilization would be allowed only with the consent of the couple or a guardian (Pomfret, 1998, p. 10). Of course, the degree to which practice changed, if at all, is another question.

From the reaction to the Chinese law, it would seem that views in the West have been transformed. Today, many people are shocked by assumptions the rightness of which their grandparents took for granted. One aim of this chapter is to explain how and why this apparent sea change in attitudes occurred. Another is to question its depth and breadth, for it appears that the public is less committed to the principle of respect for reproductive autonomy than are those "who are authorized to speak for and defend" that public in bioethical matters.[1]

2. WHAT HAPPENED TO EUGENICS?

The worldwide economic depression undermined the identification of status with genetic worth. For that reason, it is often assumed that enthusiasm for eugenics— and with it, the view that the state has a legitimate interest in who reproduces— waned in the 1930s. However, the Depression also increased pressures to sterilize individuals who would otherwise require expensive institutionalization. Thus, in 1935 George Reid Andrews could remark on the "rapid and remarkable" progress of eugenics, and note that people "who a few years ago were indifferent if not hostile to the teachings of eugenics, are now deeply interested because of the practical bearing of these teachings upon the problems of the present depression" (1935, p. 1). During the 1930s, when support for eugenics is said to have eroded, sterilization laws were passed in Germany, the Canadian provinces of Alberta and British Columbia, Norway, Denmark, Sweden, Finland, Iceland, and Estonia, while in the United States, the number of procedures performed under the older laws climbed (Paul, 1995, pp. 72-90). Japan passed a sterilization law in 1940 (Otsubo and Bartholomew, 1998).

In the immediate aftermath of World War II, eugenics did become unfashionable. Revelations of Nazi atrocities produced a general revulsion against genetic explanations of individual and group differences. The pendulum swung, at least briefly, from a hereditarian to an environmentalist perspective. As Dorothy Nelkin and Susan Lindee have noted, "nurture triumphed indisputably in both the scientific and popular rhetoric of the 1950s and much of the 1960s. The stories of biological determinism that had characterized the eugenics literature were replaced by narratives of cultural determinism" (1995, p. 34). But they also note that this cultural shift was short-lived. A variety of factors, including anxieties over long-term genetic damage resulting from atmospheric nuclear testing, well-publicized advances in medical treatment, and a perceived population explosion, soon converged to produce renewed enthusiasm for the view that it would be socially irresponsible for some kinds of people to reproduce.

Even in the 1950s and 1960s, eugenic language persisted among professionals in fertility and in human genetics (Nelkin and Lindee, 1995, pp. 34-7; see also Kevles, 1985, pp. 251-68; Wright, 1994, pp. 23-6). Indeed, something of a backlash among geneticists was evident shortly after the war's end. Classical geneticists such as H. J. Muller and Julian Huxley, who had championed programs of selective breeding before the war, now renewed their campaigns. In his 1949 Presidential address to the newly founded American Society of Human Genetics, Muller argued that the human species was deteriorating under an ever-increasing load of deleterious mutations. In his view, this burden was attributable both to expanded military and medical uses of radiation (especially atmospheric nuclear testing), and to therapeutic advances in medicine, which allowed individuals who would once have died before childbearing to survive and reproduce. To counter the threat of degeneration, he proposed a scheme under which the most burdened three percent of the population would voluntarily refrain from reproducing (Muller, 1950; Paul, 1987). A few years later, he also resurrected a proposal of the 1930s to bank the sperm of men outstanding in regards to their intellect, temperament, character, and physical traits (a program now called "germinal choice" to emphasize its voluntary character).

Muller's warnings about genetic deterioration, and hopes of genetic improvement, appear to have resonated widely; echoes of his argument are found in numerous popular articles and scientific symposia. The spell he exerted is reflected in Tracey Sonneborn's comment that Muller's contribution to the 1963 symposium on The Control of Human Heredity and Evolution "was received with such enthusiasm by the audience that, after hurried consultation with some of the other speakers, we agreed it would be anticlimactic and quite undesirable to throw it open for general discussion" (1965, p. 124).

Interest in the control of human evolution, which is reflected in the many symposia in the 1960s on the genetic future of humankind, was also spurred by the discovery of the double-helical structure of DNA and unraveling of the genetic "code" (Ludmerer, 1972, p. 181). Molecular scientists such as Francis Crick, Joshua Lederberg, Salvador Luria, Linus Pauling, and Robert Sinsheimer began to debate the pros and cons of what Muller called "genetic surgery" and Rollin Hotchkiss (in 1965) termed "genetic engineering" (see Kay, 1993, pp. 275-6; Kevles, 1985, pp. 258-68; Wright, 1994, pp. 123-4). Some scientists, such as Luria, argued for caution. Others, perhaps emboldened by their achievements in manipulating life, urged intervention.

Muller's pessimistic view of the future, should a *laissez-faire* attitude prevail, was shared by many molecular scientists. Even Lederberg, who was generally skeptical of schemes to control human breeding, asserted that "the facts of human reproduction are all gloomy—the stratification of fecundity by economic status, the new environmental insults to our genes, the sheltering by humanitarian medicine of once-lethal defects" (Lederberg, 1963, p. 264). Crick went much further. At the famous 1962 CIBA Foundation symposium on Man and His Future, Crick expressed his agreement "with practically everything Muller said" about the urgent need both to prevent further genetic deterioration and to increase the proportion of superior genotypes in the population. In place of the current *laissez-faire* system of reproduction, Crick argued, we might substitute a licensing scheme, whereby "if the parents were genetically unfavorable, they might be allowed to have only one child, or possibly two under special circumstances" (Wolstenholme, 1963, p. 275). Commenting on Crick's suggestion, N.W. Pirie asserted that, on the question of whether there is a right to have children, "in a society in which the community is responsible for people's welfare—health, hospitals, unemployment insurance, etc.— the answer is 'No'" (Wolstenholme, 1963, p. 282).

Five years later, Pauling, who was also influenced by Muller, proposed that all young people should have tattooed on their forehead symbols for any seriously defective recessive genes, such as those producing sickle cell anemia and PKU. He expressed confidence that, if this were done, carriers for the same defective gene "would recognize the situation at first sight, and would refrain from falling in love with one another." He also thought that "legislation along this line, compulsory testing for defective genes before marriage, and some form of public or semi-public display of this possession, should be adopted" (Pauling, 1968, p. 527). Pauling was concerned that one long-term effect of such a program would be a slight increase in the future incidence of disease genes. But he thought that result could be countered through an educational process aimed at convincing carriers "married to normals" to have fewer than the average number of children. (It is notable that he explicitly characterized his scheme as a form of "negative eugenics.")

Muller's scientific assumptions and policy proposals were strenuously contested, but the view that reproduction is not simply a private matter was accepted even by some of his severest critics. Among Muller's most influential antagonists were geneticist Theodosius Dobzhansky and anthropologist Ashley Montagu. In a book that took issue with virtually every other aspect of Muller's eugenics, Dobzhansky remarked that persons who carry serious genetic defects should be persuaded not to reproduce, and if persuasion should fail, "their segregation or sterilization is justified" (1962, p. 333). He continued: "We need not accept a *Brave New World* to introduce this much of eugenics." Montagu likewise wrote that, "there can be no question that infantile amaurotic family idiocy [Tay Sachs disease] is a disorder that no one has a right to visit upon a small infant. Persons carrying this gene, if they marry, should never have children, and should, if they desire children, adopt them" (1959, pp. 305-6). Sheldon Reed, who coined the expression 'genetic counseling,' also asserted that no couple has the right to knowingly produce a child with a serious genetic disorder (1964, p. 85). Montagu, Dobzhansky, and Reed, unlike Muller and Pauling, were unconcerned with the 'gene pool.' They thought the long-term would take care of itself. In their view, it was the potential short-term consequences, for the child-to-be and perhaps for the larger society, that mattered. But all the scientists agreed that reproduction was a social concern.

In theology as well, intense disagreements about the nature of the problem and the validity and utility of various eugenic solutions have obscured underlying agreement on fundamental issues. During the 1960s, Protestant thinking on bioethical matters was dominated by two theologians: liberal Episcopalian Joseph Fletcher and conservative Methodist Paul Ramsey (Jonsen, 1998, pp. 34-51). Ramsey was an influential critic of the nascent field of genetic engineering, and he sparred with both Fletcher and many scientists over the ethics of sperm banking, cloning, and other real or potential genetic manipulations. At the same time, he argued that the marriage licensing power of the state be used to prevent the transmission of grave diseases. "After all," he wrote, "it ought never to be believed that everyone has an unqualified right to have children, or that children are simply for one's own fruition...The freedom of parenthood is a freedom to good parentage, and not a license to produce seriously defective individuals to bear their own burdens" (Ramsey, 1970, pp. 98-9). Fletcher frequently made the same point in even stronger terms. Indeed, as late as 1980, he claimed that, "reproductive rights are not absolute and those who are at risk for passing on clearly identifiable, severely deleterious genes and debilitating genetic disease should not be allowed to exercise their reproductive prerogative" (Fletcher, 1980, p. 131). According to Fletcher, "testes and ovaries are communal by nature, and ethically regarded they should be rationally controlled in the social interest" (1980, p. 134).

As noted earlier, the resurgence of interest in eugenics (now focused primarily on clinical disease rather than mental defect) was fueled in part by anxieties over the potential long-term genetic damage resulting from atmospheric nuclear testing, an issue vigorously publicized by Muller, whose 1947 Nobel Prize for the discovery of the mutagenic properties of X-rays allowed him to speak with great authority. But it was also buttressed by well-publicized advances in medical treatment, especially insulin therapy for diabetes and dietary treatment for phenylketonuria (PKU).[2] In the mid-1960s, the first population-wide screening programs to identify newborns with PKU were established. It soon became clear that as a result of such screening and treatment, individuals who would ordinarily not have reproduced would enjoy near-normal fertility. Although PKU itself was rare, it was thought at the time that dietary treatment would soon be available for a large number of metabolic conditions and that, in general, the new science would result in many more treatments for genetic disease. But unless the germ line could be directly altered, such progress would come at a cost to the 'gene pool.'

Sir Peter Medawar, a severe critic of Muller, argued that the follies of many past and present eugenic proposals should not prevent us from recognizing "that rationally founded and humane procedures in the area of negative eugenics *are* possible..." (1977, p. 61). He warned of the inevitable economic implications of advances in therapeutic medicine, noting that, "if diabetics are to be kept alive and restored by medical procedures to something approaching a state of normal health, as it is right that they should be, then whatever elements of their genetic make-up may have contributed to their diseased state will for that reason be disseminated more widely throughout the population," and he suggested that carriers of the recessive gene for PKU be discouraged from marrying each other (Medawar, 1977, pp. 58-9, 63-4).

Modern medicine was also held responsible for a population explosion. At numerous conferences and in many books and articles, it was argued that, if world population were not checked, it would be impossible to maintain a minimum standard of living (e.g., Ehrlich, 1968). Proposals to lower birth rates ranged from encouragement of voluntary family planning to schemes involving direct social controls (for a summary of suggestions, see Berelson, 1969). The sense of alarm surrounding the issue is reflected in economist Kenneth Boulding's proposed system of marketable licenses to have children. He explained:

> Each girl on approaching maturity would be presented with a certificate which will entitle its owner to have, say, 2.2 children or whatever number would ensure a reproductive rate of one. The unit of these certificates might be the 'deci-child', and accumulation of ten of these units by purchase, inheritance, or gift would permit a woman in maturity to have one legal child. We would then set up a market in these units in which the rich and the philoprogenitive would purchase them from the poor, the nuns, the maiden aunts, and so on. (Boulding, 1964, p. 135)

(Boulding thought the plan would have the added advantage of reducing income inequality, since the rich would have more children, leaving them poorer, while the poor would have fewer children, leaving them richer.)

Boulding aimed to solve the population problem while preserving a maximum of individual choice. But coercive proposals were also common, justified both by the gravity of the situation and by the assumption that limiting population growth is a public good unachievable through an appeal to individual families (which have economic or emotional interests in having more children). The concept of a "tragedy of the commons"—situations in which individuals acting rationally will produce an outcome in which everyone is worse off—was popularized by Garrett Hardin, who argued that, "the only way we can preserve and nurture other and more precious freedoms is by relinquishing the freedom to breed, and that very soon" (1968, p. 1248).

It was but a short step to the conclusion that, if breeding must anyway be limited, the restrictions should be selective, particularly since the population control movement had deep roots in the pre-World War II eugenics establishment. Indeed, all the main organizations and individuals promoting population control in the 1950s and 1960s had earlier supported eugenics research and advocacy (see Gordon, 1990, pp. 386-97). It was only natural for them to ask why procreative liberty should not be limited for the purpose of improving the population if it could legitimately be limited for the purpose of reducing its growth. Vance Packard summarized a common argument of the day: "If you are going to try to control the quantity of the population, why not also control the quality?" (1977, p. 257). This argument has strong echoes in contemporary China, where eugenics has been justified by the one-child population policy. Thus, Sun Dong-Sheng (1981) of the Jinan Army Institute writes: "Eugenics can also play a considerable role in controlling population growth. If a couple gives birth to a disabled or retarded child, they will invariably want to have a second child. As a result, the proportion of our population which is of poor quality increases as does the overall birth rate."

Bentley Glass's 1970 presidential address to the American Association for the Advancement of Science exemplifies the often-asserted link between the need to restrict how many and who should reproduce. According to Glass,

> in an overpopulated world, it can no longer be affirmed that the right of the man and woman to reproduce as they see fit is inviolate. On the contrary, if my own additional child deprives someone else of the privilege of parenthood, I must voluntarily refrain, or be compelled to do so. In a world where each pair must be limited, on the average, to two offspring and no more, the right that must become paramount is not the right to procreate, but rather the right of every child to be born with a sound physical and mental constitution, based on a sound genotype. No parents will in that future time have the right to burden society with a malformed or a mentally incompetent child, just as every child must have the right to full educational opportunity and a sound nutrition, so every child has the inalienable right to a sound heritage. (Glass, 1971, p. 28)

For these (and perhaps other) reasons, proposals to limit reproductive freedom were widely discussed during the 1960s. In 1966, a writer for *Harper's Magazine* reported on a proposal, discussed at a human genetics conference, to issue licenses to reproduce only to those whose genes received a passing grade, and commented that "eugenic proposals like this are commonplace at scientific meetings nowadays. After twenty years of ill repute, eugenics is again the subject of respectable scientific investigation" (Eisenberg, 1966, p. 53).

Of course, the morality and utility of the enterprise were also strenuously criticized, and there were important countercurrents. Proponents of schemes for selective breeding often complained that they were swimming against the tide, especially if they advocated compulsion. Indeed, oppositional forces were building and would soon swamp efforts to rehabilitate both the term 'eugenics' and *any* principle of reproductive responsibility. Already in 1967, thirty nations signed a UN statement to the effect that "the Universal Declaration of Human Rights describes the family as the natural and fundamental unit of society. It follows that any choice and decision with regard to the size of the family must irrevocably rest with the family itself, and cannot be made by anyone else." In the same year, Kingsley Davis (who served as President both of the American Sociological Society and the Population Association of America) complained that "in the sphere of reproduction, complete intellectual initiative is generally favored even by those liberal intellectuals who, in other spheres, most favor economic and social planning" (Davis, 1967, p. 737). But within some professional communities, including genetics, there was little fear of characterizing favored policies as 'eugenics.' And even geneticists such as Dobzhansky or Montagu, who would have rejected the 'eugenicist' label, assumed that it was wrong to knowingly transmit a serious genetic disease.

Given the number of discussions in scientific circles about the need for reproductive restraint, it is understandable that many commentators at the time thought they had spotted a trend. For example, Harvard historian Donald Fleming wrote: "What we may reasonably expect is a continually rising chorus by the biologists, moralists, and social philosophers of the next generation to the effect that nobody has a right to have children, and still less the right to determine on personal grounds how many" (1969, p. 69). Similarly, Paul Ramsey thought it could "safely be predicted that the future will see more rather than less discussion of proposals for genetic control" (1970, pp. 1-2).

As it turns out, their predictions proved utterly wrong. Within a decade, views that seemed to Fleming, Ramsey, and many others in the 1960s to have been in the ascendancy were decidedly out of fashion. It may seem astounding that assumptions could change so quickly. But as Cass Sunstein has argued, allegiance to social norms is often weaker than it seems. What he terms a "norm cascade" can result if people with an interest in changing attitudes can exploit that fact (Sunstein, 1997, p. 36). In the 1970s, there were groups committed to changing norms in respect to reproduction. And a variety of events, described in the next section, allowed them to prevail.

3. THE RISE OF AUTONOMY

The social turmoil that began in the 1960s was one such development, particularly the closely linked patients' rights and feminist movements, which themselves followed on the civil rights and anti-war campaigns. Doctors' authority was denounced as patriarchal. "Our Bodies, Ourselves," as the book title had it, along with "autonomy," "choice," and "self-determination," became movement slogans (Gordon, 1990, pp. 400-6). To early feminists, autonomy implied "control over our bodies, our labor and economic resources, our life decisions," and as such was viewed as central to the achievement of feminist political goals (Di Stefano, 1996, p. 95).

At the same time, reproductive genetic services began a period of rapid expansion. Prior to the 1970s, few individuals made use of the limited genetic counseling services then available. Most of those who did came from families with a history of some disorder or already had an affected child. All the counselor could offer was an (often imprecise) estimate of risk. Moreover, the clients' choices were severely limited since the risk could be avoided only by refraining from childbearing. Under these conditions, genetic counseling had little to offer, and it remained a small-scale affair, mostly practiced by Ph.D. geneticists and by physicians trained in genetics (Porter, 1977, p. 23; Sorenson and Culbert, 1974).

That situation changed in the 1970s when the development of prenatal diagnosis coincided with the legalization of abortion. Amniocentesis was first developed in the 1960s, and entered clinical practice in the 1970s. In Britain, abortion was legalized by an Act of Parliament in 1967, while in the United States, the Supreme Court prohibited states from unconditionally barring the procedure in the 1973 case of Roe *v.* Wade. As a result of this convergence, the demand for genetic services, and hence counselors, exploded. The earliest professional degree program in genetic counseling, at Sarah Lawrence College, graduated its first class in 1971. Its students were trained in techniques of Rogerian therapy, according to which the role of professionals was to clarify the clients' own values, not impose their own. The education of these overwhelmingly female students also coincided with the rise of the autonomy-oriented feminist movement.

It is of particular importance that genetic services expanded in the context of impassioned controversy over the morality of abortion. That charged social context guaranteed that medical geneticists and genetic counselors would stress clients' freedom to make their own decisions. After all, these professionals did not want to be accused of fostering such a contentious practice as abortion. Given the paucity of treatments for genetic conditions, the accusation that prenatal diagnostic services promoted abortion had considerable force. Denying that there was any correct reproductive decision functioned to defuse this charge; hence the insistence that the goal of reproductive genetic services, or at least the only acceptable goal, is increasing the choices available to women.[3] According to the official view, "autonomous decision making should be the goal in prenatal diagnosis and... health professionals, society, and the state [should] be neutral on the outcome of individual reproductive choices. Reproductive genetic services should be aimed at increasing individual control over reproductive options and should not be used to pursue eugenic goals" (Andrews *et al.*, 1994, p. 103).

At least one other development is germane to the apparent shift in ethos. In the 1970s, bioethics emerged as a distinct intellectual discipline, and its practitioners became the new arbiters on a wide range of ethical issues in medicine. Although autonomy was originally proposed as one of four equally important values in bioethics, it soon came to dominate beneficence, nonmaleficence, and justice (Jonsen, 1998, p. 335; Wolpe, 1998, p. 43). And the principle of respect for autonomy was interpreted as implying that a person "should be free to perform whatever action he wishes—even if it involves serious risk for the agent and even if others consider it to be foolish" (Beauchamp and Childress, 1979, pp. 56-9; quoted in Jonsen, 1998, p. 335). The reasons for the triumph of autonomy are complex and perhaps not fully understood (see Jonsen, 1998; Wolpe, 1998). It is certainly significant that bioethics developed in the context of a series of scandals involving experiments on human subjects, most notably the Thalidomide disaster of 1961 (when testimony revealed that drug companies provided doctors with samples of experimental drugs, with doctors then paid to collect data on their patients), Henry Beecher's 1966 exposé of risky experiments conducted on patients at distinguished medical institutions without their knowledge or consent, and the 1972 exposé of the syphilis experiments at Tuskegee (Faden, 1996, pp. 80, 97-103). These and other scandals undermined the assumption that physicians could be trusted to act in their patients' best interests and therefore strengthened the case for patient autonomy.[4] Susan Wolf observes that "bioethics has embraced a liberal individualism with more vigor than it has embraced anything else. The bioethics revolution to establish patients' and research subjects' rights has been an effort to unseat both traditional physician paternalism and a societal willingness to sacrifice the individual. The central tool has been the esteem of the individual as an end not a means and as someone entitled to self-rule or autonomy" (Wolf, 1996, p. 16).

Given its emergence in the1970s, bioethics was inevitably affected by the other emergent social movements, including civil rights and feminism, and by the general cultural fracturing of the time. Albert Jonsen notes in his recent history of the field: "In a pluralistic society, where broad agreement on the content of morality seemed to be fading, a principle of autonomy, as the sole or primary moral principle, solved many a conundrum; one merely respects the wishes and choices of every person without passing judgment on further moral grounds" (1998, p. 335). (He goes on to note that "this shallowest meaning of respect for autonomy, unfortunately, seemed the most readily grasped.") That assessment is fundamentally shared by Paul Root Wolpe, who notes that the principle of autonomy (at least as understood in bioethics) is relatively easy to apply, unlike the principles of nonmaleficence, beneficence, or justice, each of which has more obviously contested meanings and implications. "Once we agree on the primary importance of letting the patient decide," he writes, "finding a way to apply ideas of autonomy in the clinical setting becomes a technical problem" (Wolpe, 1998, p. 46).

In any case, what matters for us is that the principle of respect for autonomy came to dominate bioethics at a time when bioethicists began to replace scientists as the primary spokespersons on social and ethical issues in genetics. During the 1950s and 1960s, most books on this theme were authored by distinguished scientists, such as Theodosius Dobzhansky, Ashley Montagu, Linus Pauling, H.J. Muller, L.C. Dunn, and Julian Huxley. It was scientists to whom journalists, foundations, and conference organizers typically turned for guidance on genetics-related ethical issues. For example, participants at the 1962 CIBA Foundation meeting on Man and His Future included such scientific notables as H.J. Muller, Julian Huxley, Joshua Lederberg, J.B.S. Haldane, Albert Szent-Gyorgyi, Francis Crick, Peter Medawar, Gregory Pincus, Carleton Coon, and N.W. Pirie; five of the twenty-seven participants were Nobel Prize winners (Jonsen, 1998, p. 15). Indeed, of the four major symposia on this theme held in the 1960s, nearly all the participants were scientists; the debates were primarily between geneticists who believed that genetic engineering of humans was premature and those who believed that its time had come. But by the 1970s, the discourse on issues in genetics was dominated by bioethicists.

Bioethics today includes an increasing number of scholars critical of the weight accorded autonomy (as is also true for feminism). Among the critics of an autonomy orientation are bioethicists who reject a *laissez-faire* approach to reproductive decision making. These scholars have argued that there is a responsibility not to reproduce if the parents are at high risk of transmitting a serious disease. In their view, although no one should prevent a woman or couple from reproducing and bearing a child with a debilitating condition, it may be irresponsible for them to do so, especially given the possibilities of contraception, adoption, and assisted reproduction with preimplantation genetic selection. These bioethicists insist on a distinction between ethical and legal/policy issues. They believe that some acts that may be immoral should not, for other important reasons (such as conflicts with other important moral values, "slippery slope" dangers, or practical considerations), be legally prohibited (e.g., Steinbock and McClamrock, 1994; Purdy, 1995).[5] They would agree with Thomas Murray's charge that we have an unfortunate "tendency to conflate judgments about what is morally right and wrong with judgments about what are wise and defensible public policies" (Murray, 1996, p. 97; see also pp. 108-12).

Thus, bioethicists do not speak in one voice. What had been an ideological near-consensus in the 1970s and 1980s has begun to fracture. (However, Renée Fox argues that bioethics is moving in the opposite direction, toward a more individualistic and "private-entitlement" perspective [1994, pp. 49-50]). Wherever bioethics may be headed, autonomy—especially in respect to reproduction—generally remains a trump value. Those who argue that there are social responsibilities in procreation are a minority, and have little public presence. So we need to ask: to what degree are the values of bioethicists representative of other professionals and of laypersons? Has there really been a wholesale transformation in attitudes toward reproduction or does the appearance of such a transformation result, at least in part, from the fact that a group largely committed to a new norm was authorized to speak for everyone?[6]

4. THE COMPLEXITY OF NORMS

While changes in ethos have certainly occurred, the question is how broad and deep those changes have been. In reporting on the controversy sparked by the decision to hold the 1998 genetics congress in Beijing, many commentators remarked on the existence of a gulf between the values of Asians, who are said to place a premium on the collective good (and therefore reject the concept of reproductive autonomy), and Westerners, who ostensibly take individual rights as fundamental (and therefore approve it). But this generalization is overly broad. First, the "West" is not a useful category. International surveys conducted by Wertz and colleagues demonstrate substantial national variability among European genetic professionals regarding such questions as whether it is fair to conceive a child with a serious genetic disorder or whether counseling should be non-directive. The locus of opinion in Spain is quite different from Britain. Moreover, even in the most autonomy-oriented countries, opinion is hardly unitary. Although few U.S. geneticists would restrict a woman's or couple's right to choose, twenty-six percent agree that "it is socially irresponsible knowingly to bring an infant with a serious genetic disorder into the world in an era of prenatal diagnosis" (Wertz, 1998, p. 501).

Second, we should not assume that Chinese culture is monolithic. It is certainly true that the perspective of Chinese geneticists diverges greatly from that of their counterparts in the United States and Britain. For example, whereas ninety-two percent of Chinese geneticists agree that people known to be carriers of a recessive genetic disorder should not marry each other (primarily because their children would have a one-in-four chance of inheriting the disorder), the figure for geneticists in the United States and Britain is six percent and seven percent respectively (Mao, 1998, p. 699). According to a Chinese geneticist involved with the survey, "the Chinese culture is quite different, and things are focused on the good of society, not the good of the individual. It would shock people in the West, but my survey reflects cultural common sense" (quoted in Coughlan, 1998, p. 18). In introducing the new law, China's Minister of Public Health asserted that "the state of inferior-quality births has aroused grave concern in the whole society, and their latent effects have alarmed and worried the people in various circles....[T]he broad masses of the people demand that a eugenics law be enacted and effective measures be taken to reduce inferior-quality births as quickly as possible" (Anonymous,1995, p. 699). But we should be wary of generalizing from the expressed views of geneticists or any other elites to those of "the people." While there is evidence that the Minister's statements reflect the views of China's urban elite, we know little about the opinions of rural or marginalized people. Moreover, surveys indicate that even in the coastal cities, a significant minority disagrees with the official view (Dikötter, 1998, pp. 170-2).

Recently, Chinese medical ethicist Jing-Bao Nie (1999) has argued that there exist both individualistic and communitarian strains in Chinese society and hence in its medical ethics, and he protests the simplification involved in equating Chinese culture with its classical works, official ideologies, or the views of elite spokespersons. He notes that there is no one Chinese perspective on abortion, approaches to death and dying, or any other issue in medical ethics.

In support of his argument, Nie draws on Martha Nussbaum's critique of generalizations about non-Western cultures. According to Nussbaum, *no* real culture is unitary. She writes: "Real cultures are plural, not single;" "Real cultures contain argument, resistance, and contestation of norms;" "In real cultures, what most people think is likely to be different from what the most famous artists and intellectuals think;" "Real cultures have varied domains of thought and activity," which extend beyond philosophy, religion, and literature, and include daily life and the lives of rural as well as urban people (Nussbaum, 1997, pp. 127-8). Ironically, Nussbaum thinks that we would readily "see the defects in a monolithic portrayal of 'American values,'" and urges us to be equally critical of such characterizations of China or India (1997, p. 127). But discussions of the Chinese eugenics law are rife with expansive generalizations about Western values. And they are misguided, for all the reasons she gives.

Even applied to the United States, where the tradition of liberal individualism, and its associated autonomy-centered ethic, has been especially strong, the contrast with China is overdrawn. After all, signs of discontent with the individualistic liberal ethic are everywhere evident both in scholarship and in ordinary social and political life.

Revisionist historians have challenged the once-prevailing assumption that eighteenth-century Americans were classical liberals. In the classical liberal perspective, it is expected that individuals will disagree about the nature of the good life. The state should therefore adopt a stance of neutrality on religious and moral matters. Its proper role is to provide a framework of rights within which individuals will strive to achieve their own self-chosen ends. The content of those ends is none of the government's business. People should generally be allowed to live their lives as they please (but see Dworetz, 1994; Greenstone, 1993; Bird, 1999).

However, most social and intellectual historians now agree that the fundamental ethos of that era was actually a secular republicanism, which stressed communal responsibilities, mutual obligations, and the value of active participation in the life of the polity. Indeed, Michael Sandel has noted that the individualist version of liberalism is actually a late arrival, having displaced the rival republican tradition only forty or fifty years ago (1996, p. 5). Barry Shain (1994), in a recent critique of the revisionists, argues that eighteenth-century attitudes were dominated by a narrow and constraining Protestant-inspired vision, deriving from the dogma of original sin, rather than being truly republican in the classical or Renaissance sense. Nevertheless, he agrees that a version of communalism, not liberalism, was the dominant spirit of the Revolutionary and Founding era.

The search for an alternative past reflects contemporary dissatisfaction with liberal individualism. That discontent became manifest in the 1970s, when liberals, and particularly the liberal university with its claims to intellectual and moral neutrality, came under attack from the Left student activists. Today, liberals are again on the defensive. Few politicians today apply the label to themselves; more often they try to attach it to their opponents. But now the most powerful contemporary challenge to liberalism comes from the religious Right. While promoting *laissez faire* in the economic realm, most conservatives reject autonomy elsewhere. Wishing to impose a moral orthodoxy, they advocate restricting access to abortion, allowing prayer in schools, banning assisted suicide, imposing harsher sentences for drug use, barring gays from serving in the military, and so forth. On the intellectual level, discontent is reflected in the critiques of liberalism by "communitarians" of various political stripes, including Charles Taylor, Amitai Etzioni, Jean Bethke Elshtain, Robert Putnam, Michael Sandel, and Alasdair MacIntyre (who would disavow the label). It is also reflected in contemporary feminism, which as Christine Di Stefano notes, is now marked by "a complex ambivalence toward autonomy" (1996, p. 95). The negative voices include many socialist feminists as well as feminist theorists of an "ethics of care," such as Carol Gilligan, Sarah Ruddick, Nel Noddings, Eva Kittay, and Joan Tronto. Whatever their other differences, these scholars agree that liberalism (at least in its dominant, individualist form) provides an impoverished account of persons and their relationship to community and state, and fosters anomic and selfish behavior. Feminists add that it obscures patriarchal oppression.

At the same time as these theorists challenge the philosophical adequacy of an autonomy-centered ethic, ethnographers deny that it captures the values of the poor and marginalized (see Jennings, 1998; Fox, 1994). For many poor women, in particular, such an ethic seems to have little appeal. Thus, Elizabeth Bussiere has shown how the mostly black and female rank-and-file of the National Welfare

Rights Organization lived within "a web of concrete responsibilities" that made them suspicious of the abstract and universalizing rhetoric employed by the organization's primarily white, middle-class leaders (1997, pp. 112-7). She argues that the latter's language of autonomy, individual rights, contract, and consent conflicted with the women's sense of the nature of the self and of the particular obligations of motherhood, a conflict that ultimately fragmented the movement and contributed to its decline.

Given the complexity of the American political tradition, we should not expect consensus on the view that reproductive decisions are wholly a private affair. Consider the diverse reactions to the case of Bree Walker-Lampley, a highly successful CBS television news anchor in Los Angeles. Walker-Lampley and her daughter have a heritable condition that results in the absence of digits in both hands and feet. In July 1992, when she was seven months pregnant with a second child, Jane Norris questioned the morality of Walker-Lampley's pregnancy on the radio talk show she hosts. Norris asked her listeners whether they thought it was fair to "pass along a genetically disfiguring disease to your child." She said that she asked herself whether she would make the same decision and concluded that, given the availability of adoption, surrogate parenting, and other options, "I have to say, I don't think I could do it" (Mathews, 1991). She also urged her listeners to say whether they would conceive a child knowing that the baby would have a fifty-fifty chance of inheriting Walker-Lampley's condition (Anonymous, 1991a).

Most callers were highly critical of Walker-Lampley, asserting that her decision was unfair both to the child and to society. Others argued that the pregnancy was none of Jane Norris's business. The incident was widely publicized, with nearly all the television, radio, and newspaper commentators expressing dismay at the views of Norris and the majority of her listeners. The assertion by one genetic counselor— that society has no right "butting its nose into people's [reproductive] lives"—is typical (Seligmann, 1991, p. 73; Hubbard and Wald, 1993, pp. 30-1). Walker-Lampley was honored with the 1992 National Courage Award (Baden, 1992). She and her husband, joined by about one hundred individuals and twenty-five disabled rights groups, filed a complaint before the FCC, asking it to examine whether the station's owner should be fined, reprimanded, or lose its license. Commenting on the (ultimately unsuccessful) complaint, the chairman of the Equal Employment Opportunity Commission argued that the station should not be disciplined for debating the issue but noted that he was "appalled and sickened" by aspects of the program (Anonymous, 1991b, p. 33). The show and its aftermath reveal the plurality of views on the ethics of reproductive decision making and suggest where some of the fault lines lie.

Walker-Lampley's condition was hardly disabling, and talk-show callers may be a highly unrepresentative group. However, there is evidence that, when it comes to the issue of transmitting a serious disease (genetic or otherwise), many Americans do not share the perspective of genetics professionals and bioethicists. For example, according to M. Gregg Bloche, physicians who counsel HIV-infected women about their reproductive options often advise them not to get pregnant, and if they are pregnant, to abort (1996, p. 258). He concludes that, "at the clinical level, the counseling of reproductive abstinence has been widely recommended and is probably an established practice" (Bloche, 1996, p. 260). The physicians' perspective is apparently shared by many laypersons. Surveys conducted by Wertz and her colleagues indicate that, unlike geneticists, most physicians and members of the American public think it is unfair to the child, to siblings, and to society in general to knowingly run the risk of having a child with a serious genetic disease (1997a). Indeed, eighty-one percent of patients—as opposed to ten percent of geneticists—believed that people at high risk of transmitting a genetic disease should not have children unless they use prenatal diagnosis and selective abortion (1997b). If we had a more nuanced picture of American culture, those statistics would not seem so startling.

This essay began by asking how a viewpoint that was generally taken-for-granted until the 1940s, and still seriously espoused in the 1950s and 1960s, virtually vanished from public discourse in the mid-1970s. We can now see that discourse was transformed to a much greater degree than were underlying norms. The social movements of the 1960s certainly produced changes in values, but those changes were very uneven. Respect for autonomy became a trump principle for various social and intellectual elites, particularly genetics professionals (for whom it served to defuse the conflict over abortion) and bioethicists. Since these are the groups whose opinions on reproductive issues are sought by governments and the media, the view that childbearing decisions are no one's business but that of the parents seemed to be general. But appearances are deceiving. Concern for social responsibility in reproduction has been a central theme throughout the twentieth century. There is less than meets the eye in the apparent transition to and dominance of an autonomy orientation. Given the shallowness of allegiances to the new social norm, we should not be surprised if attitudes shift again.

University of Massachusetts
Boston, Massachusetts, U.S.A.

NOTES

1. Paul Rabinow (1998) uses the phrase in reference to French intellectuals' disquiet with recent developments in science and technology.
2. Writing of the 1960s Albert Jonsen notes that geneticists worried that the gene pool was becoming polluted because the early death of persons with certain genetic conditions was now preventable (1998, p. 14). In addition to antibiotics, insulin for diabetes and diet for phenylketonuria were frequently mentioned.
3. Ruth Chadwick (1993) has cogently argued that fostering reproductive choice is not an intelligible rationale for providing genetic services, since their very existence presumes that genetic disease is undesirable. See also Paul, 1998.
4. However, recent research indicates that patients themselves generally place a low value on autonomy; see Schneider (1998, pp. 35-46).
5. Deborah Mathieu (1996) has made an analogous argument in respect to preventing other forms of prenatal harm, such as damage resulting from drug use by a pregnant woman.
6. For an analogous argument on bioethicists' stem-cell debate, see Wolpe and McGee (in press).

REFERENCES

Andrews, G.R., *Signs of Eugenic Progress*. American Eugenics Society, 1935.

Andrews, L. *et al.*, *Assessing Genetic Risks: Implications for Health and Social Policy*. Washington, D.C.: National Academy Press, 1994.

Anonymous. Talk show flap. Los Angeles Times 1991a; Dec. 31:B6

Anonymous. Official defends station's show on disabled. New York Times 1991b; Dec. 15:33

Anonymous. New Chinese law on maternal and infant health care. Population and Development Review 1995; 21:698-702

Baden P.L. Encouraging equality: disabled anchor wins award from Courage Center. Star Tribune 1992; June 6:1B

Beauchamp, T., Childress, J., *Principles of Biomedical Ethics*, 1st ed. New York: Oxford University Press, 1979.

Berelson B. Beyond family planning. Science 1969; 163(1 February):533-43

Bird, C., *The Myth of Liberal Individualism*. Cambridge: Cambridge University Press, 1999.

Bloche, M.G. "Clinical Counseling and the Problem of Autonomy-Negating Influence." In *HIV, AIDS & Childbearing: Public Policy, Private Lives*, R.R. Faden, N.E. Kass, eds. New York: Oxford University Press, 1996.

Boulding, K., *The Meaning of the 20th Century*. New York: Harper and Row, 1964.

Bussiere, E., *(Dis)entitling the Poor: The Warren Court, Welfare Rights, and the American Political Tradition*. University Park, PA: Penn State University Press, 1997.

Chadwick R.F. What counts as success in genetic counseling? Journal of Medical Ethics 1993; 19:43-6

Coughlan A. Perfect People's Republic. New Scientist 1998; 160(24 Oct.):18

Davies, S.P., *Social Control of the Mentally Deficient*. New York: Thomas Y. Crowell, 1930.

Davis K. Population policy: will current programs succeed? Science 1967; 158:730-9

Dikötter, F., *Imperfect Conceptions: Medical Knowledge, Birth Defects, and Eugenics in China*. New York: Columbia University Press, 1998.

Di Stefano, C. "Autonomy in the Light of Difference." In *Revisioning the Political: Feminist Reconstructions of Traditional Concepts in Western Political Theory*, N.J. Hirschmann, C. Di Stefano, eds. Boulder, CO: Westview Press, 1996.

Dobzhansky, T., *Mankind Evolving: The Evolution of the Human Species*. New Haven: Yale University Press, 1962.

Dong-Sheng, S., *Popularizing the Knowledge of Eugenics and Advocating Optimal Births Vigorously*. *Renkou Yanjiu [Population Research]*. Translation by M. Desilets and D. Vining, 1981. Available [Jan. 25, 2000] at: http://www.mankind.org/man22.htm.

Dworetz, S.M., *The Unvarnished Doctrine: Locke, Liberalism, and the American Revolution*. Durham, NC: Duke University Press, 1994.

Ehrlich, P.R., *The Population Bomb*. New York: Ballantine, 1968.

Eisenberg L. Genetics and the survival of the unfit. Harper's Magazine 1966; Feb.:53-8

Faden, R., ed., *The Human Radiation Experiments: Final Report of the President's Advisory Committee*. New York: Oxford University Press, 1996.

Fleming D. On living in a biological revolution. Atlantic Monthly 1969; 223:64-70

Fletcher, J.F. "Knowledge, Risk, and the Right to Reproduce: A Limiting Principle." In *Genetics and the Law II*, A. Milunsky, G.J. Annas, eds. New York: Plenum Press, 1980.

Fox, R.C. "The Entry of U.S. Bioethics into the 1990s." In *A Matter of Principles? Ferment in U.S. Bioethics*, E.R. DuBose, R.P. Hamel, L.J. O'Connell, eds. Valley Forge, PA: Trinity Press International, 1994.

Glass B. Science: endless horizons or golden age. Science 1971; 171(8 January):23-9

Gordon, L., *Woman's Body, Woman's Right: Birth Control in America*, Revised ed. New York: Penguin Books, 1990.

Greenstone, D., *The Lincoln Persuasion: Remaking American Liberalism*. Princeton: Princeton University Press, 1993.

Hardin G. The tragedy of the commons. Science 1968; 162(13 December):1243-8

Hubbard, R., Wald, E., *Exploding the Gene Myth*. Boston: Beacon Press, 1993.

Jennings, B. "Autonomy and Difference: The Travails of Liberalism in Bioethics." *In Bioethics and Society: Constructing the Ethical Enterprise*, R. DeVries, J. Subedi, eds. Upper Saddle River, NJ: Prentice Hall, 1998.

Jonsen, A.R., *The Birth of Bioethics*. New York: Oxford University Press, 1998.

Kay, L.E., *The Molecular Vision of Life*. New York: Oxford University Press, 1993.

Kevles, D.J., *In the Name of Eugenics: Genetics and the Uses of Human Heredity*. Berkeley: University of California Press, 1986.

Lederberg, J. "Biological Future of Man." In *Man and His Future*, G. Wolstenholme, ed. Boston: Little, Brown and Co., 1963.

Lepicard F. Eugenics and Roman Catholicism: an encyclical letter in context: *Casti Connubi*, December 31, 1930. Science in Context 1998; 11:527-44

Ludmerer, K.M., *Genetics and American Society: A Historical Appraisal*. Baltimore: Johns Hopkins University Press, 1972.

Mao X. Chinese geneticists' views of ethical issues in genetic testing and screening: evidence for eugenics in China. American Journal of Human Genetics 1998; 63:688-95

Mathews J. The debate over her baby: Bree Walker Lampley has a deformity. Some people think she shouldn't have kids. Washington Post 1991; Oct. 20:F1

Mathieu, D., *Preventing Prenatal Harm: Should the State Intervene?* 2nd ed. Washington, D.C.: Georgetown University Press, 1996.

Medawar, P. "Eugenics." In *The Life Science: Current Ideas of Biology*, P.B. Medawar, J.C. Medawar, eds. New York: Harper and Row, 1977.

Montagu, A., *Human Heredity*. Cleveland: World Publishing, 1959.

Muller H.J. Our load of mutations. American Journal of Human Genetics 1950; 2:111-76

Murray, T.H., *The Worth of a Child*. Berkeley: University of California Press, 1996.

Nelkin, D., Lindee, M.S., *The DNA Mystique: The Gene as Cultural Icon*. New York: W.H. Freeman, 1995.

Nie J-B. The myth of *the* Chinese culture, the myth of *the* Chinese medical ethics. Bioethics Examiner 1999; 3:1-2, 5-6

Nussbaum, M., *Cultivating Humanity*. Cambridge: Harvard University Press, 1997.

Otsubo S., Bartholomew J.R. Eugenics in Japan: some ironies of modernity, 1883-1945. Science in Context 1998; 11:545-65

Packard, V., *The People Shapers*. Boston: Little, Brown and Co., 1977.

Paul D.B. Our load of mutations 'revisited.' Journal of the History of Biology 1987; 20:321-35

Paul, D.B., *Controlling Human Heredity: 1865 to the Present*. New York: Prometheus Press, 1995.

Paul D.B. Genetic services, economics and eugenics. Science in Context 1998; 11:481-9

Paul D.B., Spencer H. The hidden science of eugenics. Nature 1995; 374:302-4

Pauling L. Reflections on the new biology: foreword. UCLA Law Review 1968; 15:267-72

Pius XI, *Pius XI On Christian Marriage*, The English Translation. New York: Barry Vail Corporation, 1931.

Pius XII. Discourse of his Holiness Pope Pius XII to the International Congress on Blood Transfusion; 1958 Sept. 5. Reprinted in the Dight Institute Bulletin 1959; 11

Pomfret J. China clarifies its law on sterilization. Washington Post 1998; Aug. 18:10

Porter, I.H. "Evolution of Genetic Counseling in America." In *Genetic Counseling*, H.A. Lubs, F. de la Cruz, eds. New York: Raven Press, 1977.

Purdy, L. "Loving Future People." In *Reproduction, Ethics, and the Law: Feminist Perspectives*, J.C. Callahan, ed. Bloomington, IN: Indiana University Press, 1995.

Rabinow, P. Life: dignity and value. Proceedings of Postgenomics? Historical, Techno Epistemic, and Cultural Aspects of Genome Projects sponsored by Max-Planck-Institute for the History of Science; 1998 July 8-11; Berlin.

Ramsey, P., *Fabricated Man: The Ethics of Genetic Control*. New Haven: Yale University Press, 1970.

Reed, S., *Parenthood and Heredity*. New York: John Wiley, 1964.

Rosenthal E. Scientists debate China's law on sterilizing the carriers of genetic defects. New York Times 1998; Aug. 16:14

Sandel, M.J., *Democracy's Discontent: America in Search of a Public Philosophy*. Cambridge: Harvard University Press, 1996.

Schneider, C.E., *The Practice of Autonomy: Patients, Doctors, and Medical Decisions.* New York: Oxford University Press, 1998.

Seligmann J. Whose baby is it, anyway? Newsweek 1991; 118(18):73

Shain, B.A., *The Myth of American Individualism: Protestant Origins of American Political Thought.* Princeton: Princeton University Press, 1994.

Sonneborn, T.M., ed., *Control of Human Heredity and Evolution.* New York: Macmillan, 1965.

Sorenson, J.R., Culbert, A.J. "Genetic Counselors and Counseling Orientation: Unexamined Topics in Evaluation." In *Genetic Counseling,* H.A. Lubs, F. de la Cruz, eds. New York: Raven Press, 1974.

Steinbock B., McClamrock R. When is birth unfair to the child? Hastings Center Report 1994; 24:15-21

Sunstein, C.R., *Free Markets and Social Justice.* New York: Oxford University Press, 1997.

Wertz D.C. Society and the not-so-new genetics: what are we afraid of? Some future predictions from a social scientist. Journal of Contemporary Law and Policy 1997a; 13:299-346

Wertz D.C. Proceedings of the Van Leer Institute, Jerusalem sponsored by the Workshop on Eugenic Thought and Practice: A reappraisal towards the end of the twentieth century; May 26- 29; Published in part as Eugenics is alive and well: a survey of genetics professionals around the world. Science in Context 1997b; 11:493-510

Wertz, D.C. "International Research in Bioethics: The Challenges of Cross-Cultural Interpretation." In *Bioethics and Society: Constructing the Ethical Enterprise,* R. DeVries, J. Subedi, eds. Upper Saddle River, NJ: Prentice-Hall, 1998.

Wolf, S.M. "Introduction: Gender and Feminism in Bioethics." In *Feminism and Bioethics: Beyond Reproduction,* S.M. Wolf, ed. New York: Oxford University Press, 1996.

Wolpe, P.R. "The Triumph of Autonomy in American Bioethics: A Sociological View." In *Bioethics and Society: Constructing the Ethical Enterprise,* R. DeVries, J. Subedi, eds. Upper Saddle River, NJ: Prentice-Hall, 1998.

Wolpe, P.R., McGee, G. "'Expert Bioethics' as Professional Discourse: The Case of Stem Cells." In *Beyond Cloning: Embryos, Ethics, and Immortality,* S. Holland, K. Lebaczq, L. Zoloth, eds. Cambridge: MIT Press, in press.

Wolstenholme, G., ed., "Eugenics and Genetics: Discussion." In *Man and His Future.* Boston: Little, Brown and Co., 1963.

Wright, S., *Molecular Politics.* Chicago: University of Chicago Press, 1994.

PART TWO:
PERSPECTIVES FROM
THE PHILOSOPHY OF SCIENCE

CHAPTER 6

FRED GIFFORD

UNDERSTANDING GENETIC CAUSATION AND ITS IMPLICATIONS FOR ETHICAL ISSUES CONCERNING MEDICAL GENETICS

In this chapter, I explore some conceptual insights about genetic causation and consider their implications for ethical issues arising in medical genetics. As these insights are rather abstract, their most important implications will be indirect ones concerning the shaping of our general conceptions and tendencies of thought, serving as a curb on inappropriate geneticization of disease. I end by critiquing the idea that the scientific and practical successes resulting from the genetic point of view should convince us to accept such geneticization.

1. INTRODUCTION

A variety of difficult ethical issues arise from our increasing powers of prediction, explanation, and control in the realm of human genetics. We have for some time been able to screen individuals for genetic diseases, estimate the probabilities of various outcomes for offspring, and detect a number of genetic diseases *in utero*. Recent and ongoing research puts us on the brink of a tremendous expansion of these powers. Using recombinant DNA methods, we can detect the existence of a precise nucleotide sequence in a sample of genetic material, thus allowing more accurate and powerful tests for a whole range of genetic diseases and susceptibilities. These and other advances in technology and knowledge, including the recent mapping of the human genome, will make all the more pressing the dilemmas concerning consent, confidentiality, stigmatization, potential misuse, and

L.S. Parker and R.A. Ankeny (eds.), Mutating Concepts, Evolving Disciplines: Genetics, Medicine and Society, 109-125.
© *2002 Kluwer Academic Publishers. Printed in the Netherlands.*

the ambiguity of the goals associated with such measures as prenatal diagnosis, carrier and susceptibility screening, and genetic therapy. New policy questions will arise concerning what sorts of techniques should be developed further, for which conditions tests should occur, and whether a whole range of traits should be block-tested together in a routinized way. Given a wider array of findings about their own genetic makeup or that of their potential offspring, individuals will face more complicated decisions regarding what sorts of findings warrant what sorts of actions.

This new knowledge expands our power by yielding greater precision, but also by increasingly moving us beyond 'single gene' conditions, such as phenylketonuria (PKU), Tay-Sachs disease, and sickle cell anemia, to include the much more common set of multifactorial conditions, such as schizophrenia, depression, alcoholism, Alzheimer disease, coronary heart disease, and various cancers (Korenberg and Rimoin, 1995). Further, the detailed picture that emerges may come to affect our general conception of genes and their influence on human traits. Clearly, we must pause and think about the nature of this new knowledge, in order to be clear about ways in which it does and does not affect the ethical issues facing us.

It is a truism that proper resolution of or even progress on ethical and policy issues concerning health care requires a careful consideration of the relevant empirical facts. Many debates are resolved not by argumentation over basic moral principles, but by the overturning of some incorrect factual assumption. For instance, it has often been pointed out that each of us carries a number of genes which, if they occurred in homozygous form, would be deleterious, thus arguing for the unfeasibility of the suggestion that we eradicate genetic diseases by identifying and discouraging the reproduction of all carriers of recessive diseases (Hirschhorn, 1972).

But it is equally important, in addressing these normative questions, to address certain *conceptual issues*. Positions on moral and policy matters can be criticized not only on the basis of unacceptable basic moral stances and a failure to be apprised of the relevant facts, but also on the basis of flaws in their implicit understandings of certain conceptual matters. In other areas of healthcare ethics, we see examples of this in debates over such concepts as autonomy, competence, and personhood, as well as the concepts of health and disease. Indeed, moral dilemmas in genetic counseling, for example, raise each of these questions. But there is a conceptual issue of particular importance in these genetics, which will be the focus here: the question of how we conceptualize genetic causation, or the causal relation between genes and phenotypic traits.

This is a topic about which a number of important things have been said in recent years by biologists and philosophers of biology (Burian, 1981; Dawkins, 1982; Gifford, 1989, 1990; Hesslow, 1984; Hull, 1979; Lappé, 1979, 1987; Lewontin, 1974; Smith, 1992; Wendler, 1996). Relevant discussions concerning causation and the language used to describe it have also occurred in more general philosophical literature (Gorovitz, 1965; van Fraassen, 1980, ch. 5; Salmon, 1984; Mackie, 1974; Schaffner, 1993, ch. 6). Yet these discussions in the philosophy of science and discussions concerning ethics and genetic technologies have been carried out for the most part in isolation from each other. One goal of this chapter, then, is to consider some philosophical themes concerning genetic causation and explore their relevance for these questions in the realms of ethics and human genetics, in order to further the project of building a bridge between these two realms of inquiry.

On the other hand, certain challenges exist regarding the applicability of these conceptual insights to these practical issues. For instance, the abstractness of these matters may make them seem irrelevant in the context of day-to-day genetic counseling. Yet surely these conceptual issues are relevant to our general understanding, which may be expected to affect individuals' decisions indirectly. I also will argue that these conceptual insights have implications for questions about the nature of the knowledge that we gain as we discover more connections between genes and diseases, especially concerning some implicit biases.

2. TWO INSIGHTS ABOUT GENETIC CAUSATION

A number of features of genetic causation are noteworthy and also relevant to the sorts of medical genetics decisions under consideration. Though these features interact with one another, we can usefully distinguish considerations having to do with causal complexity, those having to do with population-relativity, and others having to do with subtleties of description or individuation. We can understand each of these sorts of considerations in terms of how they force us to think beyond our common, simple conceptualization of the causation involved between, say, two billiard balls. When we think in a fairly unreflective manner about such causation, we are likely to imagine one cause and one effect, connected by a straightforwardly universal or deterministic causal relation, and we are likely not to deliberate about different ways of describing those factors. Examining the causal relationship between genes and traits makes clear that this conceptualization will not do. And yet the simple picture has a hold on us.

The causal relationship between genes and traits can be quite *complex*; there are many causal factors, not just one. And there is context-sensitivity; the effect that one factor (say, a gene) has on a trait will be dependent upon what other causal factors are present, either environmental factors or other genes. This can manifest itself in epistasis (where the presence of one gene prevents the expression of another gene), variable penetrance (where the probability of the trait, given this gene, is altered by the presence of some other factor), or variable expressivity (where different individuals express a quantitative trait to varying degrees) (Rimoin *et al.*, 1996, p. 89). These facts about complexity have relevance to practical and moral decisions. Variable expressivity of a trait, for example, obviously needs to be taken into account in counseling and reproductive decisions; consider the difference between decisions about Tay-Sachs disease and sickle cell anemia, which differ not only in general degree of severity but also in degree of variation of expressivity.

But it is important to see that our understanding of the causal process must go beyond this to include some more general features of genetic causation. I focus on two rather simple but powerful insights of a conceptual or philosophical sort. Each of these two insights about genetic causation concerns the fact that it is too simple, and thus can be highly misleading, to say simply that a certain gene G is the cause of some phenotypic trait P. (Similar things can be said of other statements we often make, such as that P is a genetic trait, or that G is the 'P gene.') But these two insights concern opposite ends of this causal relation, the first focusing on whether it is appropriate to specify or describe 'the cause' as G (the gene), and the second focusing on whether it is appropriate to specify or describe 'the effect' as P (the phenotype).

2.1 Population-Relativity and Taking the Environment Seriously

In the previous discussion of variable penetrance and expressivity, I emphasized the ubiquity of environmental factors that affect traits. It might be objected that many phenotypic features have only genes as their cause, and that these are the ones at issue here: the *genetic* traits or diseases, to be considered for genetic diagnosis and therapy, etc.

One response to this would be that we will focus increasingly on less straightforward cases, such as schizophrenia or depression, so such simple cases will obviously not provide adequate models for the future. But the deeper response is to insist on the generality of the thesis that all traits, even paradigmatically 'genetic' traits, have environmental as well as genetic causes. It is important to note how this tends to be masked if we fail to examine the matter carefully. For instance, the genetic disease PKU is easily seen to have environmental as well as genetic causes. The normal gene at the PKU locus produces the enzyme, phenylalanine hydroxylase, which enables the metabolism of phenylalanine into tyrosine. An individual homozygous for the PKU gene cannot produce this enzyme, and thus if the diet of such an individual contains the normal amount of phenylalanine, the serum level of phenylalanine increases, resulting in the characteristic symptoms, including mental retardation. However, these effects are avoided if a special diet low in phenylalanine is provided, and this is what is done for therapy. The screening program for PKU is often touted as a great success, for a simple test provided at birth allows effective action to be taken.[1]

From this perspective, then, having a certain level of phenylalanine in the diet is just as much a cause of the disease as is the defective gene. This environmental factor is necessary in the circumstances; it is also a potential causal lever, a factor that could be manipulated in such a way that the effect would not occur. Yet we call PKU a 'genetic' disease. We do this by implicitly ruling out of consideration those factors that are considered part of the causal background, or 'causal field' (Mackie, 1974). For a genetic disease, all environmental factors fall into this category on the grounds that they are constant throughout the population and thus do not account for any of the differences between individuals. Thus the claim that such a trait is genetic is 'population-relative' (Gifford, 1990).

Note the significance of this population-relativity: genetic diseases are not picked out as such because of some fact about the causal story in each individual, e.g., with this individual gene causing this individual trait. Rather, a certain genetic *difference* causes or explains a given phenotypic *difference*, and to determine whether we should cite genetic or environmental causes, we have to specify a comparison situation, namely the other members of the population.[2]

Identifying the gene as 'the cause,' then, involves picking out one factor from among a set of conditions each equally necessary in the circumstances. But many facts not cited as causes are just as much a part of the individual causal picture; they are not any less strong or potent than those that are cited. So at the very least, to say 'gene G is the cause of trait P' (or 'P is a genetic trait') is misleading.

2.2 Proper Trait-Individuation

To say 'gene G is the cause of trait P' (or further, 'G is the P gene') can often be misleading for another reason as well. This has to do with how we pick out or describe the *effect*. This issue thus underlies various difficulties that arise in trait definition, and is especially important in the case of more complicated, multifactorial traits.

For consider the following: if a trait has been found to vary when there is variation in a certain gene, we might infer that there is a gene 'for' that trait. But that which the gene in question *causes* may be quite different from the trait of interest, even though there will of course be some relationship between these. First, any particular gene will have lots of effects; there will be many things that can be described as the effect of a given gene. Geneticists use the concept of 'pleiotropy' to describe the situation where one gene has two or more quite separable effects, for example on different organ systems (Gelehrter *et al.*, 1997, p. 347; Rimoin *et al.*, 1996, p. 104). But one might also pick out different points or features along a single causal path. Add to this the fact that our definitions of a trait are often fluid and controversial, as in the case of alcoholism or schizophrenia (Gottesman and Shields, 1982, ch. 1), and we see that there is ample room for confusion.

In fact, the trait picked out may be only one part of the gene's effect, or, conversely, what the gene causes may be only part of the trait of interest. For example, one of a gene's effects might be on the ability to perform some specific subtask of cognitive activity, and this might be interpreted as affecting intelligence *per se*. Or the gene may affect the ability to learn language, while we, looking at the difference made in our population, might describe the effect as the inability to learn *English*.

There is a connection between these two insights of population-relativity and proper trait-individuation. The simplifications to which they are to be antidotes can conspire to encourage a sort of 'geneticizing' or 'genetic reductionism,' a tendency to see all these diseases in genetic terms, or to see genetic causes as all-important. Each of these insights has important implications for the moral issues concerning medical genetics cited at the outset. Identifying and clarifying misleading positions is surely the sort of thing that will enhance informed, rational decisions. I will focus in the balance of this chapter on the implications of the population-relativity insight.

3. IMPLICATIONS OF POPULATION-RELATIVITY

The population-relativity of our causal language and the recognition that it constitutes a narrow view of the entire causal picture have a number of important implications, each suggesting that we take seriously that genes and environmental factors are at least in some sense on a par with respect to the causation of phenotypic traits, even those we call genetic (Hesslow, 1984). First, our *classification* of traits into genetic and environmentally caused is a function not just of the causal factors involved in the production of the trait in individuals, but also of the relative frequency of the various causal factors in the population, and therefore of the individuals with which comparison is being made. The classification is, in this important sense, arbitrary.

The same is true of our assessments of *relative causal contribution*, or relative importance, of different causal factors. This general point was central in debates, motivated by controversies over the genetics of race and IQ, about the meaning and significance of heritability, and the preferability of the 'norm of reaction' conception of the effect of a given genotype (Block and Dworkin, 1976; Lewontin, 1974; Feldman and Lewontin, 1975). Indeed, the fact that degree of causal contribution is relative in this way to the population is really just a quantitative extension of the same fact about qualitative judgment of whether a trait is to be labeled 'genetic.' Also note that a measure such as penetrance, which appears initially to be some measure of 'causal potency,' i.e., some fact about how strong the causal processes are in the individual, is also actually population-relative, dependent on the frequency of various causal factors in the population as well as on the causal potency of the gene in the individual.

In addition, these ideas explain why efforts at cure or prevention of genetic diseases need not focus on the genetic components. Such manipulation is something that takes place in the individual, whereas the claim that the trait is genetic is a function of the population as well. Presumably this should encourage us to look more seriously for environmental solutions, not to view certain outcomes as inevitable, and perhaps to be less eager to assume the appropriateness (and likely effectiveness) of genetic therapy. Many genetic disorders, in fact, have non-genetic treatments, some quite successful. For example, Wilson disease (hepatolenticular degeneration), a defect of copper metabolism, is treatable via various drugs, such as penicillamine; it can also be treated by liver transplantation (Owen, 1981; Rimoin *et al.*, 1996).

While I will not explore this argument in detail the proper individuation insight also has practical implications with regard to manipulability. One might initially assume that avoiding or removing the appropriate gene would result in removing a given trait with surgical precision. But in fact by doing this one might affect additional traits as well, or only remove one aspect of the intended trait. Finally, the understanding of population-relativity also provides a challenge to the idea that research into diseases labeled 'genetic' is always most appropriately carried out by further study of the genetic factors (Edlin, 1987).

It would seem reasonable to believe that these ideas would have important implications for the moral dilemmas concerning medical genetics such as those mentioned at the outset. And thus surely those involved in or facing such dilemmas, such as policy makers, genetic counselors, and other healthcare professionals, as well as individuals facing health or reproductive decisions for themselves, must be made aware of or educated about these things.

4. CHALLENGES

In fact, however, drawing out important implications from these insights about causality for ethical issues in medical genetics is not a simple matter. A number of challenges must be considered:

(1) It might be argued that the above picture of population-relativity and the consequent arbitrariness involved in labeling a trait 'genetic' is *inaccurate*, itself an overly simple and thus misleading view;

(2) Even if the characterization is granted, it might be said not to have the expected implications, for example because it is *too abstract* a consideration to have any impact on people's practical decisions (e.g., a genetic counselee would not be helped by being told any of this);

(3) It might be held that what is correct and of relevance in this characterization is *obvious and already known* (at least, by those who need to know it), thus the philosophical points do not in fact advance the cause of making headway on questions of medical ethics in any significant way.

4.1 Inaccuracy

First, let us consider the direct challenge of our general thesis about population-relativity, that this central insight is itself highly misleading. This challenge might proceed as follows: it may be held that diseases such as PKU disease, and perhaps many diseases focused on in genetic counseling, screening, and therapy, *are* genetic simpliciter, and that it has only seemed otherwise because we have understood the traits in the wrong way. For example, it might be argued that the PKU disease state should be identified not with high phenylalanine levels or the mental retardation that results from those levels, but with the lack of phenylalanine hydroxylase or perhaps even the inability to make phenylalanine hydroxylase. That is, once we define the disease more precisely, we find that it *is* simply genetic, and in a way not subject to population-relativity.

This view raises some interesting issues, foreshadowed in our discussion of 'proper trait-individuation,' concerning what precisely is to count as the disease state in a given case: to what level of organization are we to attribute the disease state, and how much of the causal process from initial causes to symptoms is to be labeled part of the disease itself? I am not convinced that the above revision of what to count as the disease is an appropriate one. One could still insist that certain environmental factors remain relevant. But we do not need to resolve this here. For whether or not this view captures correctly how we do or ought to use the term 'PKU disease,' it will not affect what is important about our theses. Consider the manipulability thesis. If we construe PKU disease in the way suggested, as the inability to make phenylalanine hydroxylase, then we simply could restate the thesis (about what can be environmentally manipulated) to one not about the disease but about the symptoms, such as damage to the brain. It is these symptoms that we are concerned to avoid and that are the direct source of the medical complaint, and indeed they can be manipulated by alterations in diet. Thus, unless we adopt some implausibly strict account of the relation between the definition of disease and the point of appropriate medical intervention, the same implications result in terms of practical matters. (The other point to make is that no matter which way this question is decided, its evaluation requires thinking about the key conceptual question concerning proper trait description.)

But a second challenge might be posed as follows. Suppose it is admitted that being the factor that makes the difference in the population does not by itself make something the factor that is most appropriate to focus on when looking for a cure or method of prevention, and thus that environmental factors can, in principle, prevent or cure a genetic condition. It might still be said that this in principle possibility is not likely to come to fruition in practice. Consider the following: the idea that a trait is genetic does not obviate the possibility of environmental prevention or cure, since there are always other necessary, causal antecedents beside the relevant gene, as evidenced by the successful treatment of PKU disease via diet. But a closer look at the mass screening and treatment for PKU reveals that it was not in fact the complete success story initially thought (Bessman and Swazey, 1971; Rimoin *et al.*, 1996, p. 1870; Paul, 1999). For example, while IQ is significantly improved, it does not appear to be increased, on average, to 100. For various reasons, including genetic heterogeneity, other causes of high phenylalanine levels, and the difficulty of adherence to the diet, the programs of newborn screening and subsequent dietary treatment for PKU have had a number of problems. Thus it might be argued that the trait's being genetic *does* stand in the way of an effective environmental prevention or cure. The diet does not completely ameliorate the problem (no matter how the trait is defined), whereas a change in the gene presumably would. And, the claim goes, this was to be expected on the basis of its being a genetic disease. Perhaps in the future we can be more precise with dietary and other environmental interventions, and conceivably this asymmetry may disappear. If so, this would be significant. For as it stands now, it seems as though there is indeed an asymmetry here, which helps to explain why genetic diseases are not so effectively treatable by environmental inverventions, as well as why someone would say that gene therapy (preferably germline) is a better solution in the long run. A specific gene change is unlikely ever to be perfectly compensated by a particular environmental change. (Underlying this is the idea that before the genetic mutation came about historically during the evolutionary process, there was a norm of health or adaptedness. The genetic mutation disrupts this, in a discrete way—perhaps in a variety of such ways—since any changes are likely to be maladaptive. It is unlikely that some simple environmental remedy would be adequate to precisely reverse each of these disruptions.) Hence this in-principle possibility, that environmental intervention would be effective as well, is not likely to be all that helpful in practice.

This is an important challenge. Still, the basic claim being put forward here is not that environmental cure or prevention will always be just as easy, efficient, or appropriate. It is rather that the trait's being 'genetic' *does not by itself settle the matter*. If a genetic trait indeed turns out to be quite unsatisfactorily dealt with via environmental manipulation, this will have to be for reasons other than the fact that the trait is genetic in the population-relative sense described above. So, given the state of our present knowledge, it remains the case that we should not uncritically infer unchangeability from the fact that a trait is genetic.

Yet another challenge to the accuracy of the picture arising from these conceptual insights is based on the claim that the more traditional genetic picture is being successfully filled in and thus shown to be correct. I will pursue this point in the final section.

4.2 Is this Information Too Abstract to be Useful?

Second, even if the central insight is accepted, the *utility* of this information for those making real decisions might still be challenged. For it will be insisted that these sorts of insights about population-relativity are too abstract and complicated to be readily understood and thus to have practical relevance to people's deliberations; in particular, they will not yield useful advice for genetic counselees in the midst of a decision about selective abortion or genetic therapy. For instance, it is not appropriate to point out to parents of a child (or a couple considering a fetus) with what is traditionally known as a genetic disease that, when you take a careful and properly broad view of causation in the individual, the disease is really no more genetic than environmental, or that there is surely some environment conceivable in which the child would be normal.

There are, of course, important challenges involved in properly conveying the appropriate information to genetic counselees, and much care must be taken here (Kessler, 1980; Bartels *et al.*, 1993; Gelehrter *et al.*, 1997, pp. 283-7). The more information conveyed, the greater the worry that it will not all be understood or retained. Adding complicated or abstract information presents the danger of confusion and may be counterproductive. It can be argued that the main focus for counselors should be on being as sure as possible that counselees understand—in simple, standard terms that can be explained without considering these causal insights—things such as the nature and magnitude of the risks, or the likely impact of the disease on the life of the affected individual and the family.

But simply to be dismissive of the relevance of these conceptual insights is to take too narrow of a view about the way in which they could be of use. There are different sorts of dilemmas, different sorts of audiences for these insights, different points in time and contexts in which education or information can be provided, as well as more and less subtle and sensitive things to be said.

Consider the issue of different audiences. As well as counselees, there are counselors, other healthcare professionals, the general public, and so on. It is plausible that these insights about causation would be of more use to some of these other groups. Indeed, I believe that it is important to integrate these insights into the education of scientists, physicians, geneticists, and genetic counselors, those who themselves will educate the public and those who will have an impact on research priorities. We want these people to have an especially sophisticated understanding of these matters, even if they can only convey *some* of this to their clients or the general public.

But suppose we focus our attention on lay persons. Granted, this abstract information will not be appreciated if brought to their attention when they are on the verge of an important decision. But people in fact have some general background conceptions about genetic causes, which are likely to affect what options they take seriously, how they hear or process certain information, how they view the disease, and, ultimately, their tendency to make one decision or another. For instance, if one has a very geneticized view, one is more likely to put stock in genetic information and accept genetic therapies. (Note that in order to evaluate these matters more precisely—just how to present this information, and the importance of doing so—it would be important to know how such conceptions and misconceptions in fact affect people's practical decisions.) Therefore, where possible, it would be helpful to have more discussion about these issues, and a full appraisal of this topic would require examination of public education and media coverage.

4.3 Is There Already Awareness?

Finally, it might be said that these are indeed important facts to take into consideration, at least by someone, such as genetic counselors and medical genetics researchers. But it might then be claimed that these individuals *already know* these things, that these philosophical insights do not tell them anything new. Perhaps, in the past, many such individuals were ignorant of, or too complacent about, these complications. But we have learned from our mistakes, and conceptual claims about causation are not likely to help us further. For instance, we obviously already know that PKU disease and many other genetic diseases can be manipulated environmentally.

Let us put the challenge another way. Those involved intimately with these medical genetics issues will, if they are reflective, surely note that one must keep in mind such important issues as (1) false positives and false negatives in screening tests, caused in part by heterogeneity and resulting in diagnostic uncertainty; (2) variable expressivity, resulting in prognostic uncertainty; (3) at least an open possibility of environmentally-based treatments; and (4) the fact that things get much more complicated when we consider the broader set of traits that do not neatly segregate, and thus do not fit the simple paradigms of genetic disease. Do the insights discussed here concerning genetic causation add something *further* to these sorts of considerations? It might be argued that they do not.

But one response here is that these conceptual matters may aid in more precise formulations of the knowledge of the sort contained in points 1-4. Note especially that point 4, how to think about more complicated traits, is something that requires much further consideration. Such traits have a variety of both genetic and environmental causal factors, even amongst factors which vary in the human population, and their causation commonly involves complex and poorly understood causal pathways, as well as much variability between individual cases, resulting in controversy even over proper trait definition. These multifactorial traits, like hypertension, diabetes, allergies, and schizophrenia, will increasingly become important targets for genetic screening as more and more genetic markers are found. Concerning these cases, it is much more apparent that a more sophisticated understanding of the complicated causal relations involved, and thus of certain conceptual issues about causation, will be necessary.

Whether we are thinking of the lay person's conceptualization or that of a scientist or genetic counselor, it is important to note that we are still groping for clear language to use regarding such cases. We do not yet have fully adequate conceptual models for intelligibly, accurately, and helpfully representing the exact role of genes (and other factors) in such complicated cases as schizophrenia (Gottesman and Shields, 1982, ch. 11), let alone for how best to convey information to lay persons. And when one doesn't know precisely how to characterize the situation, one is likely to fall back into simple claims such as "it's genetic," or "genes give one the *potential* for the disease," truisms which may not be particularly helpful.

It may be especially promising to use some of our conceptual insights here as tools for developing models or ways of thinking both practically and theoretically about such multifactorial traits. So here plausibly is a role for some general ideas about the causal relationship between genes, environment, and phenotype in lending a hand to the ethics of medical genetics. One particular example is worth noting: the concepts of penetrance and assessments of risk are themselves dependent upon the population studied. Any lack of attention to the concept of population-relativity could lead to an unreflective reliance on such measures, so this notion should play a role in the thinking of genetic counselors and researchers.

5. THE ENTRENCHMENT OF GENETICIZATION

Finally, I would like to clarify a role these insights might be able to play by returning to the issue of geneticization. As mentioned at the outset, we are in the process of uncovering increasingly precise information about the genetic causes of phenotypes. As we locate the pieces of DNA responsible (in part) for more and more diseases, and as these causes become more concrete and even manipulable, the temptation may become overwhelming to assume that we are uncovering *the* roots of the diseases, and to adopt a 'genetic reductionism,' or a 'geneticized' view of disease, further encouraging us to focus on genes as the things to study and manipulate.

The general attitude I have been taking here is to resist such geneticization, urging that it is inaccurate and harmful. But this skepticism might be challenged. In particular, it might be argued that we should take seriously that the success this endeavor has encountered in fact counts in favor of this genetic or 'geneticized' view. After all, the (scientific) fruitfulness of the perspective of supposing traits to be substantially genetic is being shown, and thus gives evidence for the truth of its presuppositions.

One component of this 'argument from success' connects to our discussion of proper individuation. For in addition to our success in locating genes tied to various diseases, we sometimes successfully use genetic underpinnings to individuate the diseases to be explained. It might be argued that, as we learn more, we come correctly to individuate traits according to their genetic causes, and thus come to see the genes as more fundamental. For example, it is claimed that hypercholesterolemia should really be viewed as a disjunction of two traits previously conflated, familial hypercholesterolemia and non-familial hypercholesterolemia, on the grounds that these two classes of cases have different causes, the former caused by a certain gene (Brown and Goldstein, 1974). It is useful to separate such traits since they will typically have different features; the overall picture is blurred if we treat them simply as a single class, and genetic information is what allows us to clarify the situation. This is in line with what happens elsewhere in science: as we learn more, we may learn to discard certain concepts (such as humors, or phlogiston) and replace them with others. Such reconceptualization, at both the theoretical and observational levels, may be seen as progress.

So an 'argument from success' might be made against the resistance to geneticization. Indeed, it will be said that the success is really quite impressive. It is not simply that we are finding these genes (and reconceptualizing traits), but that, further, we are doing this *so quickly*, more quickly than we have ever found environmental factors, and more quickly than was expected. I suspect that many, seeing this progress, and seeing more and more media reports of more and more genes being discovered or associated with various traits, will indeed see this as evidence for the genetic view, for the view that genes are fundamental. But again, I think that this inference should be resisted.

One thing that blocks this inference is that it is not really clear what is meant by one sort of causal factor being more *fundamental* than another, being a more 'potent' causal factor, or being ontologically more important. This is part of the lesson of population-relativity, which shows the assertion that X is the cause to be a less significant sort of claim, being less about the causal story in individuals, than might be thought. I believe that this gap is indeed a serious reason to be skeptical about this 'argument from success.'

But the main point I want to consider here is that this success should not be seen as surprising or impressive. We have known the general story for some time: the genome is composed of a large number of genes, and they will match up with our traits *somehow*, as all traits are the result of an interaction between genes and their context (including the environment). We will have ways to study the effects of these genes productively (Collins, 1997). Surely we should expect that increasing numbers of genes will be discovered. Presumably *all* of them will be discovered. But we do not need the hypothesis that genes are somehow more fundamental, prior, or important to explain any of this.

The matches we find between genes and traits are, of course, not perfect. That is, even when we find not just a marker but the gene itself, we may be talking about a gene that yields, for instance, a lifetime risk of a certain sort of cancer, in a certain percentage of cases. Of course, this can be useful information in various contexts. The point here is only to emphasize that there is a tendency to count it as 'finding the gene' even in such a case. (Of course, some people are more careful than others with respect to what language they would use here.) There are no clear criteria concerning whether the connection is too probabilistic or non-specific to 'count.'

Further, concerning the claim that we are finding genes so quickly, and much more quickly than environmental factors, it should be clear that there are other explanations for this. First, genes are well understood theoretically and known to be relatively discrete and precise in their effects, and thus are inherently discoverable. Second, we have developed a set of powerful technologies that enables us to do this discovering. Third, we have focused a great deal of effort and resources on doing this.

The claim that our success has outstripped our expectations should not be based merely on an intuitive assessment, but on one that is as objective as possible. Such an evaluation would require not only gathering empirical information, but also making some difficult decisions about just what counts as progress. Further, assessing whether this progress had surpassed expectations would, of course, require that we specify what the prior expectation was. It is not altogether clear just what this baseline should be; as suggested above, I suspect the expectation has been fairly high for some time.

An additional complication arises: different people (different scientists, and especially scientists as compared to the public) may have had different expectations. Some members of the public may well have been under the impression that traits in general were mostly, primarily, or essentially environmental. For them, the new information would properly upset or revise their previous picture or conception. This raises a variety of issues concerning *whose* views count here. I submit that the more relevant baseline for assessing whether the incoming information is especially impressive is that of the scientific community, and there does not seem to be reason to believe that the outcome differs much from this baseline.[3]

Still, it must be admitted that the matter of whether expectation-changing evidence has been or will be found is a contingent, empirical one, so perhaps the matter should not be seen as closed. The above mitigating factors that would make us *expect* great progress concerning discovery do seem legitimate, but we should nevertheless be open to the possibility of real evidence turning up in favor of the 'genetic picture.' So perhaps we should be more open to looking at the evidence that exists, as well as looking to see if such evidence accrues over time.

Were we to do this, it is important that we keep in mind yet another source of bias. To see this, let us return to the other aspect of this genetic success. Suppose we do find that individuation of traits via genetic information becomes common and is scientifically productive, apparently yielding evidence that this is the right way to individuate traits. Recall our earlier example of hypercholesterolemia, which was divided into two different diseases on the basis of different causal underpinnings. In this case, we distinguished a genetic and a non-genetic trait, but often we will distinguish between two traits (or subdivide one 'earlier' trait into two) on the grounds that one is caused by one gene and the other by another. Or, as we look in a more fine-grained manner, we may find genetic heterogeneity and distinguish between trait-types on the grounds that they result from two different alleles at the same locus. The greater the precision and security of our knowledge gained at the genetic level, the more we will be able to do this, and this may seem to constitute scientific progress.

Now in principle, we can distinguish traits on the basis of different environmental causes (as we do in the case of diseases caused by different strains of bacteria or virus). But, as already suggested, due to increased theoretical and technical knowledge, the distinctness of genetic causes, and the placement of our resources and efforts, we are increasingly likely to do this on the basis of genetic differences. But this will in turn affect our assessment of the genetic influence on these traits. For, by definition, the traits so constructed will be 'more genetic.' They will have higher heritability; that is, more of their phenotypic variance will be explainable by reference to genetic variation, as compared with the traits as previously individuated. So if traits are increasingly individuated on the basis of genetic factors, then differences between them will be increasingly genetically caused.

This categorization may come to be seen as quite natural. And yet there is something quite unnecessary and contingent about it. It is not (or at least not *only*) because of the causal facts that we will have the resulting individuation scheme; rather, it will be due in part to contingent matters of what technologies have been available and which causes are more easily experimentally manipulable. We can be led by this conclusion to feel that there is increased scientific evidence for the genetic reductionist picture, for viewing genes as fundamental or of greater causal importance. The public and scientists alike, observing increasing success at explaining trait differences by reference to genes, might see this as evidence of the genetic picture being filled in. Hence, this reindividuation exacerbates the tendency to view genes as fundamental, or our traits as importantly genetic, resulting in a sort of 'entrenchment' of these ways of individuation. But if we take to heart the insights about population-relativity and the reindividuatability of traits, we see why this would be a mistaken inference. Thus even if there is some reason to take seriously the ontological significance of reindividuation by causal factors, there is also reason to believe there is a bias in how this will be done, giving undue preference to genetic individuation.

It is worth bearing in mind here a theme articulated by Ian Hacking (1992) about the contingency of research paths. Focusing on laboratory science, and extending Pierre Duhem's thesis about the role of auxiliary hypotheses, Hacking develops the thesis that the various elements of scientific practice (including theories, questions, background knowledge, instruments, methods, and data) are mutually adjusted over time in a way that brings about a consilience between them. Once a trajectory has been followed for some time, it is very unlikely to be drastically altered. But the result is by no means determined by the facts of the world, but instead is dependent on the contingent research path taken so far, and does not constitute evidence that we have really come to the truth, or to the only possible way that we could interpret our world scientifically.

I conclude that the argument from success should not move us. We already have good reason to believe that genes can make a difference in almost all traits imaginable. It is not obvious why uncovering particular genes and clarifying what their effects are should alter our views about, for example, nature and nurture. There is some danger that such alteration of views could come about via a mere psychological effect of contemplating impressive discoveries, rather than via a reasoned argument.

We need to be careful not to fall into common ways of talking that privilege the genetic perspective; we should consider whether we want to say we have found *the gene* for this trait, or only a gene that has an effect on this trait. As we learn more about the genetic bases of disease, we need to be careful about how we state or conceptualize what it is being learned, what evidence is gathered, what the nature and significance is of our new knowledge. For it is one thing (and surely correct) to say that for almost any particular disease, there are of course a range of genetic influences, as well as a range of environmental influences, which underlie it; as it turns out, and for a variety of reasons, we are finding the genetic ones. But it is another to say that we are finding the bases of these diseases, and these bases, it turns out, *happen* to be genetic. It is important that these things be kept in mind when communicating with the public.

6. CONCLUSIONS

Genetic causation and our claims about it are complex, population-relative, and dependent upon different modes of trait-description. Application of these insights to the practical decisions of medical genetics is often not straightforward, but there is reason to try to have these insights influence such discussions and decisions. Of course I have not offered a complete program for how information is to be presented in the context of genetic counseling or how these general ideas should be communicated in science education and the media. But I have provided reasons for attempting to do so, and argued that this is necessary if people are to be helped to approach ethical issues surrounding medical genetics in a fully informed way.

As is true of *all* discussions in healthcare ethics, there are a number of empirical questions the answers to which would aid in advancing the discussion here. Indeed, I think that discussion of these conceptual issues calls our attention to their importance. It would help to be clearer about just what people's conceptions of genetic causation are, how well people can understand certain kinds of information, and how best to communicate it to them so as to maximize this understanding. It also would be important to learn what effect the discoveries of various genes (and media reports of these) have on people's general conceptions about genetic causation, and on their ideas about the importance of genetic causes or the possibility of manipulation, as well as to learn how all of these factors affect the practical, clinical decisions that people make. The conceptual insights about genetic causation cause us to focus on these questions, and they should also offer guidance in constructing more appropriate ways to communicate about these issues.

Michigan State University
East Lansing, Michigan, U.S.A.

NOTES

1.   Prenatally, the individual utilizes maternal phenylalanine. There have been difficulties in determining and providing the exactly appropriate diet, both with respect to the level of phenylalanine and the level of other substances, and the success is certainly not complete. There has also been controversy concerning if and when the special diet can be terminated, and questions about exactly what dietary control is necessary for pregnant women with PKU so as to prevent the mental retardation of their offspring (Guttler and Lou, 1986; Rimoin *et al.*, 1996, pp. 1867-70).
2.   Various philosophical tools can aid in analyzing this comparison: reference class (Salmon, 1984, ch. 2) and contrast class (van Fraassen, 1980, ch. 5).
3.   Note that differences between scientists and the public here raise important questions concerning the responsibility of scientists to make clear what they are doing and what its significance is. We should also be mindful of the effect that portrayals in the media of exciting scientific progress may have on the public's impression in this regard (Nelkin, 1987, ch. 10).

REFERENCES

Bartels, D. *et al.*, eds., *Prescribing Our Future: Ethical Challenges in Genetic Counseling*. New York: Aldine De Gruyter, 1993.
Bessman, S.P., Swazey, J.P. "Phenylketonuria: A Study of Biomedical Legislation." In *Human Aspects of Biomedical Innovation*, E. Mendelson *et al.*, eds. Cambridge: Harvard University Press, 1971.
Block, N., Dworkin, G., *The IQ Controversy*. New York: Random House, 1976.
Brown M., Goldstein J. Familial hypercholesterolemia. Proceedings of the National Academy of Sciences 1974; 71:73-7
Burian R. Human sociobiology and genetic determinism. The Philosophical Forum 1981; 13(2):43-66
Collins F. Sequencing the human genome. Hospital Practice 1997; 32:35-53
Dawkins, R., *The Extended Phenotype*. Oxford: Oxford University Press, 1982.
Edlin G. Inappropriate use of genetic terminology in medical research: a public health issue. Perspectives in Biology and Medicine 1987; 31:47-56
Feldman M., Lewontin R. The heritability hang-up. Science 1975; 190:1163-8

Gelehrter, T. *et al.*, eds., *Principles of Medical Genetics*, 2nd ed. Baltimore: Williams and Wilkins, 1997.

Gifford F. Complex genetic causation of human disease: critiques of and rationales for heritability and path analysis. Theoretical Medicine 1989; 10:107-22

Gifford F. Genetic traits. Biology and Philosophy 1990; 5:327-47

Gorovitz S. Causal judgments and causal explanations. Journal of Philosophy 1965; 62:695-711

Gottesman, I., Shields, J., *Schizophrenia: The Epigenetic Puzzle*. New York: Cambridge University Press, 1982.

Guttler F., Lou H. Dietary problems of phenylketonuria: effect on CNS transmitters and their possible role in behaviour and neuropsychological function. Journal of Inherited Metabolic Disease 1986; 9(Suppl. 2):169-77

Hacking, I. "The Self-Vindication of the Laboratory Sciences." In *Science As Practice and Culture*, A. Pickering, ed. Chicago: University of Chicago Press, 1992.

Hesslow, G. "What is a Genetic Disease: On the Relative Importance of Causes." In *Health, Disease and Causal Explanation in Medicine*, L. Nordenfelt, B.I.B. Lindahl, eds. Dordrecht: D. Reidel Press, 1984.

Hirschhorn K. Practical and ethical problems in human genetics. Birth Defects 1972; 8(4):17-30

Hull, R. "Why 'Genetic Disease?'" In *Genetic Counseling: Facts, Values and Norms*, A. Capron *et al.*, eds. New York: Alan R. Liss, 1979.

Kessler S. The psychological paradigm shift in genetic counseling. Social Biology 1980; 21:167-85

Korenberg J., Rimoin D. Medical genetics. Journal of the American Medical Association 1995; 273:1692-3

Lappe, M. "Theories of Genetic Causation in Human Disease." In *Genetic Counseling: Facts, Values and Norms*, A. Capron *et al.*, eds. New York: Alan R. Liss, 1979.

Lappe M. The limits of genetic inquiry. Hastings Center Report 1987; 17:5-11

Lewontin R. The analysis of variance and the analysis of causes. American Journal of Human Genetics 1974; 26:400-11

Mackie, J.L., *The Cement of the Universe*. Oxford: Clarendon Press, 1974.

Nelkin, D., *Selling Science: How the Press Covers Science and Technology*. New York: W.H. Freeman and Co., 1987.

Owen, C. A., *Wilson's Disease: The Etiology, Clinical Aspects, and Treatment of Inherited Copper Toxicosis*. Park Ridge, NJ: Noyes Publications, 1981.

Paul, D. "PKU Screening: Competing Agendas, Converging Stories." In *The Practices of Human Genetics: Sociology of the Science Yearbook 1997*, M. Fortun, E. Mendelsohn, eds. Dordrecht: Kluwer Academic Publishers, 1999.

Rimoin, D. *et al.*, eds., *Emery and Rimoin Principles and Practice of Medical Genetics*, 3rd ed., Vol. II. New York: Churchill Livingstone, 1996.

Salmon, W., *Scientific Explanation and the Causal Structure of the World*. Princeton: Princeton University Press, 1984.

Schaffner, K., *Discovery and Explanation in Biology and Medicine*. Chicago: University of Chicago Press, 1993.

Smith K. The new problem of genetics: a response to Gifford. Biology and Philosophy 1992; 7:331-48

Wendler D. Innateness as an explanatory concept. Biology and Philosophy 1996; 11:89-116

van Fraassen, B., *The Scientific Image*. Oxford: Clarendon Press, 1980

CHAPTER 7

RACHEL A. ANKENY

REDUCTION RECONCEPTUALIZED: CYSTIC FIBROSIS AS A PARADIGM CASE FOR MOLECULAR MEDICINE

1. INTRODUCTION

Information regarding the genetic basis of disease is being gathered at an ever-accelerating rate, particularly because of the use of molecular techniques to map and sequence the human genome. In the last decade, genes related to numerous genetic diseases have been identified, including cystic fibrosis (CF), neurofibromatosis 2, amyotrophic lateral sclerosis (Lou Gehrig disease), and Huntington disease. In the earliest years of the U.S. Human Genome Project (HGP), it was said that molecular sequencing would provide the basis for diagnosis, treatment, and even cure of the various disease entities encountered in clinical genetics. This approach relied on the assumption that explanations in terms of properties of clinical genetic disease entities (the phenotype as evidenced by abnormal or variant cellular and/or physiological entities and processes) would be reducible to explanations in terms of properties of molecular entities (the genotype), which on the surface seemed plausible. In this era of genetic sequencing, George W. Beadle and Edward L. Tatum's (1941) 'one gene-one enzyme' hypothesis has mutated into what some term the 'OGOD' hypothesis, 'one gene-one disease,' so that the goal is to find *the* gene for this cancer, that type of arthritis, and even for more complex diseases such as alcoholism. Mapping and sequencing thus is said to constitute a new form of clinically relevant human anatomy, with the gene as object (McKusick, 1989). Progress in genetics has "...given all of medicine a new paradigm. Specialists in all medical areas approach the study of their most puzzling diseases by first mapping the genes

L.S. Parker and R.A. Ankeny (eds.), Mutating Concepts, Evolving Disciplines: Genetics, Medicine and Society, 127-141.
© 2002 Kluwer Academic Publishers. Printed in the Netherlands.

responsible for them" (McKusick, 1991, p. 17). However, critics have argued that reductionist genetic approaches to medicine have many limitations (e.g., see Strohman, 1993; Tauber and Sarkar, 1992).

This chapter is motivated by the epistemic gaps present between the concept of 'reduction' as commonly invoked in the goals underlying the HGP and the concept as it has been developed in the philosophy of science and particularly philosophy of biology. My argument addresses the difficulties raised by the claims of proponents of the HGP by developing a new model of explanatory reduction called the 'SUPER' model. I argue that the explanations sought in this area must conform to the SUPER model if the optimism of their proponents is justified. Then by focusing on recent research in medical genetics on cystic fibrosis (CF), I explore the extent to which it will be possible to reduce explanations about the properties of clinical genetic disease entities as they are currently identified to molecular explanations in a manner that fulfills the SUPER model's criteria. Using CF as a paradigm case, I argue that explanatory reduction of the sort initially proposed by researchers in this area is an inappropriate model for scientific progress, and discuss what methodological position for future research in medical genetics is entailed by this conclusion.

2. HISTORICAL BACKGROUND: THE HUMAN GENOME PROJECT

The HGP is part of an international research effort that is the culmination of the search for causes of genetic disease and human disease in general. By far the largest and most expensive project undertaken to date in biology, the HGP set out to provide high-resolution maps of—and ultimately to sequence—the entire human genome, a task that was completed in 2000.[1] Although individual projects within the HGP are somewhat diverse in their aims, the HGP's explicit goals generally shape the agenda for most research in medical genetics since a large percentage of funding in the field is under the auspices of the HGP's governing bodies.

The idea of sequencing the human genome gained support in the mid-1980s due to a number of factors (see Cook-Deegan, 1994 for a general history). Basic techniques for analyzing DNA had been developed, such as electrophoresis separation and the use of linkage markers and cloned DNA fragments. In addition, the Department of Energy (DOE) was interested in investigating methods of detecting inherited molecular mutations in humans, including those found among atomic bomb survivors, and began its own research program (DeLisi, 1988). Finally, Renato Dulbecco proposed that the best way to accelerate research toward finding a cure for cancer was to sequence the entire human genome, since "the sequence of the human DNA is the reality of our species, and everything that happens in the world depends on those sequences" (1986, p. 1056). The association with a cure for cancer helped to focus and accelerate implementation of plans already under discussion by various governmental agencies and scientific bodies, and even to excite public interest in the mapping and sequencing of the human genome (Watson and Cook-Deegan, 1991).

Early mapping of the chromosomal position of genes was predicted to hold immediate payoffs for medicine because it would allow analysis of the structure, function, and disease relevance of genes in molecular terms (Roberts, 1988; Berg, 1990; Rossiter and Caskey, 1991). According to HGP proponents, mapping and sequencing would permit direct diagnostic testing as well as identification of the molecular lesions that produce phenotypic disorders (McKusick, 1993; Collins, 1991a). In the early 1990s, it was thought by many scientists that mapping and sequencing would also lead to therapy to interrupt the phenotypic expression of deleterious molecular sequences not only for diseases associated with single gene mutations, but also for more complex polygenic or multifactorial disease phenomena (Yager, Nickerson, and Hood, 1991). Medicine eventually would become preventive, proponents of genomics argued, since genetic mutations that predispose individuals to disease would be identifiable and treatable at the molecular level before clinical disease was evidenced (Collins, 1991b). Even more generally, it was claimed that "when finally interpreted, the genetic messages encoded within our DNA molecules will provide the ultimate answers to the chemical underpinnings of human existence" (Watson, 1990, p. 44). Therefore, implicitly in its historical development and in the beliefs of its planners and supporters, as well as explicitly in its goals, the HGP was aimed at identifying causes of disease at the molecular level and reducing explanations of properties of clinical genetic diseases to explanations of properties of molecular sequences to facilitate diagnosis, treatment, and prevention.

When interpreting these claims, it must be noted that clinical genetics uses a rather broad definition of 'genetic disease' to delineate its domain. Most non-genetic diseases are thought to have genetic components that give a higher probability of developing the disease to persons possessing them than to persons lacking them. Most so-called genetic diseases typically have environmental and developmental components that are necessary for the disease to occur. In its current usage, application of the adjective 'genetic' reflects an emphasis on a genetic component over all other factors because the disease has been observed to be transmitted in certain familial patterns or associated with a particular chromosomal defect. A working definition will be adopted here that relies on the present state of medicine: a certain disease phenotype will be considered "genetic" if it is currently thought to be primarily caused by the genetic makeup of an individual (see Gifford, 1990). In biomedicine, causes are those conditions that can be manipulated to prevent the occurrence of a disease. Which particular cause (or related causes) from among these is emphasized often depends on the research interests of the speaker, author, or researcher (Whitbeck, 1977).

A number of interrelated questions that arise in clinical genetics help to illustrate some of the types of explanations sought in the area: what causes some genetic disease X; what sorts of people develop X; why do these patients have these symptoms; what can be done for X; and what will happen to people with X (McKusick, 1993). The first question is primarily relevant to research in human genetics, while the rest concern the practice of clinical genetics; the explanations provided in the process of clinical practice are greatly dependent on the information gathered in response to the first question type. In order for explanations in clinical genetics to be successful, information must be provided that can ground accurate predictions concerning which individuals are at increased risk for developing certain diseases.

Questions of the first type traditionally were answered using explanations following the Mendelian paradigm: a certain disease was said to be caused by inheritance of a single copy of a dominant gene or two copies of a recessive gene (e.g., sickle cell anemia was caused by inheriting two copies of the defective sickle cell gene). Observation of hereditary metabolic disorders, together with the Beadle-Tatum hypothesis (1941) of "one gene-one enzyme" (eventually modified to "one gene-one polypeptide" and, as noted previously, to "one gene-one disease" in recent years), altered the type of explanation being sought. The same question type then came to be considered as a request for a biochemical explanation with a basic causal-mechanical form: a gene (inherited in a specific Mendelian pattern, information that became part of a background package[2]) causes a change in the normal biochemical pathway that results ultimately in a set of clinical symptoms, the disease's phenotype. In the case of sickle cell disease, a single amino acid change results in production of unstable hemoglobin protein, which causes the sickling phenomenon.

There are a number of clinical genetic diseases for which fairly well-developed molecular characterizations are currently available. Using the paradigmatic example of CF, the question of what it would mean to claim that higher level disease explanations can be reduced, or are in principle reducible, to explanations at the molecular genetic level will be examined.

3. MODELS OF REDUCTION

To assess whether it will be possible and useful to reduce explanations in clinical genetics to molecular genetic explanations as early proponents of the HGP advocated, it is necessary to explicate a working model of explanatory reduction that reflects their aims. Many models of reduction have been proposed by philosophers of science (for reviews, see Hooker, 1981; Sarkar, 1992; and Schaffner, 1993). The applicability and adequacy of these models have been hotly debated, especially with regard to their relevance to biology and medicine, which are claimed to be notoriously difficult, or in principle impossible, to formalize in terms of laws.[3] In general, explanatory reduction occurs when some "reduced" entity can be explained by a "reducing" entity, where the entities in question can be, for example, theories, laws, or empirical generalizations. Models of reduction in this literature (e.g., Kemeny and Oppenheim, 1956; Nagel, 1949, 1961) fail to capture the methodology inherent in biomedicine and especially in the modern push aimed at molecular understandings of genetic diseases.

As a partial response, Kenneth Schaffner (1967, 1969a, b) proposed his "general reduction paradigm." In this model, a corrected, reduced theory that is "strongly analogous" to the original theory to be reduced, but that provides "more accurate experimentally verifiable predictions," can be used in the derivation. Early criticisms of Schaffner's paradigm drew attention to the fact that his revisions blurred reduction and replacement (Ruse, 1971; Hull, 1974). Schaffner's latest model (1993), called the 'general reduction-replacement' (GRR) model, explicitly allows for cases where the original reduced theory is not correctable, but is instead replaced by a reducing theory, or even a corrected, reducing theory; hence a continuum of reduction relations is possible, with replacement at one end.

By allowing replacement to count as a form of reduction, however, the GRR model lacks clarity about what is being "reduced." If only the weakest versions of the formal conditions of the GRR model are satisfied, the corrected reducing theory may bear little resemblance to the original reduced theory. Further, due to the weakness in the conditions of the GRR model, the corrected reducing theory does not necessarily have to make more accurate predictions or help to show why the reduced theory worked as well as it did, but simply must "explain" the domain of the reduced theory. The explanation may be expanded in complexity rather than reduced, and need not be explanatorily stronger in any way. Clearly those who are seeking to reduce complex phenotypic presentations into explanations at the level of the genotype are seeking a reduction resulting in an explanation which is more adequate or stronger, in virtue of its leading to improved diagnosis, prognosis, treatment, or even cure. In addition, the GRR model is broad enough to allow unilevel "reduction" relations and does not require the reduction to have any particular directionality, unlike what the proponents of the HGP are advocating. Therefore in its generality, the GRR model allows almost any type of theory change, whether it be progressive or reductive in any sense, to count as a "reduction," Thus the GRR model fails to reflect the epistemic goals underlying the HGP which focus on progressive changes through explanations at the lower, molecular level.

In light of the criticisms of the GRR model reviewed above, the following 'SUPER' (Stringent Unidirectional Progressive Explanatory Reduction) model will be adopted here as a more precise and rigorous summary of the epistemic framework in play in molecular medicine. The formal conditions for this model are that:

(1) the terms of the reducing explanation must be associated via synthetic identity with the terms of the original, reduced explanation, and the reducing explanation must contain terms that make reference to a lower level than those in the reduced explanation (revised connectability condition);

(2) the entire domain of the reduced explanation must be derivable from the reducing explanation together with the identity relations (revised derivability condition); and

(3) the reducing explanation should have greater explanatory power than the reduced explanation.

Condition (1) restricts reduction to a specific case of the GRR model particularly appropriate for the biomedical sciences and molecular medicine—

namely, reduction of an explanation of higher-level properties to an explanation in terms of lower-level properties; hence it is termed 'unidirectional.' There must be some function that expresses a synthetic identity (it is 'stringent' in this sense) of the properties at a higher level with the properties at a lower level. One important problem with this model is that the idea of 'levels' is used commonly in the biomedical sciences, but is very difficult to define unambiguously. Much analysis has occurred regarding 'parts' and use of part-whole relations, where one entity is said to be at a "higher level" than another if the former has the latter among its parts.[4] This part-whole concept is not particularly useful when the higher level entities are clinical phenotypes. It is not clear what would constitute proper parts of a phenotype, and an analysis of phenotypes in terms of parts and wholes thus seems inadequate; this conclusion is not surprising, since although many natural systems are hierarchical, they are not decomposable into parts (Bechtel and Richardson, 1993). Therefore, a very general definition of 'levels' is adopted here: $level_1$ will be considered lower than $level_2$ if the entities at $level_1$ are relatively similar in size, or some other generalizable measure, and smaller than the entities at $level_2$.

In condition (2), 'domain' is used to indicate a complex of experimental results that can be explained by the original, reduced theory (or should be able to be explained once the reduced theory is fully developed; see Shapere, 1974). This condition is similar to the corresponding condition in the GRR model; however, the SUPER model is strictly reductive (hence I term it 'stringent') since it does not explicitly allow substitution of corrected explanations in place of either the original or the reducing explanation. This avoids problematic replacements that bear little resemblance to reductions, as was discussed above.

Condition (3) requires that for a reduction to be successful, explanatory power with regard to the phenomena of most interest must be gained by using the reducing explanation rather than the reduced explanation (thus the P for 'progressive'). For example, the reducing explanation should make more accurate predictions than the reduced explanation did, should indicate why the reduced explanation worked as well as it did historically, or should serve as the basis of fruitful future research. This condition helps to rule out trivial decompositions; moreover, it requires that some scientific progress occur through the reduction process, even if the reducing explanation is more complex, which it is likely to be if all details are made explicit. Against these conditions, consider the following brief case study focused on cystic fibrosis.

4. THE CONFUSIONS SURROUNDING CF

Cystic fibrosis, which was first recognized in the 1930s (Andersen, 1938), is the most common potentially lethal autosomal recessive disorder among whites. In 1989, a deletion (ΔF506) in the CF transmembrane conductance regulator (CFTR) gene on chromosome 7 was identified as the mutation responsible for nearly 70% of CF chromosomes (Rommens et al., 1989; Riordan et al., 1989; Kerem et al., 1989).

The discovery of the CF gene and its most common mutations has been cited by proponents of the HGP (Collins, 1991a) and others (Harris and Schaffner, 1992) as an example of resoundingly successful applications of molecular genetic techniques and reductionistic strategies, rendering the recent history of molecular approaches to CF a particularly good (and fair) example to examine in the context of explanatory reduction.

CF affects the exocrine glands, causing production of thick mucus typically resulting in chronic respiratory infections, pancreatic insufficiency, malabsorption, and positive sweat tests (elevated chloride concentrations). Males with CF are often infertile because they have congenital bilateral absence of the vas deferens (CBAVD), the excretory duct of the testis responsible for sperm transport. Since ΔF506 was discovered, the situation has become much more complex; over 900 different mutations related to CF have been identified (Monaghan $et\ al.$, 2000), and the number continues to grow. Although ΔF506 accounts for a large percentage of the CF mutations in the U.S. white population, its frequency is much lower in other CF populations. Patients with CF can be either homozygotes (with two copies of the same mutation) or heterozygotes (with two different mutations associated with CF).

Complexities of this sort, however, do not rule out the in-principle reducibility of higher level explanations about the causation of CF to explanations at the molecular level. Assume that an exhaustive list of the mutations $(d_1 ... d_n)$ associated with the clinical CF phenotype will become available, and that the amino acid sequence related to CF that results from carrying two of these mutations and produces the CF phenotype will be articulated. The phenotypic explanation for CF could be formulated as follows: "defective chloride transport across membranes results in dehydrated secretions, which in turn lead to persistent mucus in the lungs and pancreas, leading to infections...." Then for the explanation "Mendelian genotype aa (inheritance of two copies of gene a) causes the clinical CF phenotype," identities can be established between the genotype and the various mutations. The reduction functions would be "Mendelian genotype aa = DNA sequence $(d_1 d_1$ or $d_1 d_2$ or $d_2 d_1$ or...or $d_n d_n)$" and "clinical CF phenotype = amino acid sequence CF." Thus the following explanation at the molecular level, which fulfills the revised connectability condition, would be obtained: "DNA sequence $(d_1 d_1$ or $d_1 d_2$ or $d_2 d_1$ or...or $d_n d_n) \rightarrow$ amino acid sequence CF," where the arrow represents a promissory note for a telescoped causal chemical account (as described by Schaffner, 1993).

However, it is not obvious whether the other two conditions for achieving reduction are fulfilled by this molecular explanation. Although the reduced explanation is derivable from the reducing explanation together with the identity relations, it is not immediately clear whether the $entire$ domain of the reduced explanation is thereby explicable. Further, it remains uncertain whether greater explanatory power has been gained. To answer these questions, some additional research about CF must be considered.

Identification of the great variety of mutations related to CF has drawn attention to the highly variable expression of the CF phenotype and even resulted in establishment of the Cystic Fibrosis Genotype-Phenotype Consortium. This group found that on average, those patients with a certain heterozygote combination (ΔF506/R117H) have less severe disease, including, for example, a lack of pancreatic insufficiency, than those with other common heterozygote combinations or those who are homozygous for ΔF506 (see Cystic Fibrosis Genotype-Phenotype Consortium, 1993). Nonetheless, the researchers concluded that predictions could not be made regarding the occurrence, severity, or course of the common CF complications for any of the genotypes studied (Hamosh and Corey, 1994). Others have obtained similar results concerning the correlations between less severe CF phenotypes and heterozygosity and between more severe phenotypes (especially pancreatic insufficiency) and homozygosity for ΔF508 as well as the correlation of phenotypic differences with genetic heterogeneity (McColley, Rosenstein, and Cutting, 1991).[5] However, many researchers report conflicting conclusions, including that patients who are ΔF506 homozygotes do not have more severe phenotypes (e.g., in terms of pulmonary and nutritional status) as compared with those with various heterozygote combinations (McColley et al., 1991; Santis et al., 1990a; al-Jader et al., 1992; Burke et al., 1992; Borgo et al., 1993; Lester et al., 1994) and that wide phenotypic variation exists even among patients who carry the same mutation combination (Burke et al., 1992; Santis et al., 1990; Campbell et al., 1991; Liechti-Gallati et al., 1992). One study even concluded that there is no correlation between the structural region of the protein affected by the CF mutation and the clinical phenotype (Santis et al., 1990a, b). Further, at least in some cases, truncation or even absence of CFTR is not particularly correlated with severe pulmonary disease, contrary to expectation (Cutting et al., 1990).

Even more puzzling results subsequently were noted. Three decades ago, it was hypothesized that males with CBAVD might have a mild form of CF (Holsclaw et al., 1971). When relatively large numbers of otherwise healthy men with CBAVD began presenting at infertility clinics in the early 1990s to undergo sperm aspiration for in vitro fertilization, researchers aware of the known association between CBAVD and CF performed genetic testing and found that many carried CF mutations (Dumur et al., 1990a; Rigot et al., 1991; Anguiano et al., 1992; Gervais et al., 1992; Oates and Amos, 1993; Patrizio et al., 1993). They were typically heterozygotes at the CF locus, with one mutation that causes traditional CF symptoms when inherited in two copies. Most importantly, some of the men were found to be carrying two mutations usually found in CF patients but did not have any clinical symptoms of CF except CBAVD. CF mutations also have been found in siblings with mild disease and normal sweat tests (Strong et al., 1991). Finally, a greater than expected frequency of mutations normally associated with phenotypic CF has been found in patients with other, milder pulmonary problems (e.g., allergic bronchopulmonary aspergillosis and chronic pseudomonas bronchitis), nasal polyps, and neonatal transient hypertrypsinemia (Dumur et al., 1990b; Burger et al., 1991; Laroche and Travert, 1991; Lucotte et al., 1991).

Given all of this evidence, the reduced explanation constructed above, "DNA sequence (d_1d_1 or d_1d_2 or d_2d_1 or...or d_nd_n) $\rightarrow$ amino acid sequence CF," appears to be currently invalid. The domain of the clinical CF phenotype in the reduced explanation is not derivable from the reducing explanation using the posited identity relations, since the traditional CF phenotype does not occur in many cases where the reducing explanation predicts that it should, given the DNA sequence. Thus there seems to have been either falsification of the direct causal link between the mutation and the amino acid sequence related to CF or of the assumed identity between the amino acid sequence and the clinical CF phenotype, or both.

Accordingly, a number of additional hypotheses have been proposed by researchers to account for these sorts of findings. First, some claim that careful examination of the various mutations will allow establishment of a "spectrum" of forms of CF, including CBAVD and milder pulmonary diseases, where different mutation combinations result in different phenotypes (Oates and Amos, 1993). However, this theory fails to explain the lack of pulmonary disease in those with CBAVD who carry two copies of the mutation normally associated with the traditional CF phenotype, for example, and has not been borne out by more recent research. Another hypothesis frequently proposed is that a modifying gene exists at another locus that mediates the expression of mutations at the CF locus. A related option is that the "chromosomal background" of the CF gene (hitherto undetected mutations in the immediate genetic context in which the main CF mutations occur) determine the phenotype (Stuhrmann *et al.*, 1991; Kiesewetter *et al.*, 1993), as has been suggested by twin studies (Santis *et al.*, 1992). All of these hypotheses attempt to preserve the assumption that higher level explanations about the causation of the CF phenotype will be reducible to explanations at the molecular level. To assess this assumption, it is necessary to discuss how to evaluate additional hypotheses such as these and to review more general evidence regarding genetic and epigenetic phenomena.

All of the additional hypotheses in the case of CF have a similar form: one of the DNA sequences associated with CF *plus* another DNA sequence, either at the same or a different locus, is said to cause an amino acid sequence that results in the traditional CF phenotype: "DNA sequence$_1$ (d_1d_1 or d_1d_2 or d_2d_1 or…or d_nd_n) + DNA sequence$_2$ (details to be determined) $\rightarrow$ amino acid sequence CF." It may be that not all of the previously identified DNA sequences (d_1d_1 or…d_nd_n) together with the modifying sequence or sequences produce the traditional phenotype, but that each produces a different but fairly consistent phenotype. In that case, a series of molecular explanations would be available for the various CF phenotypes. The revised explanation thus is clearly an extension of the original explanation and is likely to be empirically testable. The remaining question is whether there is any theoretical evidence in favor of such an explanation, given the current lack of specific empirical evidence.

The theoretical background of current research in molecular genetics generally assumes that variable expressivity or even variable penetrance (whether a mutation is expressed at all) of mutations is due to modifying effects of mutations at other loci, as has been proposed in the case of CF. However, there is theoretical evidence that expression can also be affected by phenomena at higher levels than the molecular, including three-dimensional chromosomal position, genomic imprinting, environmental exposure, fetal environment, and so on. Another theory put forward to explain differential phenotypic expression holds that chance plays a significant developmental role along with genetic and environmental effects. With CF, recent genotype-phenotype studies have tracked the extent to which various CFTR alleles contribute to clinical variation in CF, and showed that the degree of correlation between CFTR genotype and CF phenotype varies depending on the phenotypic presentation. The correlation is highest for those with pancreatic symptoms and lowest for those with pulmonary disease (the classic symptom associated with CF). The poor correlation between the CFTR genotype and severity of lung disease indicates an influence of *both* environmental and secondary genetic factors (for a review, see Zielenski, 2000).

Thus it is likely that some combination of these sorts of mechanisms is responsible for the differential phenotypic expression of the genetic mutations associated with CF. Consequently, the assumptions of modifying effects due to specific molecular sequences that underlie the hypotheses typically proposed by researchers to account for the causation of the various CF phenotypes can be faulted for being *ad hoc* and premature. Of course it cannot be concluded that reduction of the higher level explanations about CF to the molecular level is in principle impossible, but such a reduction is clearly far from being achieved currently despite efforts over the past decade. However, there are more significant philosophical and methodological lessons to be learned from the paradigmatic example of CF that do not require decisive evidence about precise genotype-phenotype correlations, as will be discussed in the closing section.

5. CONCLUSIONS

As this brief review of CF research has indicated, phenotypes thought previously to be unrelated have been found to be associated with mutations at the same locus. Research in this area has made very uncertain what seemed to be a clear set of diagnostic criteria for CF, including certain phenotypic manifestations and particular detectable mutations. For example, although it was known that many males with CF also have CBAVD, it was not clear that the occurrence of CBAVD without CF was also associated with mutations at the CF locus. CBAVD is now being called a 'primarily genital form' of CF (Anguiano *et al.*, 1992). Similarly, CF formerly was thought to be invariably severe and fatal; recent research suggests that a number of pulmonary diseases, some of which may be quite mild, are related to mutations at the CF locus.

A new, broadened disease class is likely to emerge that will replace the traditional, relatively limited definition of CF. Similar significant reclassifications will likely occur with most other genetic diseases, resulting in subdivision (or perhaps unification) of many current disease entities. Some diseases previously thought to be primarily genetic, for example various congenital malformations and chromosomal disorders, likely will be found to have more significant environmental or developmental components. Conversely, some disease entities that currently are not believed to be genetic undoubtedly will be found to be associated with molecular mutations. Finally, conditions formerly thought to be within the normal range of variation will be associated with newly detected genetic mutations and hence potentially reclassified as diseases. Identification of such new disease entities and reclassification of diseases will have even broader methodological, ethical, social, and economic consequences than those raised by the traditional genetic disease entities, especially given the high amount of variation present in the human genome.

Reduction of higher level explanations of *current* genetic diseases to lower level, molecular explanations is thus highly unlikely, since corrected explanations that reflect the revised classificatory systems will need to be substituted for the original explanations. Such a replacement would be permissible for reductions under the GRR model; however as was argued earlier, it is not clear why this type of replacement should qualify as a *reduction*. More importantly, although such replacements may refer to terms at the molecular level, explanations will not be completely reducible to the molecular level; instead, the resulting explanations will need to be interlevel[6] in order to retain explanatory power. Obtaining an explanation of a particular phenomenon likely will require a constant shifting between levels, and not a unidirectional, reductionist explanatory flow.

The extended example of CF has shown resistance to complete reduction of higher level explanations to explanations at the molecular level in the strict sense of the SUPER model. This lack of explanatory reducibility is evident because of the failure to date of efforts to correlate mutations (and even protein alterations produced by them) with phenotypic outcomes. The strategies employed in this area show that research continues at both higher and lower levels because of the need to shift constantly between these levels to generate new, accurate information as explanations are being constructed. For instance, by identifying higher level patterns of inheritance, affected individuals, and sequences of phenotypic development, clinical geneticists help to circumscribe research in molecular genetics. Conversely, molecular geneticists find particular mutations at certain loci and other genetic data that can help to identify relevant research subjects as well as phenotypic traits and possible inheritance patterns for clinical geneticists to investigate.

There will be inevitable reclassification of disease entities (see also Temple *et al.*, 2001), which will result in replacement of previous explanations as improved clinical and molecular data are gathered in this manner. Such replacement explanations may indeed reflect focus on the molecular level. However, by adopting a strictly molecular explanation and relegating the remaining information to the initial conditions and definitions, it is not clear that there would be any gain in explanatory power. As has been seen in the case of CF, attempts to provide strictly molecular explanations have resulted in inaccurate predictions. Similarly, some interlevel causal details likely will need to be part of valid future explanations, and as in the case of CF, some of the most interesting and useful information for the purposes of continued research probably will be contained in these interlevel relations.

These conclusions should not be surprising, since medical genetics is basically a complex amalgam of a number of related fields, including classical genetics, clinical genetics, cytogenetics, and molecular genetics. Hence, identity relations should more properly be seen as interlevel causal sketches that present solutions to problems that are shared by these fields and reflect the interactions between them. Research in molecular genetics has furnished an explanatory extension of clinical genetics by explicating many causal connections between genotype and phenotype. However in addition, research in clinical genetics and other areas of medicine has helped to explicate causal connections between phenotype and genotype. Therefore, a model such as the SUPER model—implicit in promises made in many of the goals and probably much of the research of the HGP (and consequently, of modern genetics) that advocates reduction to a single, lower level—appears to be an inappropriate characterization of the most fruitful manner of achieving progress in clinical and even molecular genetics. Instead, the research advances and difficulties that have been outlined in the paradigmatic case of CF more properly reflect the need for using a multilevel explanatory model to achieve the goals of modern biomedicine.

University of Sydney
Sydney, New South Wales, Australia

ACKNOWLEDGEMENTS

I wish to thank Jim Lennox, Jason Grossman, Fred Gifford, Eric Juengst, and audiences at the American Philosophical Association Pacific Meeting 2000 and at the Unit for HPS, University of Sydney for comments on various versions of this chapter. A much earlier draft was awarded honorable mention in the 1995 Dwight J. Ingle Memorial Writers Award, sponsored by *Perspectives in Biology and Medicine,* and I am grateful to the anonymous referees who provided comments in that process.

NOTES

1.   For the original planning documents that reflect the changing goals of the project, see U.S. Department of Health and Human Services and the Department of Energy, 1990; Collins and Galas, 1993; and Collins *et al.*, 1998. A joint announcement was made in June 2000 by the public human genome sequencing consortium and a private corporation, Celera Genomics, that the sequence had been completed, to great international fanfare (see Macilwain, 2000). Their respective versions of the sequence were published separately in the following year (see Lander *et al.*, 2001 and Venter *et al.*, 2001).
2.   Schaffner (1993) refers to the procedure of assuming some background processes ("packages") as a sort of "information hiding." The amount of detail necessary will differ from science to science, and all sciences probably will allow explanations used in research, teaching, or publication to make some assumptions as part of a larger explanation.
3.   This literature on reduction in biomedicine is enormous and highly varied: for instance see Zucker, 1981; Robinson, 1986; Shapiro, 1989; Harris and Schaffner, 1992; Heim, 1992; and Tauber and Sarkar, 1992. For general approaches to reduction, see Wimsatt, 1976a; Kitcher, 1984.
4.   Discussions on how to conceptualize and formalize part-whole relations in biology are provided by Simon, 1962; Grobstein, 1969; Oppenheim and Putnam, 1968; Kauffman, 1976; Maull, 1977; Wimsatt, 1986; and Bechtel and Richardson, 1993.
5.   For the details of these studies, see Curtis *et al.*, 1991; Férec *et al.*, 1993; and Gan *et al.*, 1994.

6.   On interlevel explanations, see Wimsatt, 1976b; Maull, 1977; Culp and Kitcher, 1989; Darden 1991; and Schaffner, 1993.

REFERENCES

al-Jader L.N. *et al*. Severity of chest disease in cystic fibrosis patients in relation to their genotypes. Journal of Medical Genetics 1992; 29:883-7

Andersen D.H. Cystic fibrosis of the pancreas and its relation to celiac disease. American Journal of Diseases of Children 1938; 56:344

Anguiano A. *et al*. Congenital bilateral absence of the vas deferens: a primarily genital form of cystic fibrosis. Journal of the American Medical Association 1992; 267:1794-7

Beadle G.W., Tatum E.L. Genetic control of biochemical reactions in *Neurospora*. Proceedings of the National Academy of Science 1941; 27:499-506

Bechtel, W., Richardson, R.C., *Discovering Complexity: Decomposition and Localization as Strategies in Scientific Research*. Princeton: Princeton University Press, 1993.

Berg P. All our collective ingenuity will be needed. FASEB Journal 1990; 5:75

Borgo G. *et al*. Cystic fibrosis: the ΔF508 mutation does not lead to an exceptionally severe phenotype: a cohort study. European Journal of Pediatrics 1993; 152:1006-11

Burger J. *et al*. Genetic influences in the formation of nasal polyps. Lancet 1991; 337:974

Burke W. *et al*. Variable severity of pulmonary disease in adults with identical cystic fibrosis mutations. Chest 1992; 102:506-9

Campbell P.W. *et al*. Cystic fibrosis: relationship between clinical status and F508 deletion. Journal of Pediatrics 1991; 118:239-41

Collins F., Galas D. A new five-year plan for the U.S. Human Genome Project. Science 1993; 262:43-6

Collins F.S. Medical and ethical consequences of the Human Genome Project. Journal of Clinical Ethics 1991a; 2:260-7

Collins F.S. The genome project and human health. FASEB Journal 1991b; 5:77

Collins F.S. *et al*. New goals for the U.S. Human Genome Project: 1998-2003. Science 1998; 282:682-9

Cook-Deegan, R., *The Gene Wars: Science, Politics, and the Human Genome*. New York: W. W. Norton & Company, 1994.

Culp S., Kitcher P. Theory structure and theory change in contemporary molecular biology. British Journal for the Philosophy of Science 1989; 40:459-83

Curtis A. *et al*. Association of less common cystic fibrosis mutations with a mild phenotype. Journal of Medical Genetics 1991; 28:34-7

Cutting G.R. *et al*. Two patients with cystic fibrosis, nonsense mutations in each cystic fibrosis gene, and mild pulmonary disease. New England Journal of Medicine 1990; 323:1685-9

Darden, L., *Theory Change in Science: Strategies from Mendelian Genetics*. New York: Oxford University Press, 1991.

DeLisi C. The Human Genome Project. American Scientist 1988; 76:488-93

Dulbecco R. A turning point in cancer research: sequencing the human genome. Science 1986; 231:1055-6

Dumur V. *et al*. Abnormal distribution of cystic fibrosis ΔF508 allele in adults with chronic bronchial hypersecretion. Lancet 1990a; 335:1340

Dumur V. *et al*. Abnormal distribution of CF ΔF508 allele in azoospermic men with congenital aplasia of epididymis and vas deferens. Lancet 1990b; 336:512

Férec C. *et al*. Genotype analysis of adult cystic fibrosis patients. Human Molecular Genetics 1993; 2:1557-60

Gan K.H. *et al*. Correlation between genotype and phenotype in patients with cystic fibrosis. New England Journal of Medicine 1994; 330:865-6

Gervais R. *et al*. High frequency of the R117H cystic fibrosis mutation in patients with congenital absence of the vas deferens. New England Journal of Medicine 1992; 328:446-7

Gifford F. Genetic traits. Biology and Philosophy 1990; 5:327-47

Grobstein C. Organizational levels and explanation. Journal of the History of Biology 1969; 2:199-221

Hamosh A., Corey M. Cystic correlation between genotype and phenotype in patients with cystic fibrosis. Fibrosis Genotype-Phenotype Consortium. New England Journal of Medicine 1993; 329:1308-13

Hamosh A., Corey M. Correlation between genotype and phenotype in patients with cystic fibrosis. New England Journal of Medicine 1994; 329:866-7

Harris H.W., Schaffner K.F. Molecular genetics, reductionism, and disease concepts in psychiatry. Journal of Medicine and Philosophy 1992; 17:127-53

Heim S. Is cancer cytogenetics reducible to the molecular genetics of cancer cells? Genes, Chromosomes, and Cancer 1992; 5:118-96

Holsclaw D.S. et al. Genital abnormalities in male patients with cystic fibrosis. Journal of Urology 1971; 106:568-74

Hooker C.A. Towards a general theory of reduction. Dialogue 1981; 20:38-59, 201-36, 496-529

Hull, D., Philosophy of Biological Science. Englewood Cliffs, NJ: Prentice-Hall, 1974.

Kauffman, S.A. "Articulation of Parts Explanations in Biology and the Rational Search for Them." In Topics in the Philosophy of Biology, Boston Studies in the Philosophy of Science. Vol. 27, M. Grene, E. Mendelsohn, eds. Dordrecht: D. Reidel, 1976.

Kemeny J.G., Oppenheim P. On reduction. Philosophical Studies 1956; 7:6-19

Kerem B.-S. et al. Identification of the cystic fibrosis gene: genetic analysis. Science 1989; 245:1073-80

Kiesewetter S. et al. A mutation in CFTR produces different phenotypes depending on chromosomal background. Nature Genetics 1993; 5:274-8

Kitcher P. 1953 and all that: a tale of two sciences. Philosophical Review 1984; 93:335-73

Lander E.S. et al. Initial sequencing and analysis of the human genome. Nature 2001; 409:860-921

Laroche D., Travert G. Abnormal frequency of ΔF508 mutation in neonatal transitory hypertrypsinaemia. Lancet 1991; 337:55

Liechti-Gallati S. et al. Genotype/phenotype association in cystic fibrosis: analyses of the F508, R553X, and 3905insT mutations. Pediatric Research 1992; 32:175-8

Lucotte G. et al. Transient neonatal hypertrypsinaemia as a test for F508 heterozygosity. Lancet 1991; 337:998

Macilwain C. World leaders heap praise on human genome landmark. Nature 2000; 405:983-6

Maull N.L. Unifying science without reduction. Studies in History and Philosophy of Science 1977; 8:143-71

McColley S.A. et al. Differences in expression of cystic fibrosis in blacks and whites. American Journal of Diseases of Childhood 1991; 145:94-7

McKusick V.A. Mapping and sequencing the human genome. New England Journal of Medicine 1989; 320:910-5

McKusick V.A. Current trends in mapping human genes. FASEB Journal 1991; 5:12-20

McKusick V.A. Medical genetics: a 40-year perspective on the evolution of a medical specialty from a basic science. Journal of the American Medical Association 1993; 270:2351-6

Monaghan K.G. et al. Frequency and clinical significance of the S1235R mutation in the cystic fibrosis transmembrane conductance regulator gene: results from a collaborative study. American Journal of Medical Genetics 2000; 95:361-5

Nagel, E. "The Meaning of Reduction in the Natural Sciences." In Science and Civilization, R.C. Stauffer, ed. Madison, WI: University of Wisconsin, 1949.

Nagel, E., The Structure of Science. New York: Harcourt, Brace & World, 1961.

Oates R.D., Amos J.A. Congenital bilateral absence of the vas deferens and cystic fibrosis: a genetic commonality. World Journal of Urology 1993; 11:82-8

Oppenheim, P., Putnam, H. "Unity of Science as a Working Hypothesis." In Concepts, Theories, and the Mind-Body Problem, H. Feigl, M. Scriven, G. Maxwell, eds. Minneapolis, MN: University of Minnesota, 1968.

Patrizio P. et al. Aetiology of congenital absence of vas deferens: genetic study of three generations. Human Reproduction 1993; 8:215-20

Rigot J.-M. et al. Cystic fibrosis and congenital absence of the vas deferens. New England Journal of Medicine 1991; 325:65-6

Riordan J.R. et al. Identification of the cystic fibrosis gene: cloning and characterization of complementary DNA. Science 1989; 245:1066-73

Roberts L. Academy backs genome project. Science 1988; 239:725-6

Robinson J.D. Reduction, explanation, and the quests of biological research. Philosophy of Science 1986; 53:333-53

Rommens J.M. *et al.* Identification of the cystic fibrosis gene: chromosome walking and jumping. Science 1989; 245:1059-65

Rossiter B.J.F., Caskey C.T. Molecular studies of human genetic disease. FASEB Journal 1991; 5:21-7

Ruse M.E. Reduction, replacement, and molecular biology. Dialectica 1971; 25:39-72

Santis G. *et al.* Linked marker haplotypes and the ΔF508 mutation in adults with mild pulmonary disease and cystic fibrosis. Lancet 1990a; 335:1426-9

Santis G. *et al.* Independent genetic determinants of pancreatic and pulmonary status in cystic fibrosis. Lancet 1990b; 336:1081-4

Santis G. *et al.* Genotype-phenotype relationship in cystic fibrosis: results from the study of monozygotic and dizygotic twins with cystic fibrosis. Pediatric Pulmonology 1992; 14(suppl 8):239

Sarkar S. Models of reduction and categories of reductionism. Synthese 1992; 91:167-94

Schaffner K.F. Approaches to reduction. Philosophy of Science 1967; 34:137-47

Schaffner K.F. Theories and explanations in biology. Journal of the History of Biology 1969a; 2:19-33

Schaffner K.F. The Watson-Crick model and reductionism. British Journal for the Philosophy of Science 1969b; 20:325-48

Schaffner, K.F., *Discovery and Explanation in Biology and Medicine*. Chicago: University of Chicago, 1993.

Shapere, D. "Scientific Theories and Their Domains." In *The Structure of Scientific Theories*, F. Suppe, ed. Urbana, IL: University of Illinois, 1974.

Shapiro B.L. The pathogenesis of aneuploid phenotypes: the fallacy of explanatory reductionism. American Journal of Medical Genetics 1989; 33:146-50

Simon H. The architecture of complexity. Proceedings of the American Philosophical Society 1962; 106:467-82

Strohman R.C. Ancient genomes, wise bodies, unhealthy people: limits of a genetic paradigm in biology and medicine. Perspectives in Biology and Medicine 1993; 37:112-45

Strong T.A. *et al.* Cystic fibrosis gene mutation in two sisters with mild disease and normal sweat electrolyte levels. New England Journal of Medicine 1991; 325:1630-4

Stuhrmann M. *et al.* Genotype phenotype correlations in cystic fibrosis patients. Advances in Experimental Medicine and Biology 1991; 290:97-103

Tauber A.I., Sarkar S. The Human Genome Project: has blind reductionism gone too far? Perspectives in Biology and Medicine 1992; 35:220-35

Temple L.K.F. *et al.* Defining disease in the genomics era. Science 2001; 293:807-8

U.S. Department of Health and Human Services and Department of Energy, *Understanding Our Genetic Inheritance. The U.S. Human Genome Project: The First Five Years FY 1991-1995*. Washington, D.C.: Department of Energy, 1990.

Venter J.C. *et al.* The sequence of the human genome. Science 2001; 291:1304-51

Watson J.D. The Human Genome Project: past, present, and future. Science 1990; 248:44

Watson J.D., Cook-Deegan R.M. Origins of the Human Genome Project. FASEB Journal 1991; 5:8-11

Whitbeck C. Causation in medicine: the disease entity model. Philosophy of Science 1977; 44:619-37

Wimsatt W. Reductive explanation: a functional account. Boston Studies in Philosophy of Science 1976a; 32:671-710

Wimsatt, W. "Reduction, Levels of Organization, and the Mind-Body Problem." In *Consciousness and the Brain*, G. Globus, G. Maxwell, I. Savodnik, eds. New York: Plenum, 1976b.

Wimsatt, W. "Forms of Aggregativity." In *Human Nature and Natural Knowledge*, A. Donagan, A.N. Perovich, Jr., W.V. Wedin, eds. Dordrecht: D. Reidel, 1986.

Yager T.D. *et al.* The Human Genome Project: creating an infrastructure for biology and medicine. Trends in Biochemical Sciences 1991; 16:454-61

Zielenski J. Genotype and phenotype in cystic fibrosis. Respiration 2000; 67:117-33

Zucker A. Holism and reductionism: a view from genetics. Journal of Medicine and Philosophy 1981; 6:145-63

CHAPTER 8

JOSEPH L. GRAVES, JR.

SCYLLA AND CHARYBDIS: ADAPTATIONISM, REDUCTIONISM, AND THE FALLACY OF EQUATING RACE WITH DISEASE

1. INTRODUCTION

In the Odyssey of Homer, the goddess Circe gives Odysseus instructions to aid him in his journey home. He is warned that only one sea road can take him to Ithaca. That road runs between an impassable whirlpool created by the monster Charybdis and along the cliffs of the horrible six-headed monster Scylla. Here are Circe's instructions:

> Don't let your ship draw near when she [Charybdis] is gulping brine; no one—not even he who makes the earth tremble—could save you then. Hold closer to the cliff of Scylla: better far to mourn six men than to lament the loss of all of them. (Mandelbaum, 1990, pp. 245-6)

One lesson learned from this example is that given two horrors, it is better to choose the lesser. However, even that choice carries with it a price. Modern biology is now faced with Odysseus' choice. Advancement cannot occur without a journey involving the ideologies of adaptationism and reductionism, although it is not clear which takes on the role of Scylla or Charybdis in any particular research problem. This chapter examines the role of adaptationism and reductionism in how medical research historically has viewed the biological category of 'race' and its relationship to disease prevalence. The errors that result from this confusion are not trivial and have major practical importance for the progress of biomedical research relevant to "minority" populations that historically have been underserved by the medical and scientific community.

L.S. Parker and R.A. Ankeny (eds.), Mutating Concepts, Evolving Disciplines: Genetics, Medicine and Society, 143-164.
© 2002 Kluwer Academic Publishers. *Printed in the Netherlands.*

2. CONFUSED DEFINITIONS AND RESEARCH PROGRAMS

Modern evolutionary genetics recognizes that no "races" exist within the human species (e.g., see Graves, 1993, 2001, 2002; Cavalli-Sforza, Menozzi, and Piazza, 1994, pp. 19-20; Templeton, 2002). What generally is not known is that this recognition dates back to Charles Darwin's discussion of the problem in *The Descent of Man* published in 1871. Properly defined, 'biological' races refer to subspecies; these are populations that have a considerable amount of genetic distance between them or could be described as populations that have maintained unique evolutionary lineage. Neither of these conditions currently exists for human beings. What are popularly described as 'races' are really populations that have considerable genetic overlap across the genome and have maintained relatively high levels of gene flow throughout human history (Cavalli-Sforza, Menozzi, and Piazza, 1994; Templeton 2002). Nor can human races unambiguously be defined using phenotypic characters (Montagu, 1974). Attempts to do so utilizing general anthropomorphic characters lead to phylogenetic trees that do not match evolutionary genetic history. For example, L. Luca Cavalli-Sforza and Anthony W. F. Edwards (1964) attempted to construct a tree of human relatedness using general anthropometric characters such as anatomical measurements of body parts and skin color. Their resultant tree gave population groupings that made no genetic or evolutionary sense. For example, Swedes and French were on the same branch as Eskimos and North American Indians. However, North American Indians were on a different branch from Northeast Asians. All these groups diverged at an original branch point separating them from Papuans, Australoids, and Bantu (meaning most of the people who are native inhabits of sub-Saharan Africa). Of course we now know that Australoids and 'Bantu' are the most genetically divergent groups in the human species and that Eskimos and North American Indians belong on the same branch as Northeast Asians. Africans are the most genetically divergent population and contain more genetic variation than all of the rest of the modern human populations combined (see Figure in Cavalli-Sforza, Menozzi, and Piazza, 1994, p. 80). The failure of the phenotypic trees to reveal genetic history results from two factors: first, phenotypic variation is discordant. That is, genetic forces that determine the frequency of alleles impacting various phenotypic traits do not vary consistently over time and space. Second, the interaction of genes and environments influences the expression of genetic material. Neither genetic backgrounds nor environments can be expected to be expressed uniformly with regard to any particular trait.

What is the impact of this on the determination of the genetic aspects of disease in any particular population or "race?" First we must recognize that human genetic variation can be partitioned in loci that are considered monomorphic as well as those that are considered polymorphic. These loci may be involved in either the cause of an inborn genetic disease, regulating fundamental features of life history such as fecundity or longevity, or in the resistance to disease-causing agents, such as environmental toxicity or pathogenic organisms. Human populations may vary with regard to the frequency of alleles related to either of these categories.

Monomorphic loci are those that can be said to contain a dominant allele at a frequency of greater than ninety-nine percent (Roychoudhury and Nei, 1988, p. 19). Monomorphic loci occur generally where any genetic variation at such a locus is highly detrimental to the individual's fitness. For example, the frequency of the allele for the mutation that causes hemophilia-A is extremely small (2.0×10^{-4}), and thus the frequency for the normal allele(s) is 99.98 percent (Gelehrter, Collins, and Ginsburg, 1998, p. 45). Hemophilia-A is caused by a defect in factor VII clotting protein, and the gene for it is located at the end of the X chromosome. In reality at the molecular level, there is a variety of hemophilia-A mutations. These include large deletions removing all or part of the gene, and small deletions and insertions that cause frame-shift mutations (78 large deletions and 223 point mutations are known; see Gelehrter, Collins, and Ginsburg, 1998, p. 140). Thus it is likely that unrelated families actually carry distinct mutations that cause the hemophilia-A syndrome. The frequency of the hemophilia-A mutations is determined by mutation-selection balance, which is the relationship between the allele's negative impact on fitness (which will drive its frequency to zero) and its tendency to recur by spontaneous mutation. Genetic drift also may have produced populations with frequencies of these diseases above that which is predicted by the mutation-selection balance theory.

Polymorphic loci, on the other hand, will have numerous alleles present at various frequencies, with no single allele at a frequency greater than ninety-nine percent. Phenylketonuria (at 1.0 frequency), Tay Sachs disease A (1.7%), cystic fibrosis (2.2%), and sickle cell anemia (5.0%) are examples of genetic diseases occurring at polymorphic loci. Selection and genetic drift also determine the frequency of these alleles, as they do for all polymorphic loci. For example, the alleles mainly responsible for the ability of mammals to identify self versus foreign tissue are located in the major histocompatibility complex (MHC). In humans the MHC loci are located on the short arm of chromosome 6, and comprise the HLA-A, HLA-B, and HLA-C type I loci, and the HLA-DQ type II loci (Gelehrter, Collins, and Ginsburg, 1998, p. 207). A question that can be reasonably asked is whether variation at these polymorphic MHC loci is concordant with an individual's identification with a particular "race."

Table 1 shows variation at the HLA-A locus by socially-defined "race":

Allele	Whites	Blacks	American Indians
A1	0.169	0.083	0.044
A2	0.263	0.174	0.369
A3	0.151	0.136	-
A11	0.038	0.046	-
A23	0.040	0.091	-
A24	0.102	0.046	0.335
A25	0.021	-	-
A26	0.027	0.046	0.044
A28	0.053	0.046	0.087
A29	0.027	0.023	0.022
A30	0.029	0.068	-
A31	0.038	-	0.087
A32	0.027	0.046	-
AW33	0.011	-	-
AW34	-	0.046	-
AW36	-	0.105	-
AW43	-	-	-
AW66	-	-	-
AX	0.003	0.046	0.013
n (sample)	362	44	46
H (heterozygosity)	85.8	90.5	73.2

Table 1: North America HLA-A Histocompatibility Type.[1]

An examination of variation at this locus shows that many of the alleles are found in all three of the so-called human races. In fact it is solely the rare alleles that seem to be found in only one or two of these groups. Variation at the HLA-B, HLA-C, and HLA-DQ loci follow the same general pattern (Roychourdhury and Nei, 1988, p. 235-9). Linkage disequilibrium between specific alleles at the HLA loci has been observed (Hedrick et. al., 1991). This means that certain alleles will associate with each other more than we expect by chance alone. Population subdivision and isolation by distance will mean that populations that are closer together geographically are more likely to share HLA haplotypes. For example, Gambian populations are more closely related to Nigerians than they are to Europeans, and African- Americans are more closely related to Gambians than they are to Nigerians or Europeans (Allsopp et al., 1992).

However, one must recognize that the existence of rare alleles limited to any given population is not sufficient to guarantee identification of an individual as belonging to a particular "race." For example, in the United States different states utilize various formulas to certify "racial" identity. In 1963, Arkansas defined anyone with a "drop of Negro blood" in them as Negro. In Florida, if you were less than 1/8th Negro blood, you ceased to be Negro, and in Oklahoma, anyone that was not of Negro "blood" was classified as "white." Currently under Louisiana law, anyone with less than 1/32th African descent is considered "white." Ironically, Homer Plessy, who was removed from a "whites-only" railroad car and was the subject of the landmark Plessy v. Ferguson Supreme Court decision in 1872 validating "separate but equal accommodations," had only 1/8th African ancestry (Wilkinson, 1979, p. 18). In 1982-83, Susie Guillory Phipps, classified by the state of Louisiana as "black" sued to have her racial status reversed to "white" (she was found to be of 1/32th Negro blood and classified as black; Omi and Winant, 1986, p. 57). Thus depending on the phenotypic formula one uses, the allele frequencies in these "racial" categories could be completely redefined. The African-American genome is, on average, at least twenty percent European, and as much as ten percent Asian (American Indian) in its origin (Cavalli-Sforza and Bodmer, 1971; Hartl and Clark, 1989, pp. 308-9). Another fact also follows from the overlap in allele frequencies at the MHC loci; 'race' is not a means by which one should limit a search for tissue compatibility. This has important implications for organ transplantation. Only individuals who have the rare alleles at these loci would be precluded from finding a match in the general pool, and in fact all individuals, regardless of "race," have the highest probability of finding a match with close relatives. In the American South, this takes on a particular irony. Due to the widespread practice of concubinage between African-American women and their Euro-American slave masters, many Southern African- and Euro-American families share common descent. Hence plantation records could be used to find possible MHC matches between these families, as compared to a search in the general pool (Jackson, 1997).

The amount of overlap in gene frequencies between African- and Euro-Americans raises an obvious, but traditionally overlooked question. Does the amount of genetic distance between African- and Euro-Americans match their general health disparities? The traditional medical research literature treats genetic difference as if it is the obvious relevant factor for health status. For example, Michael A. Winkelby et al. suggest that after normalizing some socioeconomic differences between ethnic groups at risk for cardiovascular disease (CVD), the residual difference in risk might be genetic (1998, p. 356). However if one asks how large the disparity in CVD occurrence is when compared to the amount of genetic distance at polymorphic loci, this proposal seems absurd. Current data show that African-Americans have a thirty-seven percent greater age-adjusted rate of death from CVD as compared to Euro-Americans.[2] The genetic distance at polymorphic loci is, however, on average only five percent between Africans and Europeans, and African-Americans are even closer with regard to shared alleles. This means the CVD disparity is 7.4 times greater than one would predict based on genetic distance alone. Anthony Polednak (1989) showed that African-Americans had a higher rate of age-adjusted incidence in twenty-two of twenty-four categories of mortality. Genetic distance alone cannot account easily for such disparities. If such a difference were mainly genetically-based, a sophisticated genetic explanation would be required (such as pleiotropy). However such explanations generally are not advanced when the disease disparity between "races" is discussed. One must wonder what intellectual program could dismiss such obvious inconsistencies. The answer lies in an irrational faith in the reality of the race concept, and in the twin pillars of "racial" science, adaptationism, and reductionism.

3. THE WHIRLPOOL OR THE MONSTER?

Some of the greatest advances in biology since the middle of the twentieth century have occurred in molecular genetics. These advances, however, have not occurred without a price. That price has been the rise of molecular reductionism. Part of this price arises from the necessary growth of specialization as the knowledge base grows within the discipline. The difficulty with specialization is that it almost always entails a lack of communication among allied fields. The lack of communication between evolutionary and molecular genetics has resulted in failures in both disciplines. Lack of understanding of molecular mechanisms has often fostered a naive allegiance to adaptationism in evolutionary biology ("naive adaptationism" is discussed in Rose and Lauder, 1996), while lack of understanding of evolutionary genetics has often fostered misinterpretation of the origin of genetic variation in molecular biology (e.g., see Graves, 1993, 1997). Here I shall examine the ideological commitment to reductionism in molecular genetics and some of its applications to the problem of "race" and disease incidence.

Molecular biology has allowed us to get a greater handle on the underlying nature of phenotypic variation by examining differences in allelic structure at the DNA level. In cancer research, for instance, particular allelic variants (such as p53 mutations and BRCA1 and BRCA2 mutants) have been demonstrated to be associated with specific cancer phenotypes. For some researchers the discovery of these new mutations (some revealed in small populations such as the Ashkenazi Jews) suggests that long-standing differences in supposed "racial" cancer incidence rates might be related to underlying allelic variation between supposed human races. Such an assumption is unwarranted, and in fact is not supported in theory; it stands in opposition to evolutionary and population genetic theory for a number of reasons. The first set of reasons concerns our understanding of the processes by which gene frequencies get determined in populations, particularly age-associated mechanisms (see Graves, 1997). The second is related to the already measured differentials in both gross level phenotypic characters and gene frequencies (as measured by nuclear protein methods and mitochondrial DNA techniques at multiple loci; e.g., see Cavalli-Sforza, Menozzi, and Piazza, 1994).

3.1 Age-related Population Genetic Mechanisms

The incidence of biological sources of mortality in human populations shows the classical Gompertzian age-related increase given by equation 1: $Mr = Ao * ebt$. This equation can be rewritten to reveal more readily the initial and age-specific mortality rate in the form of equation 2: ln (Mr) = ln (Ao) + bt. That is, the age-specific mortality rate is an exponentially increasing function in which (b) is the mortality rate coefficient, (t) is time, and (Ao) is the initial mortality rate. An examination of Euro-American and African-American mortality rates suggests that throughout the twentieth century African-Americans always have had about 1.6 times the age-adjusted rate of mortality as compared to "whites." For example, in 1963 the total ratio was 1.68, and the specific rates are included in Table 2.

| Age | *Rate per 100,000 in population* | | |
	"White"	"Black"	Ratio ("White"/"Black")
Under 1	2230	4170	1.87
1-4	90	180	2.00
5-14	40	60	1.50
15-24	100	160	1.60
25-34	130	320	0.46
35-44	260	650	2.50
45-54	680	1320	1.94
55-64	1620	2790	1.72
65-74	3750	5290	1.41
75-84	8580	7490	0.87
85+	21580	14570	0.67

Table 2: Age-Specific Death Rates by "Race" in 1963 (Knowles and Prewitt, 1969, p. 97).

These same data in 1980 appear in Table 3.

| Age | *Rate per 100,000 in population* | | |
	"White"	"Black"	Ratio ("White"/"Black")
Under 1	-	-	-
1-4	-	-	-
5-14	-	-	-
15-24	92	116	1.26
25-34	108	235	2.18
35-44	181	444	2.45
45-54	472	936	1.98
55-64	1219	2008	1.65
65-74	2773	3759	1.36
75-84	6407	7407	1.16
85+	15757	13076	0.83

Table 3: Age-Specific Death Rates by "Race" in 1980 (Polednak, 1989, p. 63).

The total ratio was 1.61. Infant mortality data from 1984 suggest that this pattern was robust over this time period also. The value adjusted to 100,000 per population for under 1 year of age was 1880 and 3670 for Euro-Americans and African-Americans, respectively, yielding a ratio of 1.95. If these data are included with the 1980 data then the total mortality ratio approximates 1.65.

A simple genetic hypothesis of mortality differences might predict that the specific diseases associated with mortality are different for "whites" and "blacks." For example, in 1963 the frequency of death from tuberculosis was 3.4 in "whites" and 12.8 in "blacks," but in 1985 these figures were 13.2 and 56.5, respectively (Knowles and Prewitt, 1969, p. 98; Polednak, 1989, p. 61). These ratios of 3.76 in 1963 and 4.28 in 1985 are hard to explain using a simple genetic hypothesis. The problem becomes particularly acute given the high allele frequency overlap between African- and Euro-Americans at the HLA loci. Table 4 shows that both rates for mortality from all cancers are very low before age 40 and that "black" rates climb faster than "white" rates after age 40 (Polednak, 1989, p. 61).

| | *Rate per 100,000 in population* | | |
| Age | "White" | "Black" | Ratio |
			("White"/"Black")
15-24	5.5	5.4	0.98
25-34	12.8	15.9	1.24
35-44	43.0	70.0	1.62
45-54	161.0	255.0	1.58
55-64	438.0	618.0	1.41
65-74	829.0	1014.0	1.22
75-84	1272.0	1479.0	1.16
85+	1595.0	1636.0	1.02

Table 4: National Health Survey Total Cancer Mortality Rates (1985).

The initial mortality rate as calculated by equation 2 is higher for "blacks" by 0.128, and the age-specific mortality rate parameters are nearly identical (0.106 for "whites" versus 0.108 for "blacks"). The initial mortality rate could be differentiated using either genetic or environmental sources, or the interaction of both. At this point no intellectually credible program exists that allows unambiguous measurement of the contributions of either component to the resulting cancer phenotypes, simply because neither genetic nor environmental sources of this variance can be standardized.

Other biological mortality sources also show the general Gompertzian pattern, including tuberculosis, diabetes, hypertensive heart disease, ischemic heart disease, hypertension, cerebrovascular diseases, intracerebral hemorrhage, thrombosis, pneumonia, chronic obstructive pulmonary disease, ulcer, chronic glomerulonephritis, and renal failure (Polednak, 1989, pp. 61-2). Until the mid-twentieth century the explanation for the age-specific increase in biological sources of mortality was obscure. However, Peter Medewar (1952) and George C.Williams (1957), drawing on previous work by Robert A. Fisher (1930) and Sewell Wright (1931), described the conditions under which alleles that have negative impacts on survival late in life accumulate in populations. This theory was developed mathematically by William D. Hamilton (1966), Brian Charlesworth (1980), Michael R. Rose (1984a, b) and others, and substantially corroborated experimentally in *Drosophila* (Rose, 1984b; Luckinbill *et al.*, 1984; and others; see review in Graves, 1997). The implications of this theory for biomedical research have been widely disseminated (see Rose and Graves, 1989; Rose, 1991; Rose and Finch, 1995; Graves, 1993, 1997; Neese and Williams, 1994; Jazwinski, 1996). Unfortunately, although widely disseminated, the theory also has been widely ignored or misunderstood in most biomedical research related to aging (as discussed in Rose, 1991; Neese and Williams, 1994; Graves, 1997). It is apparent to those examining this controversy that molecular reductionism has played an important role in stimulating the neglect of evolutionary genetic theory in aging research.

This is unfortunate, particularly since evolutionary theory is the only scientific program able to explain metazoan aging. It has conclusively shown that there are only two types of population genetic mechanisms that can account for the existence in a population of any allele that has a deleterious late age effect on fitness (such as cancer-producing alleles). These mechanisms are antagonistic pleiotropy and mutation accumulation.

3.2 Antagonistic Pleiotropy

Pleiotropy is the condition under which an allele has multiple effects on the phenotype and is widespread in biological systems. The antagonistic pleiotropy mechanism predicts that if an allele has a positive effect on early life fitness, it will increase in a population despite its deleterious late life effect. The mechanism has been tested extensively in experimental systems such as *Drosophila melanogaster*, and both the predicted phenotypic trade-offs between components of early versus late life fitness and corresponding changes in gene frequencies have been shown (Rose, 1984a, 1991; Jazwinski, 1996; Graves, 1997). It has been suggested elsewhere that this mechanism may explain the widespread occurrence of cancer in mammals. Many late life acting cancer alleles have positive effects in early life (Graves, 1993) due to their role in regulating normal growth and differentiation of cells in development (Bishop, 1983; Klein, 1988; Gelehrter, Collins, and Ginsburg, 1998, p. 252). Their positive role in promoting early life fitness means that such alleles will be fixed (that is, found at a very high frequency) in all populations. Therefore we would expect no "racially"-based differences at loci such as these. Given that such loci must exist, it is problematic to suggest that the differential in cancer rates by "race" results mainly from genetic differences. The strongly corroborated evolutionary theory of aging predicts that some loci should be differentiated consistently between subpopulations, but certainly not all! This is in sharp contrast to the distribution of alleles that might be responsible for the early onset of cancer. These would be governed by the mutation-selection balance mechanism, and could be extremely different in frequency due to genetic drift. Again, however, given what we know about the broader distribution pattern of genetic variation, there is no *a priori* reason to suspect "racially" consistent variation for loci such as these.

The case of diseases related to skin color is an interesting example of weak selection resulting in clinal geographic variation. Skin color shows a clear cline in the Old World related to sunlight intensity (Cavalli-Sforza, 1983, p. 3). This cline, however, must result from very weak selection, since populations that arrived in the New World approximately 13,000 to 11,000 years ago have not re-established the gradient (for recent estimates of the time of arrival see Sandweiss *et al.*, 1998). In addition, the distribution of alleles related to melanin phenotypes in the skin follow that cline (e.g., see the distribution of Vitamin D binding protein in Table 104, Roychoudhury and Nei, 1988). Here is an example where population variation and resistance to various diseases show a well-established mechanistic relationship. Remember, however, that all so-called races show high melanin concentrations in the skin in tropical latitudes (Caucasians in India, Australoids in the South Pacific, Asians in Central America and Indochina, and Africans in tropical latitudes all have melanic skin), but even in these cases, the incidence of skin melanomas is not directly correlated with skin color. Australoids, who have very high melanin content in their skin, also show some of the highest rates of melanoma (Polednak, 1989, pp. 194-5). In addition, the frequency of the two major alleles at the Vitamin D binding protein locus range between 0.795 and 0.702 for allele 1, and 0.205 and 0.298 for allele 2 in Europe. In central Africa, the same locus shows 0.963 to 0.907 for allele 1, and 0.037 to 0.093 for allele 2. However, in Australoids (which are the group farthest removed by genetic distance from Africans), the frequencies at this locus in the tropics run between 0.900 and 0.824 for allele 1, and 0.100 and 0.152 for allele 2. Vitamin D binding protein factor may be important in at least two different diseases, rickets and diabetes mellitus (Polednak, 1989, pp. 42, 240). The lesson to be learned from this example is that one cannot assume specific variation in alleles that either predispose or protect individual "races" from a disease. Genetic variation at this locus in Australoids is discordant with their remaining genetic variation that places them closer in the human phylogenetic tree to Caucasoids as compared to Negroids (Cavalli-Sforza, Menozzi, and Piazza, 1994, p. 99). Thus the fatal flaw of the reductionist program relating disease to 'race' is its failure, in general, to establish mechanistic relationships between differentials in gene frequency and differentials in disease prevalence between the populations in question. Such relationships can be found for a few loci of major penetrance. Yet even these do not map effectively onto 'race,' as I shall demonstrate below.

One of the most interesting recent discoveries related to clinical variation for disease resistance has to do with the CCR5 molecular polymorphisms that confer resistance to some strains of the HIV virus (Quillent, 1997; Michael, Louie, and Sheppard, 1997; O'Brien and Dean, 1997). Individuals who have the CCR5 mutant allele are resistant to the action of the HIV virus (either through resisting infection or development of AIDS). The mutation is a deletion of thirty-two base pairs in the codon that produces the CCR5 protein on the surface of human macrophages. The loss of thirty-two base pairs results in altered structure of the protein, which most likely interferes with the ability of the HIV virus to bind that particular protein. Homozygotes for the mutation have greater resistance than heterozygotes. This mutation is found at frequencies as high as eight percent in Northern European populations, and it shows clinal variation south and east toward Africa and Asia (with frequencies less than one percent). It is virtually absent in African, East Asian, and Native American populations (Martinson, Chapman, and Clegg, 1997). It is rare in African-Americans, most likely as a result of admixture during slavery. Given the fact that HIV did not occur in Europe in an appreciable frequency until the early 1980s, it is impossible for the virus to have been the selective agent that increased the frequency of this mutation. Thus it is hypothesized that either this is a case of founder effect, or some other viral infection faced by European populations as they migrated into Europe may have selected for this particular mutant. The fact that the mutant is not found in Africa or Asia suggests that the specific selective agent may have been unique to Europe. It is also clear that the distribution of this gene alone is not sufficient to explain the difference in the frequency of HIV infection and AIDS in different geographic regions, although it may be playing a role. John Caldwell and Pat Caldwell (1996) have reviewed the biological and sociological data concerning the African AIDS epidemic. They conclude that the single variable that best correlates with the spread of AIDS in Africa is the lack of male circumcision. This lack of circumcision, accompanied by widespread poverty, is in turn associated with high rates of chancroid sores (open wounds on the male penis that can spread HIV virus; see Caldwell and Caldwell, 1996, p. 66). This analysis stands in opposition to single-minded adaptationist explanations purporting genetic differentials in sexual behavior and aggression by African males as the source of the AIDS epidemic in Africa (e.g., see Rushton, 1995, pp. 165-84).

The only mechanism that could explain wholesale differences between "racial" populations for loci contributing to later life survival is some hypothesis related to whole scale life history trade-offs between them. For example, a trade-off between reproductive capacity and later life survival could account for such a differential between "races." At the turn of the century, social Darwinists felt that the low reproductive potential of Negroes and their lower survival rates meant that they were doomed to extinction on the North American continent (Tucker, 1994, pp. 32-3). However, by the end of World War II, it was clear that the African-American population was increasing faster than the Euro-American population (between 1905 and 1909 the intrinsic rate of increase was 0.0100 and 0.0204, and in 1950 it was 0.0220 and 0.0139 for African- and Euro-Americans, respectively; see Graves, 2002, pg. 761). This lead to a wave of cries of dysgenesis for the American population (Tucker, 1994, p. 198-204). J. Phillipe Rushton (1995) presents a supposed life history-based analysis of the differential reproductive and survival rates of the three major human "races" (Negroid, Caucasoid, and Mongoloid). He utilizes r- and K-selection theory to explain the position of the three "races" along the r- and K-selection continuum. Negroids are the most r-selected, with higher birth rates and lower investment in somatic tissue (including survivorship and intellect), while Mongoloids are the most K-selected (including lower birth rates and higher investment in somatic tissue). If this theory were true, it could explain phenomena such as higher mortality due to disease across the "races." Unfortunately, the Rushton program is an example of the worst form of adaptationist reasoning relying on a series of untestable just-so scenarios (as discussed in Graves, 2002).

3.3 Mutation Accumulation and Mutation-selection Balance

The theory of mutation accumulation states that there are alleles that have no detectable influence on fitness early in life, but that have a deleterious impact on fitness later in life. For these alleles there should be no specific pattern of gene frequencies, and they are able to accumulate differentially by geographic origin. Their frequency is determined by the random process of genetic drift as originally described by Sewall Wright (1931). Thus this mechanism predicts a pattern where the frequency of disease-associated alleles could be claimed to be "racially" differentiated at a particular locus, but certainly not at all such loci. We can examine this problem using a binomial equation: if we allow the rates to be higher or lower in "blacks" versus "whites" at some particular frequencies p and q, then the probability that all such independent loci would be at higher frequency in "blacks" would be p^n, where n is the number of loci in question. It is easy to show that the probability that all such loci (and their phenotypes) are always higher in "blacks" rapidly approaches zero as the number of loci increases. For instance if the original probabilities are balanced, then if only four disease phenotypes are examined, $p = (0.5)^4$, five phenotypes $(0.5)^5$, and so on, resulting in approximately 6.25 percent and 3.1 percent probabilities of higher frequencies respectively in "blacks." Again, the mutation accumulation mechanism does not predict that there will be a one-sided disease incidence result. The only mechanism that could account for all categories being uniformly distributed is either some form of close genetic linkage (however, the mapping evidence for alleles known to contribute to various diseases shows that this is not true), or the existence of some sort of massively pleiotropic, late age-impacting locus found in "blacks" that is not found in "whites." There is no reason to support the latter mechanism, nor has any credible reason been suggested.

Medical researchers are often concerned with rare mutations at loci with frequencies approaching 10^{-5} to 10^{-6}. At such low frequencies, we would expect the rates in local populations to deviate at random, particularly if the populations are small. Thus, the identification of the rare BRCA1 and BRCA2 alleles at high frequency in women of Ashkenazi Jewish descent was not a surprise for population geneticists. A similar case involved the identification of high frequencies of the APP-1 mutants (involved in the Alzheimer phenotype) in the Volga Germans, a population that was culturally isolated from surrounding Russian populations. South African "whites" and "coloreds" both have an elevated frequency of Huntington disease, as compared to the native African population (2×10^{-5} versus 1×10^{-7}), because all cases of Huntington disease in these "white" and "colored" populations can be traced back to one individual, Jan van Riebeeck, who led the first group of Dutch settlers to South Africa in 1658 (Edlin, 1984, p. 367). Here the social system of rigid race separation, known as apartheid, kept the native and Dutch populations relatively separate, thus maintaining the gene frequency differential. From these examples, however, it would be incorrect to conclude that such differential allele frequencies should be expected in large panmixtic and also admixed populations (such as those identified as "black" in the United States), or that such allelic frequency differences might be responsible for the historically large differences in disease incidence in the United States. As a case in point, the sickle cell anemia allele originated in the Middle East, and was transported to Western Africa via people who moved along the caravan trade. The allele is in high frequency in a number of Mediterranean populations that faced high incidences of malaria. The allele therefore cannot be used as a "racial" identifier. Its seeming association with "race" simply reflects the fact that this allele was in high frequency among slaves who survived the middle passage from Western Africa, and thus it is identified in the United States as a "black" disease. Frederica P. Perera (1997) also suggests that only about five percent of the differential in "racial" cancer incidence can be attributed to genetics.

These arguments do not deny that there are geographically-based polymorphisms that contribute to disease incidence. What they do indicate is that these genetic polymorphisms are distributed in the same way that the remaining genetic variation in the human species is organized (that is, in a discordant fashion). Thus it simply is not possible to create an unambiguous classification scheme that relates the "race" of individuals with the probability of disease incidence. We should expect each subpopulation, due to its geographic history, to contain unique frequencies of alleles related to disease incidence. These frequencies are determined by the interaction of mutation-selection balance, genetic drift, antagonistic pleiotropy, and mutation accumulation. For example, the alpha-thalessemias are in high frequency among Southeast Asians and Africans, while the beta-thalessemias are found in high frequency among Italians, Southeast Asians, Greeks, and Africans. Diabetes mellitus is found at high frequency among Ashkenazi Jews, Polynesians, members of American Indian Nations, and Mexicans (Friedman *et al.*, 1996, p. 75). The simple point here, developed eloquently by Jared Diamond (1994), is that if we would choose different genetic criteria, then we could reclassify all the "races." We could create a beta-thalessemia race just as easily as a race that contains whorls for fingerprints rather than loops, though neither of these would match the classic races of nineteenth-century anthropology (although there was never consensus even then about race; see Darwin, 1981 [1871], p. 226).

4. DO DISEASE PHENOTYPE DIFFERENTIALS REVEAL UNDERLYING GENETIC VARIATION?

The data show that the incidence of specific forms of cancer are differentiated between the so-called "races." This has led some researchers to presume differences in gene frequencies. Perera (1997) summarizes data by cancer category. Some cancers are more frequent in "blacks" as compared to "whites": esophageal (3-4 times); multiple myeloma, liver, cervical, and stomach (2 times); oral, pharynx, larynx, lung, prostate, and pancreatic (1.5 times); and lymphocytic leukemia and pre-menopausal breast (higher by a rate significantly greater, but not equal to 1.5 times). Other cancers are more frequent in "whites": melanoma, leukemia, lymphoma, endometrial, thyroid, male bladder, ovarian, testicular, and brain (all higher by a rate significantly greater, but not equal to 1.5 times). These data suggest no uniform pattern of cancer incidence. However, "blacks" show greater incidence rates in fourteen of twenty-four categories, and in no category are "whites" likely to have an order of magnitude greater incidence of a specific cancer type. Some genetic polymorphisms have been found that are consistent with these "racial" differentials. For example, the enzyme NAT2 deactivates carcinogenic aromatic amine groups through N-acetylation. The frequency of the slower form of this enzyme differs between African- and Euro-Americans (30-40% versus 50-60% respectively). The impact of this locus on the production of a cancer phenotype seems entirely dependent on smoking; however, the enzyme seems to only have a protective impact where exposure to tobacco smoke is relatively low, indicating its action is strongly influenced by environment (Perera, 1997, p. 1071). In addition, other polymorphisms have been reputedly associated with "racial" variation in cancer incidence (e.g., Reichardt *et al.*, 1995). However, these studies do not establish a causal link between the genetic variation and any particular cancer phenotypes. The typical method of these sorts of studies is to identify genetic variation at the molecular level between already socially-defined groups that also differ in the incidence of a cancer phenotype (as in Devgan *et al.*, 1997). Allelic variants are defined such that they are differentiated between the groups (often involving the number of short tandem repeats). Such a method does not *establish* genetic causality; it assumes that an observed genetic difference must have *functional* importance. That is not to say that the insertion of multiple triplet repeats in the coding portion of a protein is not likely to alter its function and cause disease (as in the well-established case of Huntington disease; see Gelehrter, Collins, and Ginsburg, 1998, p. 219). Yet many researchers err in a fundamental requirement of genetic analysis; namely, they fail to control for environmental variation between the groups in question, thus vitiating any claim of genetic causality. This is a common mistake associated with those assuming a program of molecular reductionism.

Many medical researchers begin their analysis of genetic data with the false assumption that any genetic variability they examine is racially-based and causally-related to disease incidence differentials. Consider Perera:

In a California population serum concentrations of the organochloride 1,1´-dichloro-2,2´ bis (p-chlorophenyl) ethylene (DDE), a metabolite of 2,2´-bis(p-chlorophenyl)-1,1,1 - trichloroethane (DDT) was found to be higher in black women than among white women. In black smokers, urinary concentrations of metabolites of the tobacco specific carcinogen   4-(methyl-nitrosamino)-1-(3-pyridyl)-1-butanone   (NNK)   and   serum concentrations of cotinine (a metabolite of nicotine) exceeded those of white smokers. *Although unmeasured differences in exposure cannot be ruled out, these findings are consistent with biologic variation and with the higher rates of various smoking related cancers in blacks.* (Perera, 1997, p. 1071; emphasis mine)

The author assumes that a measured phenotype in these populations necessarily can be applied to an unknown genotype—namely, a genotype she suspects may be related to the ability of each population to detoxify compounds with potential mutagenic effects. Yet her caveat seems to vitiate such an analysis. She does not report any information relating to the nature of exposure rates, yet it is well-known that "minority" and poor populations do face differential environmental exposure to toxic materials (Bryant and Mohai, 1992). The working assumption here again results from genetic reductionism. It is, however, impossible to clearly define genetic contributions to a phenotype without the ability to control for and equalize environmental inputs (Falconer and Mackay, 1996, pp. 122-46; see Graves and Johnson, 1995 for discussion of 'toxic racism' and intelligence phenotypes). The sources of the variance for any phenotype resulting from multiple genes are given by equation 3: $V_p = V_g + V_e + V_{gxe} + Cov\ (G,E) + V_{error}$, where V_p is the total variance in phenotype; V_g is the variance resulting from genetic sources $(V_a + V_d + V_i)$—$(V_a$ is the additive genetic variance resulting from multiple loci of small but quantitative impact, V_d are dominance effects, and V_i are epistatic effects resulting from differential genetic backgrounds); V_e is the variance due to environment; V_{gxe} is the variance due to gene/environment interaction; $Cov\ (G,E)$ is the covariance of genes and environment; and V_{error} is the variance that results from error in identifying a given phenotype (equation adapted from Falconer and Mackay, 1996).

Disease incidence (cancer in particular) is clearly a polygenic phenomenon, with the interaction of several genetic and environmental effects necessary for causation. Thus to examine the differential in cancer incidence, it is necessary to examine all of these sources of the variation in rates. For example, African-derived populations in the Dominican Republic have some of the lowest age-specific cancer rates in the world (Klug and Cummings, 1997, p. 607). Yet these populations have about the same genetic admixture as African-Americans (although probably fewer European genes by admixture). In fact, if we examine specific cancer incidence by "race" internationally, no appreciable pattern emerges. In 1987 the World Health Organization (WHO) data for breast cancer showed the following values (per 100,000 age-adjusted mortality).

Country	Rate

Caribbean

Bahamas	25.0
Barbados	22.5
Cuba	14.3
Dominica	39.1
Martinique	22.3
Puerto Rico	10.5
St. Vincent	28.8
Trinidad	16.6

Scandinavia

Denmark	27.6
Norway	18.2
Sweden	18.1
Finland	16.7

Table 5: Age-Adjusted Mortality of Breast Cancer, Selected Countries
(Extracted from Polednak, 1989, p. 71).

The Caribbean populations listed here are mainly admixtures of various amounts between African, American Indian, and European populations. Yet it is clear that the Caribbean and Scandinavian populations overlap in their breast cancer mortality rates.

Cancer incidence statistics show clear correlation with socioeconomic data, correlations that are larger in magnitude than reputed genetic differentials in "races" (Wingo *et al.*, 1996). Equation 3 shows that it is extremely difficult to estimate the true genetic components of variance in incidence rates due to the impossibility at present in estimating the gene/environment interaction and covariance terms. Thus simply identifying gene frequency differences between populations is not a sufficient means of assigning genetic causation to a locus. Nor is this problem entirely solved by utilizing techniques such as charting resemblance among relatives or quantitative trait locus mapping. These techniques require large samples of relatives and detailed pedigrees. These are often not present for human diseases, and are still not entirely free from complications associated with gene/environment interactions. For example, heritability estimates for systolic blood pressure for a range of populations vary widely from 0.13 to 0.42 (Weiss, 1995, p. 105). Most interestingly, the estimate reported by P.P. Moll *et al.* (1983) for African- and Euro-Americans was quite different (in Detroit, heritability was 0.13 for "blacks" and 0.32 for "whites"). This supports the point that African-Americans are not genetically predisposed to hypertension. It has been shown that the majority of the hypertension differential between African- and Euro-Americans is focused in the African-American professional class and is related to differential psychological stress, and also that genetic markers associated with hypertension in Europeans are not present in populations of Western African origin (Neser *et al.*, 1986; Broman, 1989; Calhoun, 1992; Haynes, 1992; Rotimi *et al.*, 1994; Cooper and Rotimi, 1994; Light, 1995; Kaufman *et al.*, 1996).

In another example, R.A. Irvine *et al.* (1995) demonstrate how *not* to do genetic analysis of variation and correlation to disease incidence. Their paper purports to examine microsatellite variation in an androgen receptor gene that they claim is statistically associated with prostate cancer. Once again, the first flawed assumption of this study is the "reality" of the racial categories utilized. They divide their sample study group into three "racial" categories: African-American, Asian, and non-Hispanic white. They then attempt to calculate microsatellite allele frequencies in their population (composed of 44 African-Americans, 39 Asians, and 39 non-Hispanic whites). This procedure is immediately suspect because an accurate estimation of allele frequencies in the populations they are attempting to examine is impossible with such limited sample sizes and sampling regime. Thus we cannot say anything definite about variation within and between these groups. Worse yet, they then attempt to partition the size of microsatellite alleles in the receptor gene and associate that statistically with prostate cancer incidence in a population of fifty-seven affected "white" males. The partitioning of the alleles occurred in the following way: short CAG repeat alleles were defined as less than or equal to twenty-two repeats, and long were greater than twenty-two repeats. The observed repeat numbers varied from nine to twenty-nine in the CAG microsatellites. Their rationale for dividing repeat numbers into long or short alleles was simply to use the median size in the Asian population (22) as the dividing point between long and short. There is no statistical or functional rationale provided for such a procedure. Using their assumption, they found the following distribution of short CAG repeats: African-Americans (75%), Asians (39%), and non-Hispanic whites (62%). This distribution was determined to be highly significant by chi-squared test of association at $p < 0.0005$. If one were to choose a different criterion (and this option was not examined in the paper), one could arrive at very different percentages. For example, the results using the median size of the repeat number of the African-American population (20) to divide long and short alleles are considerably different: African-Americans (58.8%), Asians (33%), and non-Hispanic whites (53.8%). Using this criterion the distributions of the repeats in the African-American and "white" populations are more similar, raising the question as to whether any significant difference could be supported by this analysis. Most importantly, there is absolutely no functional rationale for grouping the alleles into long and short in this analysis. The CAG repeat codes for the amino acid glutamine. Microsatellite repeats of this kind are thought to alter protein function by altering binding properties (as is suggested by the authors). Thus it would seem that the number of repeats here quantitatively varies within populations, and that the proper analysis of allelic number revealed by this study was not two, but twenty-one (the number of different repeats observed in the sample). The authors produced no experimental validation of the procedure of separating the allelic variation into large versus small size.

Interestingly enough, the explanation given for the functional significance of repeat number is that it might alter transmembrane properties of the resultant protein. There is no *a priori* way to know how small or large repeat number will alter such properties; for example, in the case of the CAG repeats associated with Huntington disease, it is large repeat numbers that cause the problem (and influence age of onset), and the impact is quantitative, not discrete (see Lewis, 1997; Gelehrter, Collins, and Ginsburg, 1998). Thus this study fails because it assumes correlations between genetic variation, disease, and "race" that are not substantiated by the techniques utilized.

5. CONCLUSION: THE FALLACY OF ADAPTATIONISM, REDUCTIONISM, AND RACE

Heeding the warning of Circe, Odysseus steered his ship closer to the cave of Scylla. As she predicted, the fierce monster seized six of the crew. Odysseus commented that their struggle was the saddest thing that he had ever seen while at sea, and he hoped that at least he had saved the rest of the crew. However, having escaped the cliffs of Scylla and Charybdis, Odysseus still lost his entire crew to the vengeance of the Sun God. Our hero had made the decision to not warn his crew of the dangers of the cliff, yet when he warned them of the danger on the island of Helios, they paid no heed and were still lost (Mandelbaum, 1990, p. 249-59). Here I choose to render a warning about the pitfalls of both adaptationism and reductionism.

Charles B. Davenport of the Eugenics Record Office (ERO) orchestrated what might be considered one of the great medical frauds of the twentieth century. Davenport and his coworkers 'proved' that pellagra was a genetically-inherited disease. To do so they had to purposefully suppress the work of Joseph Goldberger. Goldberger had demonstrated how to produce and cure pellagra by withholding a balanced diet to voluntary convict populations. As a result of the ERO fraud, at least 55,000 Americans, the majority of which were African-American children, died between 1917 to 1941 from a disease that was entirely treatable (Chase, 1977). In the 1930s, Euro-American physicians felt that African-Americans were a "notoriously syphilis soaked race" (Jones, 1981). The Tuskegee Study (carried out by the U.S. Public Health Service at Tuskegee) attempted to test the existence of genetic differences in syphilis progression in African-American men. There were more than 500 subjects involved in this experiment, and most died before the details of the study came to the public eye (Jones, 1981, p. 218). Both of these cases reveal the failure of adaptationist logic. In both cases these researchers assumed that the phenotypic variation they examined, whether real or not, was the result of underlying genetic causes, when obvious and apparent environmental explanations also were available. The importance of genotype/environment interaction was already well-established in evolutionary and quantitative genetics by this time. The history of genetics has shown that with each advance in theory and technique, there is a rush to apply those results to explain both human variation and condition (Chase, 1977; Provine, 1971; Bowler, 1989; Adams, 1988; Graves, 2001). Often this is done recklessly, and that recklessness is directly related to the longevity of the adaptationist assumption. The new molecular reductionism also suffers from this historical tendency. The future for genetic analysis of disease and human genetic variation can only be rescued by the integration of sound evolutionary and population genetic thinking into the research paradigm. One aspect of this integration will be a focus on individual and population variation, as opposed to reliance on nineteenth century descriptions of socially-defined "racial" categories. For example, a number of studies have shown that there are significant socioeconomic impacts on cancer susceptibility in various "races" (Baquet et al., 1991; Roach et al., 1992; Whittemore et al., 1995; Yu and Whitted, 1997). Mack Roach et al. (1992) found that uniform diagnosis and similar treatment of prostate cancer removed all significance of "'ace" as a prognostic variable in survival of prostate cancer patients. Evelyn Yu and John Whitted (1997) recommended that the Office of Management and Budget should abandon the term 'race,' and that the federal government needed to recognize subpopulations with more accurate ethnic identifiers in all its health-related research institutes, such as the National Institutes of Health (NIH) and the Centers for Disease Control (CDC). Furthermore, they recommended that the NIH and CDC increase their recruitment of minority researchers and staff, to be achieved by a long-term commitment by academe to encourage and support more minority researchers in basic biomedical research. All of these interventions would help to address the issues that I have raised in this chapter.

Furthermore, biomedical research needs to explicitly examine its allegiance to adaptationism and reductionism. Such allegiance occurs because neither physicians nor molecular biologists are strongly grounded in higher-order genetics (such as population and quantitative genetics). The schism between these fields of genetics has been costly to medical research and practice (e.g., see Neese and Williams, 1994). Sadly to say, there is still much resistance to integrative thinking in biomedical research. Until this is achieved, we always may be faced with the choice between the monster and the whirlpool.

Arizona State University West
Phoenix, Arizona, U.S.A.

ACKNOWLEDGEMENTS

This chapter is dedicated to the memory of my brother, Dr. Warren Graves, M.D. (December 14, 1956-March 6, 1998). He might have had a better chance if biomedical research had truly integrated evolutionary perspectives.

NOTES

1. Original data taken from Baur *et al.*, 1984; Baur and Danilovs, 1980; cited as Table 169 in Roychoudhury and Nei, 1988, p. 234.
2. Centers for Disease Control and Prevention, National Center for Health Statistics, National Vital Statistics System; available at: http://raceandhealth.hhs.gov.

REFERENCES

Adams, M., ed., *The Wellborn Science: Eugenics in Germany, France, Brazil and Russia.* Oxford: Oxford University Press, 1988.
Allsopp C. *et al.* Interethnic genetic determination in Africa: HLA class 1 antigens in the Gambia. American Journal of Human Genetics 1992; 50:411-21
Baquet C., Horm J.W., Gibbs T., Greenwald P. Socioeconomic factors and cancer incidence among blacks and whites. Journal of the National Cancer Institute 1991; 83:551-6
Baur, M.P. *et al.* "Population Analysis on the Basis of Deduced Haplotypes from Random Families." In *Histocompatibility Testing*, M.P. Baur and J. A. Danilovs, eds. New York: Springer Verlag, 1984.
Baur, M.P., Danilovs, J.A., *Histocompatibility Testing.* Los Angeles: UCLA Tissue Typing Laboratory, 1980.
Bishop J.M. Cellular oncogenes and retroviruses. Biochemistry 1983; 50:301-54
Bowler, P., *The Mendelian Revolution: The Emergence of Hereditarian Concepts in Modern Science and Society.* Baltimore: Johns Hopkins University Press, 1989.
Broman C.L. Social mobility and hypertension among blacks. Journal of Behavioral Medicine 1989; 12:123
Bryant, B., Mohai, P., eds., *Race and the Incidence of Environmental Hazards: A Time for Discourse.* Boulder, CO: Westview Press, 1992.
Caldwell J., Caldwell P. The African AIDS epidemic. Scientific American 1996; 274:62-9
Calhoun D. Hypertension in blacks: socioeconomic stress and sympathetic nervous system activity. American Journal of Medical Sciences 1992; 304:306
Cavalli-Sforza, L.L., *The Genetics of Human Races.* Burlington, NC: Carolina Biological Supply, 1983.
Cavalli-Sforza, L.L., Bodmer, W.F., *The Genetics of Human Populations.* San Francisco: W.H. Freeman Co., 1971.
Cavalli-Sforza, L.L., Edwards, A.W.F. "Reconstruction of Evolutionary Trees." In *Phenetic and Phylogenetic Classification*, V.E. Heywood, J. McNeill, eds. London: The Systematics Association, 1964.
Cavalli-Sfroza, L., Menozzi, P., Piazza, A., *The History and Geography of Human Genes.* Princeton: Princeton University Press, 1994.
Charlesworth, B., *Evolution in Age-Structured Populations.* London: Cambridge University Press, 1980.
Chase, A., *The Legacy of Malthus: The Social Costs of the New Scientific Racism.* New York: A. Knopf, 1977.
Cooper R.S., Rotimi C.N. Hypertension in populations of West African origin: is there a genetic predisposition? Journal of Hypertension 1994; 12:215-27
Darwin, C., *The Descent of Man and Selection in Relation to Sex.* Princeton: Princeton University Press, 1981 [1871].
Devgan S.A. *et al.* Genetic variation of 3 beta-hydroxysteroid dehydrogenase type II in three racial/ethnic groups: implications for prostate cancer risk. Prostate 1997; 33:9-12
Diamond J. Races without color. Discover 1994; 15:82-91
Edlin, G., *Genetic Principles: Human and Social Consequences.* Boston and Portola Valley: Jones and Bartlett Publishers, 1984.
Falconer, D.S., Mackay, T., *Introduction to Quantitative Genetics*, 4th ed. London: Longman Press, 1996.
Fisher, R.A., *The Genetical Theory of Natural Selection.* Oxford: Clarendon Press, 1930.
Friedman, J.M., Dill, F.J., Hayden, M., McGillivray, B. "Genetics." In *National Medical Series for Independent Study*, 2nd ed. Baltimore: Wilkins and Wilkins, 1996.

Gelehrter, T.D., Collins, F.S., Ginsburg, D. *Principles of Medical Genetics*, 2nd ed. Baltimore: Wilkins and Wilkins, 1998.

Graves J.L. Evolutionary biology and human variation: biological determinism and the mythology of race. Sage Race Relations Abstracts 1993; 18:4-33

Graves, J.L. "General Theories of Aging: Unification and Synthesis." In *Principles of Neural Aging*, S. Dani, A. Hori, G. Walther, eds. Amsterdam: Elsevier Press, 1997.

Graves, J.L. "The Misuse of Life History Theory: J.P. Rushton and the Pseudoscience of Racial Hierarchy." In *Understanding Race and Intelligence*, J. Fish, ed. Northvale, NJ: Jason Aronson Publishers, 2002.

Graves, J.L., *The Emperor's New Clothes: Biological Theories of Race at the Millennium*. New Brunswick, NJ: Rutgers University Press, 2001.

Graves J.L., Johnson A. The pseudoscience of psychometry and the bell curve. Journal of Negro Education 1995; 264:277-94

Hamilton W.D. The moulding of senescence by natural selection. Journal of Theoretical Biology 1966; 12:12-45

Hartl, D., Clark, A.G., *Principles of Population Genetics*, 2nd ed. Sunderland, MA: Sinauer and Associates, 1989.

Haynes K. Why hypertension strikes twice as many blacks as whites. Ebony 1992; 47(11):36

Hedrick, P. *et al.* "Evolutionary Genetics of HLA." In *Evolution at the Molecular Level*, R.K. Selander, A.G. Clark, T.S. Whittam, eds. Sunderland, MA: Sinauer, 1991.

Irvine R.A., Yu M.C., Ross R.K., Coetzee G. The CAG and GGC microsatellites of the androgen receptor gene are in linkage disequilibrium in men with prostate cancer. Cancer Research 1995; 55:1937-40

Jackson F.L.C. Taxonomic implications of the human genome project (HGP). American Journal of Human Biology 1997; 9:130

Jazwinski M. Longevity, genes, and aging. Science 1996; 273:54-9

Jones, J., *Bad Blood: The Tuskegee Syphilis Experiment*. New York: The Free Press, 1981.

Kaufman J.S. *et al.* Obesity and hypertension prevalence in populations of African origin: results from the international collaborative study on hypertension in blacks. Epidemiology 1996; 7:398-405

Klein G. Oncogenes and tumor suppressor genes. Acta Oncology 1988; 27:427-37

Klug, W., Cummings, M. "Genetics and Cancer." In *Concepts of Genetics*, 5th ed. Upper Saddle, NJ: Prentice Hall, 1997.

Knowles, L.L., Prewitt, K., *Institutional Racism in America*. Englewood Cliffs, NJ: Prentice Hall, 1969.

Lewis, R., *Human Genetics: Causes and Applications*, 2nd ed. Dubuque, IA: W.C. Brown, 1997.

Light R. Job status and high-effort coping influence work blood pressure in women and blacks. Hypertension 1995; 25:1

Luckinbill L.S. *et al.* Selection for delayed senescence in *Drosophila melanogaster*. Evolution 1984; 38:996-1003

Mandelbaum, A., *The Odyssey of Homer: A New Verse Translation*. Berkeley: University of California Press, 1990.

Martinson J.J., Chapman N.H., Clegg J.B. Global distribution of the CCR5 32 basepair deletion. Nature Genetics 1997; 16:100

Medewar, P., *An Unsolved Problem in Biology*. London: H.K. Lewis, 1952.

Michael N.L., Louie L.G., Sheppard H.W. The role of CCR5 and CCR2 polymorphisms in HIV-1 transmission and disease progression. Nature Medicine 1997; 3:1160

Moll P.P. *et al.* Heredity, stress, and blood pressure, a family set approach: the Detroit project revisited. Journal of Chronic Diseases 1983; 36:317

Montagu, A., *Man's Most Dangerous Myth: The Fallacy of Race*, 5th ed. London: Oxford University Press, 1974.

Neese, R.M., Williams, G.C., *Why We Get Sick: The New Science of Darwinian Medicine*. New York: Times Books, 1994.

Neser W.B. *et al.* Obesity and hypertension in a longitudinal study of Black physicians: the Meharry cohort study. Journal of Chronic Disease 1986; 39:105-13

O'Brien S., Dean M. In search of AIDS resistance genes. Scientific American 1997; 227:44-53

Omi, M., Winant, H., *Racial Formation in the United States: From the 1960s to the 1980s*. New York: Routledge & Kegan Paul, 1986.

Perera F. Environment and cancer: who are susceptible? Science 1997; 278:1068-73

Polednak, A., *Racial and Ethnic Differences in Disease*. Oxford: Oxford University Press, 1989.

Provine, W., *The Origins of Theoretical Population Genetics*. Chicago: University of Chicago Press, 1971.

Quillent C. HIV-1 resistance phenotype conferred by combination of two separate inherited mutations of CCR5 gene. The Lancet 1997; 351:14

Reichardt J.K. *et al.* Genetic variability of the human SRD5A2 gene: implications for prostate cancer risk. Cancer Research 1995; 55(18):3973-5

Roach M. *et al.* The prognostic significance of race and survival from prostate cancer based on patients irradiated on radiation therapy oncology group protocols (1976-1985). International Journal of Radiation Oncology Biology and Physics 1992; 24:441-9

Rose M.R. Genetic covariation in Drosophila life history: untangling the data. American Naturalist 1984a; 123:565-9

Rose M.R. Laboratory evolution of postponed senescence in Drosophila. Evolution 1984b; 38:1004-10

Rose, M.R., *The Evolutionary Biology of Aging*. Oxford: Oxford University Press, 1991.

Rose M.R., Finch C. Hormones and the physiological architecture of life history evolution. Quarterly Review of Biology 1995; 70:1-52

Rose M.R., Graves J.L. What evolutionary biology can do for gerontology. Journal of Gerontology 1989; B44:27-9

Rose, M.R., Lauder, G., eds., *Adaptation*. New York: Academic Press, 1996.

Rotimi C.N. *et al.* Angiotensinogen gene in human hypertension: lack of association of the M235T allele among African Americans. Hypertension 1994; 24:591-4

Roychoudhury, A.K., Nei, M., *Human Polymorphic Genes: World Distribution*. New York: Oxford University Press, 1988.

Rushton, J.P., *Race, Evolution, and Behavior: A Life History Approach*. New Brunswick, NJ: Transaction Publishers, 1995.

Sandweiss D. *et al.* Quebrada Jaguay: early South American maritime adaptations. Science 1998; 281:1830-2

Templeton, A.R. "The Genetic and Evolutionary Significance of Human Races." In *Understanding Race and Intelligence*, J. Fish, ed. Northvale, NJ: Jason Aronson Publishers, 2002.

Tucker, W.H., *The Science and Politics of Racial Research*. Urbana and Chicago: University of Illinois Press, 1994.

Weiss, K., *Genetic Variation and Human Disease: Principles and Evolutionary Approaches*. Cambridge: Cambridge University Press, 1995.

Whittemore A. *et al.* Prostate cancer in relation to diet, physical activity, and body size in Blacks, Whites, and Asians in the United States and Canada. Journal of the National Cancer Institute 1995; 87:652-61

Wilkinson, J.H., *From Brown to Bakke: The Supreme Court and School Integration: 1954-1978*. Oxford: Oxford University Press, 1979.

Williams G.C. Pleiotropy, natural selection and the evolution of senescence. Evolution 1957; 11:398-411

Wingo P.A. *et al.* Cancer statistics for African Americans, 1996. California Cancer Journal Clinics 1996; 46:113-25

Winkleby M.A., Kraemer H.C., Ahn D., Varady A.N. Ethnic and socioeconomic differences in cardiovascular disease risk factors. Journal of the American Medical Association 1998; 280:356-62

Wright S. Evolution in Mendelian populations. Genetics 1931; 16:97-159

Yu E., Whitted J. Task group I: epidemiology of minority health. Journal of Gender, Culture and Health 1997; 2:101-12

CHAPTER 9

HELEN E. LONGINO

BEHAVIOR AS AFFLICTION: COMMON FRAMEWORKS OF BEHAVIOR GENETICS AND ITS RIVALS

1. INTRODUCTION

The ascendancy of behavior genetics in the last decade has revived the nature-nurture debate: advocates both of the social and environmental approaches in the behavioral sciences and of developmental systems approaches have taken issue with the claims to explanatory self-sufficiency made on behalf of behavior genetics. Drawing on an ongoing comparative study, I want to throw in relief some features of behavior genetics and its rivals that remain obscured by the polemics.

My intention is not to criticize the research approaches I discuss, or to defend one against another, but to understand their presuppositions and limits. In this respect my aspirations differ from those of some other philosophers who have recently discussed similar material (cf. Griffiths and Gray, 1994; Schaffner, 1998). While these philosophers have been concerned to identify the correct approach to take regarding the biological study of behavior, I am more interested in understanding the different methodologies and metaphysics involved in the different approaches. My hope is to understand why the debates take the shape they do rather than to bring them to closure. Thus part of my analysis is directed to the identification of the commonalities against which differences are articulated in debate, in particular to the way in which (some) behavior comes to be conceptualized as pathology—that is, as affliction.

While the debates among the scientific adversaries help to clarify some issues, they serve to obscure others. In particular, the overall framework of investigation shapes the conception of behavior employed in these studies, which in turn

L.S. Parker and R.A. Ankeny (eds.), Mutating Concepts, Evolving Disciplines: Genetics, Medicine and Society, 165-187.
© *2002 Kluwer Academic Publishers. Printed in the Netherlands.*

influences the questions asked. In other words, the persistent oscillations between genetic and social approaches to studying behavior which are present in scientific discussions and in general public dialogue limit or are associated with limits on approaches to conceptualizing and understanding behavior. We direct our attention to the debates that divide genetically- and environmentally-oriented approaches rather than to the conceptual framework that is a condition of those debates' possibility. Here I focus on research on two kinds of behavior—aggression and sexual behavior—reviewing first the kinds of empirical studies being done and then looking at the theoretical and polemical literature to provide a guide to the intellectual contexts of these studies.[1] While their proponents often write and speak as though the issue is which of these methods provides the correct way to understand behavior, another way to understand the debates regarding these different approaches is as debates about what subfields (if any) of the biological sciences are best adapted or most appropriate for the study of human behavior, or, to put it another way, as a way of sorting out what biological tools are most useful for such studies.

2. STUDYING BEHAVIOR

It is not possible to study behavior in general. Empirical research on behavior must identify particular behaviors about which knowledge is sought: sexual behavior, aggressive behavior, parenting behavior, laboratory-elicited behaviors. A particular behavior that is the subject of research can be of either primary or secondary interest. When it is the primary subject of investigation the empirical question may concern its frequency, distribution, or etiology. Behavior is a secondary subject of investigation in studies aimed primarily at understanding some factor thought to be involved in the production or performance of some kind of behavior. Here behavior is a tool rather than a subject. For example, researchers may be interested in understanding the scope of effects of a given neurotransmitter, hormone, or other biological structure or physiologically active substance. Much research on behavioral effects of manipulating serotonin uptake or various gonadal hormones is of this nature. The behavior of an organism is used as a clue to the activity of the substance in the organism's nervous system. Similarly researchers may be interested in the cascade of effects of a given allelic mutation. In these cases, the research problem is formulated around the substance or gene in question, not around behavior *per se*. Measuring some change in behavior is important to identifying more proximate organic effects, but the behavior itself is of less interest than the gene or substance. Although this summary seems to capture the internal epistemological structure of the research, the choice of behavioral effect to be studied can be very important pragmatically in terms of attracting funding, being adopted by other research programs, and attracting external interest—what Bruno Latour (1987) calls enrolling allies —which can reach into the work to pull it in one direction rather than another and whose exigencies the internal structure can rearrange itself to satisfy.

If, as I suggest, one treats the current state of biologically-informed approaches to behavior as a collective sorting out of the issues, then there are several questions being addressed. Some are ontological questions: what is behavior? What is the particular behavior being investigated? These questions are often asked in connection with a related question and conflated with it: what components of behavior are stable enough to be treated as the outcomes of stable causal pathways? The other main question is causal: what are the causal pathways leading to the identified states, traits, or episodes? The turn to biology, and in particular to intra-organismal biology (i.e., the study of biological processes internal to individual organisms), means first that these questions are articulated with respect to the individual human or nonhuman organism. Behavior is treated primarily as a change in state of an individual organism. Correlatively, the work resisting the biological and especially the genetic turn is also articulated with respect to the individual organism, precisely to the extent it purports to rebut or refute the genetic approach.

Research on both causal and compositional questions can proceed in a top-down or bottom-up direction. 'Top-down' research identifies a behavior of interest (e.g., aggression), characterizes the behavior in a way that makes it amenable to study, finds appropriate measures of the behavior, and then seeks correlates in lower level behaviors or temperamental characteristics, or in biological factors such as neural states, non-neural physiological states, or genes. 'Bottom-up' research identifies a possible temperamental component (e.g., 'impulsivity'), a physiological state or substance, or a genetic mutation or marker, and seeks correlates of these at the behavioral level. While animal studies can be experimental, almost all human studies are of necessity correlational. The challenge to researchers is to design measurement and observational protocols sensitive enough to detect significant relationships, and to differentiate between possible causal factors. Some researchers also seek to identify interactions among causal factors. Although it looks as though there is one question—what is the cause of x?—that multiple approaches are asking, on closer examination there are quite different kinds of causal questions that are asked within the different approaches. The debates among proponents of different approaches, even when articulated as debates about different causal theories, are at least as much, if not more, about the point or value of different *kinds* of knowledge. I first describe the approaches in a general way, noting whatever explicit theoretical framework is adopted, as well as the questions addressed and the methods used to address them. I then look more closely at the criticisms each approach directs at the others in order to bring out what really differentiates them and, more importantly, what unites them. My intention is to show that the structure of the debates and the character of the public interest in the behaviors studied narrow the range of alternative representations of behavior in general and of the behaviors under study in particular.

2.1 Behavior Genetics

Traditionally, behavior genetics is the application of quantitative population genetics to behavior. Granting that behavior is causally influenced by both the genes and the environment of an organism, the behavior geneticist asks what the genetic contribution to a given behavior is. While this overall question motivates the inquiry, the tools of population genetics require it to be reframed as a question about a population. The tool used in this context is analysis of variance—that is, identification of how much of the difference in expression of a trait in a population is correlated with genetic difference in that population. This methodological approach requires being able to separate or distinguish differences in environment from differences in genotype. In humans, the methods traditionally used have been twin studies and adoption studies, in which one or another of the contributing factors purportedly is held constant in order to measure differences in the other. Researchers have compared the degree of similarity in expression of a given trait by monozygotic (MZ) twins who share an identical genotype and by dizygotic (DZ) twins who share only half of their genes. They have also compared twins reared apart with twins reared together. In adoption studies, researchers compare adoptees with both their biological and adoptive parents and sometimes with their biological and adoptive siblings.

Both sexual orientation and varieties of aggression have been investigated via twin studies and adoption studies. Dehryl Mason and Paul Frick (1994) identified seventy such studies of aggressive behavior published between 1975 and 1991 that did not involve confounding issues such as alcohol abuse, and developed a set of selection criteria for studies to be included in a meta-analysis. These included requirements that adoption studies compare concordance rates of adoptees and biological parents with those for adoptees and adoptive parents; that all studies exclude confounding forms of mental disturbance (e.g., schizophrenia); and that studies compare subjects based on the same type of behavior. Applying their criteria to the original seventy studies left them with fifteen whose data were suitable for inclusion in a meta-analysis. They then sorted the behaviors reported in the studies into three categories: criminal offending, aggression, and other anti-social behavior. Each of these was then divided into severe and non-severe, and they found stronger genetic effects for severe anti-social behavior than for non-severe. A number of similar studies have been performed for homosexuality (Bailey et al., 1993; Bailey and Pillard, 1991; Pillard and Weinrich, 1986; Weinrich, 1995). In studies conducted on gay men, the concordance for homosexuality among MZ twins was more than twice that for DZ twins and more than four times that for adoptive brothers. In a study on women, the concordance rate of homosexuality among MZ twins was three times that of DZ twins and eight times that of adoptive sisters. On the basis of such studies, researchers infer heritability of homosexuality at roughly 0.30 to 0.76 depending on assumptions about the base rate in the population. Studies such as those conducted by Michael Bailey and his coworkers or analyzed by Mason and Frick have come under severe criticism. These criticisms, as I will show later, are directed both at individual studies and at the entire enterprise, and are both methodological and ideological.

Medical genetics has provided another tool to take behavior research to the molecular level: linkage analysis. Behavior genetics is no longer limited to analysis of variance in populations, but can begin to address causes of difference at the individual level. Linkage analysis looks for common genetic markers (known multi-allelic loci on the same chromosome) in biological relations identified as sharing the same phenotype, in this case, behavior. The basic presupposition of linkage analysis is that genes closer to one another on a chromosome are more likely to retain their positions relative to one another than to genes further away through the multiple cell divisions and genetic recombinations involved in gamete formation. Thus, if biological relatives share an identical allele at a given locus, they are likely also to share genes on either side of that locus (with a probability inversely correlated with distance). The strategy is to associate allelic variation at a locus with behavioral variation. If such an allelic association can be found, it is taken as evidence that a nearby gene is involved with the behavior. Recently fourteen male volunteers from a single Dutch family, all of whom experienced episodes of aggressive behavior, were found also to share genes for the monoamine oxidase A (MAOA) enzyme of the X chromosome (Brunner *et al.*, 1993). In 1993, Dean Hamer and colleagues announced that pedigree analysis showed a higher incidence of homosexuality in brothers of gay men (13.5%) and in maternal uncles and sons of maternal aunts of gay men (7.5%) (Hamer *et al.*, 1993).[2] The pedigree analysis suggested X chromosome involvement. Linkage analysis revealed shared alleles at the q28 locus of the X chromosomes in thirty-three of forty sibling pairs selected from the larger sample. As with twin and adoption studies, the interpretation of these sorts of results is controversial. Behavior geneticists interpret the twin and adoption studies supplemented by linkage studies as indicating that there is a significant genetic component in the behaviors studied.

Molecular genetics has been an especially powerful tool in studying the genetics of non-human animal behaviors. Jeffrey Hall, a fruitfly geneticist, has studied the behavioral effects of dozens of mutations in the *Drosophila* genome, and found associations between quite precise variations in behavioral routines and specific mutations (1994). Shang-Ding Zhang and Ward Odenwald (1995) identified a peculiar variation in sexual behavior of fruitflies produced by subjecting genes to heat stress. The behavioral variation they observed, male fruitflies forming a closed chain of individuals each mounting the one in front of it, is one of the many behaviors that Hall also observed when he systematically introduced mutations into the fly genome. Zhang and Odenwald went so far as to propose this might be a model of human homosexuality.

In addition to studying behaviors through these kinds of methods, behavior geneticists also have been engaged in both empirical refinement and conceptual development. Developmental or longitudinal genetic analysis adds a temporal dimension to the above types of studies, monitoring and recording change in concordances in twin samples or adoptee/parent samples over time. These studies are used to support hypotheses about the increasing or decreasing effects of genotype as a person ages. Multivariate genetic analysis, another refinement, compares concordances for several phenotypic traits. Thus researchers can ask about relative degrees of heritability among phenotypes, the relative magnitude of common and unique genotypes and of environmental variability among phenotypes, and the sources of stability and change in single or multiple phenotypes over time. Multivariate analysis is also used to assess the heritability of (and presumed genetic contribution to) covariance among traits (e.g., among sub-traits identified as related to cognitive ability). This technique would facilitate compositional analysis of a broadly defined property such as intelligence or aggressivity.

Conceptual development includes the articulation of theoretical concepts for use in characterizing genetic influence. The concept of 'gene-environment interaction,' for example, is used to characterize the joint effects of a genotype and a given environmental condition. Different genotypes are said to produce different sensitivities to the environment. The genotype is presumed to be causally relevant to a dispositional state which, when an individual is exposed to some environmental stimulus or situation, expresses itself in a particular behavior. The associated concept of a 'reaction-range' is the span of possible phenotypic effects of a given gene across the span of actual or possible environments. These notions are invoked in studies of adoptees whose biological parents were criminal, which by differentiating between criminal and non-criminal adoptive parents compare the joint effect of criminality in both adoptive and biological parents with the effect of criminality in only one set of parents. 'Gene-environment correlation,' on the other hand, is used to characterize systematic covariance between a genotype and its environment. Behavior geneticists treat this as a matter of genotypes selecting or creating their environments, for example, different talents, temperaments, and so on, evoking different responses from parents and teachers, and thus individuals with these different talents and temperaments seeking out appropriate environments.

In sum, proponents of behavior genetics claim on its behalf that (1) it can help elucidate the mechanisms underlying behavior (McGue, 1994); (2) it represents the appropriate extension of Darwinian evolutionary concepts to behavior (Scarr, 1992, 1993); (3) it can help show the extent of the role of non-genetic, environmental factors in behavior (Plomin, Owen, and McGuffin, 1994); and (4) it can indicate the limits of various kinds of intervention strategies and the populations which may or may not benefit from such strategies (Fishbein, 1990).

2.2 *Social/Environmental Approaches*

The social/environmental approach emphasizes the environmental contribution to the development and expression of various behaviors. Socially- or environmentally-oriented studies seek to establish the role of socialization patterns, familial environments, and/or parental attitudes and interactions with their children. The methods used include correlational studies and direct observation of behavior and interactions in standardized settings. Typically the subjects in these studies are young children or adolescents, although in some cases the adult behaviors of persons identified as having had a certain kind of experience in childhood are also studied. Path analysis (which measures the strength of associations among pairs of factors in a set of multiple factors and attempts thereby to identify serial dependencies within the set) is used when many variables are being examined. The young and adolescent subjects in these studies are identified through schools or clinics. Interestingly, most studies of aggression include only boys.[3]

In a typical school study, students in several grades are investigated. In some cases, researchers conduct longitudinal studies, following individuals for six or more years. The children are identified as 'aggressive' using combinations of teacher ratings, peer ratings, parent ratings, and self-report. These are obtained using standardized inventories, checklists, questionnaires, or interview schedules. Some of these simply differentiate between aggressive and non-aggressive youths, others differentiate degrees of aggressivity, and others employ psychiatric diagnostic categories (e.g., Childhood Conduct Disorder or Oppositional Defiant Disorder; see (American Psychiatric Assn., 1997). Researchers then either seek to correlate the behavior with parental attitudes and practices or peer relations identified by the same means (questionnaires, interviews, etc.), or with directly observed interactions. In the latter case, children and parents are brought into a clinical or academic psychological setting and asked to perform a series of activities. The interactions between parents and between parents and children as they perform these activities are observed and recorded using a standardized protocol. Thus studies have attempted to establish the relation of parental disciplinary practices and the levels of harmony or discord (ascertained via self-report) within a family with behavioral problems in children, or to ascertain the relation of positive and negative interactive behaviors (e.g., endearments or verbal attacks recorded during observation) to behavioral problems (Lavigueur et al., 1995; Speltz et al., 1995). Other studies seek to coach parents in alternative interactive styles or disciplinary practices and to measure the effect of such changes in parental behaviors on child and adolescent behavior (McCord et al., 1994). Others investigate the relation between aggression and peer relations (Bierman and Smoot, 1991; Bierman et al., 1993). It is obviously much more difficult to establish any relationship between late adolescent or adult behavioral patterns and early childhood experiences, since it is difficult, unless one is doing a genuinely long-term longitudinal study, to establish with any reliability what the childhood experiences were. Several studies have used criminal justice system records to get around this obstacle, using court records to identify victims of childhood abuse and neglect and either interviews or later court records to ascertain the incidence of antisocial personality disorder (Luntz and Widom, 1994) or adult criminal offending (Widom, 1989; Rivera and Widom, 1990) in these populations.

Until recently, most literature in psychology and sociology on social/environmental determinants of homosexual orientation was written using a disorder or deviance model of homosexuality. Thus this literature explored such factors as alleged overprotective mothering of boys. In addition, this literature focused almost entirely on sexual behaviors in Western societies. Anthropological cross-cultural studies offer another perspective. In a review of a number of studies, Barry Adam (1996) proposes that in certain social configurations, homosexuality is an outgrowth of particular combinations of age, gender, and kinship structures. He argues that its institutionalized expression in a society is a function of gender meanings and kinship structures and meanings in that society, as well as of apparently more distal factors such as the degree of complexity of the division of labor (see also Ortner and Whitehead, 1981; Herdt and Stoller, 1990).

Practitioners of social/environmental approaches have various views about what can be concluded from such research and what the goal of it is. Some researchers clearly assume a causal effect of parental behavior on children (Haapasalo and Tremblay, 1994). Thus, correlations of parental attitude or behavior with offspring behavior are interpreted as evidence that the parental behavior plays a causal role in the expression of the child's behavior. Others acknowledge that in the networks of association their studies uncover, causal relations might not be as straightforward as assumed (Bierman and Smoot, 1991). Diana Baumrind (1993) takes something of an equivocal (and perhaps strategic) position in concluding that the correlational data produced in these studies do not license causal inferences. Such data are valuable for eliminating, but not confirming, hypotheses. Nevertheless, she claims, it is reasonable to conclude on the basis of a number of studies that parental behavior does influence children's development. Therefore, these environmentally-oriented studies are valuable because they can (help to) identify children 'at risk' for anti-social behavior or delinquency and to identify points of intervention in family and school dynamics. In a clearly meliorist articulation of faith, Baumrind claims that even if heritability accounts in large part for behavioral outcomes, it does not follow that social interventions cannot affect these outcomes. This point seems to follow quite straightforwardly from the apparently shared premise of these approaches that genes and environment interact, but as I will argue below, this conclusion depends on whether the meaning of the terms 'genes' and 'environment' is also shared, and on how interaction is understood.

2.3 Developmental Systems Theory

The developmental systems approach has its theoretical base in embryology and developmental biology. Here the central question is how the organism develops from a single fertilized cell into a mature individual characterized by multiple and specialized organs and tissues. Differentiation, the process of specialization of cells, is one of the key problems for developmental biologists. For the systems approach, genetic and environmental contributions to development are not separable. The relation between them is non-additive and non-linear. Nor can behavioral development be separated from other dimensions of development. Because of its emphasis on the role of environmental factors in development, the systems approach is sometimes seen as an ally of the social/environmental approach, and it is certainly a rhetorical ally in the debates with proponents of behavior genetics.

Gilbert Gottlieb (1991), one of the principal researchers in this tradition, characterizes development as coactional, emergent, and hierarchical. By using the term 'coactional,' Gottlieb draws attention to the interaction of the factors involved in development. Not only do the factors not act independently, but they can modify one another, thus altering their respective contributions in subsequent phases of development. By 'emergent,' Gottlieb means the increase in structural and functional complexity at all levels of organization of the individual as a result of interactions within and between these levels. By 'hierarchical,' Gottlieb seems to mean the multi-level character of development and the inter-level as well as intra-level character of coaction. 'Hierarchy' is not used to convey dominance or causal priority, but rather the unity of genetic, physiological, neurological, behavioral, and environmental aspects of the developmental system. It is being used, therefore, much as Niles Eldredge and Marjorie Grene (1992) use it to represent something like degrees of embeddedness or enclosure of systems within each other. These three features contribute to the description of the developmental process as 'epigenetic.' The structure of the genome is not static in the organism, but is conceptualized as the 'effective genotype,' which is equivalent to patterns of gene activation and is mutable in response to both intra- and extra-organismal environmental changes.

There are no human studies conducted explicitly using this approach, although it is used to provide an alternative interpretive framework for the behavior genetic and social/environmental data, and, as we shall see below, a platform from which to critique the interpretations given these data by their authors (Wahlsten and Gottlieb, 1997). Gottlieb (1991) describes studies that he and colleagues have performed on laboratory animals. These are intended to demonstrate either the canalization (i.e., establishment and fixation) of behaviors by experience (rather than by genes) or the principle of coaction. An experiment with ducklings showed that ducklings required *in utero* exposure to their own or their mother's vocalizations in order to respond to species-specific calls after birth. Genes were insufficient to canalize auditory recognition. Coaction is demonstrated by an experiment with rat pups which showed that pups bred to be spontaneously hypertensive (SHR) only developed hypertension if suckled by SHR mothers. They did not develop hypertension if suckled by non-SHR mothers, nor did non-SHR pups develop hypertension if suckled by SHR mothers. Other studies of behavioral plasticity in non-human animals are cited as showing the complex coactional nature of development. Gottlieb and colleagues argue that an understanding of the complex relations among the factors influencing development is only possible by conducting "controlled experiments that vary potentially influential factors in a systematic way" (Wahlsten and Gottlieb, 1997, p. 165). Although such experiments are both unethical and impractical in humans, these researchers claim the similarity of molecular, physiological, neural, and cognitive processes between humans and selected non-human animals is such that some general conclusions about developmental processes can be drawn, specifically that development has an epigenetic character.

One point emphasized by these writers is that their concept of a reaction-norm is distinct from the behavior genetic concept of a 'reaction range' (Gottlieb, 1995). The reaction range concept is implied in certain behavior genetic articulations of the genotype-environment interaction. Some behavior geneticists have described the reaction range as the upper and lower bounds set by the genotype on the expression of a trait across a range of environments. In contrast, the reaction-norm concept holds that each genotype is "associated with a characteristic pattern of phenotypic changes in response to alterations in the environment...[but] the rank order of individuals [regarding degree of expression of a trait] can change appreciably and uncontrollably under novel conditions" (Wahlsten and Gottlieb, 1997, p. 172). The point here is that while in one environment a given set of genotypes will characteristically sort into one order from lowest to highest expression, this ordering will not necessarily be preserved in another environment.[4]

Conversely, while one might be tempted to rank environments along degrees of nutritiveness or of nurturance for example, genotypes will not respond as though to a continuum of reinforcement, but may instead be differentially activated by the different environments. Instead of all allelic variants increasing (or decreasing) the intensity of the expression of a trait, some may while some may not. For example, while genotypes aa and aA are associated respectively with higher and lower frequencies of a given behavior (e.g., random hitting) in a given environment (e.g., permissive parenting), these frequencies will not necessarily both move in the same direction in a different environment (e.g., parental discipline), but may change in different directions. The implications are as dire for social-environmentalists as for behavior geneticists. Not only can neither genes nor environment be thought of as acting independently in the production of a given effect, but there appears to be no means of partitioning the separate effects of genetic differences or of environmental differences on the expression of a phenotype. Thus the assumption of the social-environmentalist that, other (environmental) things being equal, improving parenting skills will diminish the likelihood of delinquency in offspring is as problematic as the inference from heritability of a given measure of aggression in a population in a given environment (or set of environments) to genetic causality.

In an article primarily critical of exclusively biological approaches to sexual orientation, William Byne and Bruce Parsons (1993) have advanced an interactive model for the development of sexual orientation that is compatible with and may even be viewed as a form of developmental systems thinking. They note the lip service given to the role of both biological (genetic, endocrinological, and neuroanatomical) and environmental factors in various etiological studies, but insist on the importance of understanding how these factors interact. Such an interactive view is open to a multiplicity of developmental pathways leading to homosexuality. Some biological, environmental, or temperamental factors may be somewhat more frequently found in the developmental matrix of (subgroups of) homosexual men or women (e.g., the markers at Xq28). Their causal role, however, will depend on what other factors are interacting in that matrix. They suggest examining how biological factors, such as genotype or certain dimensions of personality (e.g., childhood gender nonconformity) which precede expression of sexual orientation and have been more strongly associated with homosexuality than other factors, interact with different environments to influence erotic choice and orientation.

Developmental systems theorists aim to understand the mechanisms of development. For them, the core questions are those of organismic differentiation: the development of the coordinated components of a complex organism from a single cell. In the realm of behavior, answering these questions means understanding how some behaviors are canalized while others remain malleable. Researchers seem less concerned about practical applications such as interventions, although some argue that there are obvious implications of taking the systems view. Richard Lerner (1991) argues that a systems or contextual approach, because it involves a finer attention to differences between individuals and between environments, requires abandoning notions of the 'generic child' (which tends to be a white, middle-class child) in the design of social interventive strategies. Instead, researchers and clinicians should take contextual variability into account (including both variation in developmental contexts and variation in the interactions of different individuals in the same context). This insistence on variability and context dependence means that if generalizations are to be obtained within this approach, they will not be generalizations about populations, individuals, or their properties, but generalizations about processes. Such generalizations will be about more abstractly conceived entities such as canalized behavior than about aggression or sexual orientation *per se*.

3. POLEMICS

3.1 The Critiques

The critical interactions of proponents of these different approaches with each other open windows onto the frameworks within which they operate, helping not only to make visible the limitations of specific methodologies, but also to reveal both differentiating and shared assumptions. I will summarize the criticisms advocates of these different approaches address to one another, as well as the distinctive questions each asks, to help make out what they hold in common and what divides them. My aim is to show that they collaborate in the construction of a common object of inquiry, developing a shared concept of behavior shaped by shared, but implicit, metaphysical and moral presuppositions.

The behavior geneticists' twin and adoption studies are criticized for failing to take gene-environment interactions into account in their calculations of heritability and thus in their inferences from a given heritability estimate in a particular environment to the same estimate of genetic influence (Gottlieb, 1995). The particular degree of parent-offspring similarity in a trait ascertained in one environment, they point out, may not hold in another. And if a trait is polygenic, studies on MZ twins will give an overestimate of the heritability of that trait in the general population, since only MZ twins share all their genes. The twin and adoption studies are also criticized for relying on two problematic assumptions about environments (Lewontin, 1991; Billings, Beckwith, and Alper, 1992). One is that twins reared apart are reared in significantly *dis*similar environments. This is affirmed in the absence of agreed on measures of environmental similarity and dissimilarity, and in the face of placement practices of finding similar homes for twins and homes for adoptees that resemble what were (or would have been) their natal home environments. It also ignores that the degree of environmental influence (and hence of estimated heritability) may be a function of age at adoption. The second and correlative assumption is that adoptive and biological siblings are reared in similar environments. This treats environmental characteristics of the home setting as gross (or shared), rather than as individualized (or non-shared) aspects. Effects of birth order (or age difference) and different parental attitudes and feelings regarding adoptive versus biological children or first-born versus later-born children are not taken into account. Developmentalists also like to point out that twin studies and adoption studies have yet to identify a single gene associated with behavior. This, however, may be more of a clue to the assumptions of developmentalists, rather than those of the behavior geneticists, and can be rebutted by showing the link between behavior genetics and molecular genetics, whose aim is precisely such identification.

Molecular genetics is criticized less for its approach than for its claims that it has already produced illuminating results. Linkage analysis, claim critics (such as Billings, Beckwith, and Alper, 1992), has proven useful only for traits with simple Mendelian or X-linked modes of inheritance (e.g., Xq28 and homosexuality among male relatives on an individual's maternal side). Most behavioral traits will be polygenic (i.e., involve many genes), so the association of any single gene with a behavioral phenomenon will be difficult to confirm. Initial reports of findings of genetic associations for bipolar disease, alcoholism, and schizophrenia have not been replicated, and replications may in general be difficult to obtain given the small samples generally used. The small samples are further problematic in that one or two misidentified alleles can affect the significance (positively or negatively) of a candidate association. Thus, even if there is a genetic component, it may be too ephemeral for current tools to isolate it.

Environmentalists are accused of valuing political correctness over knowledge and of permitting their concerns for social justice to interfere with their science (Scarr, 1995). This leads them, say critics, to select possible causes for investigation based on their perceived manipulability and to ignore genetic issues out of fear of their social implications. They also fail to extract or to include genetically relevant information from or in their samples, as they use only biological families or do not distinguish between biological and adoptive families (DiLalla and Gottesman, 1991). This leads them, according to their behavior genetics critics, to confound genetic transmission with socialization. Where biological causes are separable from social ones, the behavior geneticist claims against the environmentalist that biology is a better predictor of similarity and dissimilarity in behavior than social factors (presuming, of course, that the causes in fact are separable). A number of authors reanalyze social/environmental studies to demonstrate this point (Gottesman and Goldsmith, 1994; McGue, 1994).

Developmentalists are also accused of being nonscientific, not because of political or social motivations but due to the reintroduction of vitalism (Scarr, 1995).[5] The more substantive criticisms concern the possibility of acquiring knowledge at all within the developmental approach, as well as the multi-level, anti-reductionism of the developmentalists. Regarding the first of these, Sandra Scarr (1995) argues that it is not possible to design studies that will identify causes of phenotypic outcomes if one is committed to treating all causes as mutually modifying one another. Regarding the second, Robert Burgess and Peter Molenaar (1995) argue that reductionism is not incompatible with acknowledging multiple levels of organization. They recognize multiple levels, but nevertheless claim that the most 'general,' and hence most explanatorily basic, propositions will be found at the lowest level of organization. All explanation, they say, must ultimately invoke genes, and hence is reductionistic. Some of the critical responses to the developmentalists are defensive in character, addressed to the objections they raise to the behavior geneticist program. Scarr (1995) claims that one does not need mechanistic knowledge (i.e., knowledge of the process of development or even of gene action) to identify causes of a phenomenon. She also suggests that developmentalists are closet determinists. This is an odd accusation given developmentalists' insistence that even full knowledge of the genome will yield only probabilistic estimates of the phenotype properties. Scarr seems to think that an understanding of the mechanisms, the 'how' of development, involves the specification of necessary and sufficient conditions, and that the developmentalists' rejection of behavior genetics is a rejection of the probabilistic character of genotype-phenotype associations. The more temperate critics Eric Turkheimer and Irving Gottesman (1991) stress that behavior geneticists and developmentalists are asking different questions: developmentalists are seeking to understand proximate mechanisms, while behavior geneticists want to know about variation in a population. Furthermore, by focusing their criticisms only on single variable genetic analysis, developmentalists underestimate the kinds of information that behavior geneticists can provide.

Two features of these disputes seem most to animate the polemics. One is the shared assumption that the approaches are all asking the same question. The other is the associated assumption that there is one correct way to represent the domain, the causal landscape, in and about which the question is asked. At some very general level, they *are* asking the same question: what causes behavior? But this question is both too broad and too vague to admit of any answer. By 'behavior,' for example, do we mean particular episodes in the history of an individual, patterns of behavior, or dispositions to respond to situations in one way rather than another? Furthermore, to get a grip on this as a causal question, behaviors must be distinguished from one another and assigned criteria of identification along with strategies for determining when these are satisfied. This requirement presupposes that behaviors, at least those susceptible to causal explanation, constitute, if not a natural kind, a phenomenon that is at least stable enough to permit re-identification. The question of causality has moved far beyond the simple nature-nurture dichotomy, as it is recognized that nature *and* nurture are causally implicated both in the evolution of particular behaviors and in their expression. And while the different approaches differ on the precise implications of this for research, they do agree that, for any causal factor, one can only think in terms of its contribution to a behavior relative to that of others. The disputes, therefore, are about the weight or strength of one type of factor as compared to that of others. The supposition that there could be a straightforward resolution of these disputes is undermined by attending to the underlying dimensions of difference and commonality among these approaches. The questions that constitute their research programs increasingly differentiate and distance them from each other, and in addition they represent behavior in a way that encourages a perpetual oscillation between the different approaches.

3.2 The Questions

The critical interactions, especially the responses to criticism, reviewed above help to make clear just what questions the different approaches are asking. While each is characterized by one overarching question, those questions must themselves be made more specific in order for observational or experimental methodologies to be brought to bear on them. Here is a small sample of the fracturing process:

In behavior genetics, "What role(s) do genes play in behavior B?" becomes:

- How much of B is genetically influenced?
- To what degree is B heritable? (How much do differences in parents influence differences in offspring?)
- How much of the difference in expression of B in a population is associated with genetic difference?
- Does the degree of genetic influence on B change over time?
- How can the methodologies used to study the genetic influence on behavior (twin and adoption studies) be refined and extended?
- Can any genetic markers be associated with the incidence of B in a given pedigree? Can any genetic mutations be so associated?
- Can linkage analysis be refined and extended to strengthen claims of genetic influence?

In social/environmental approaches, "What role do environmental and other exogenous factors play in behavior B?" becomes:

- What role do gross or macro-level social variables (such as social class, ethnicity, racial and cultural identity, urban/suburban/rural setting, immigrant/native status) play in the expression/frequency of B?
- What role do micro-level variables within family, school, or peer group play in the expression of B?

- How does the influence of micro-level variables in the expression of B vary in relation to macro-level variables?
- How do differences within a family influence the expression of B by its members?
- How can familial interactions relevant to B be studied?

In developmental systems approaches, "How does B come to be expressed in individuals?" becomes:

- What developmental trajectories can be identified that culminate in B?
- What developmental factors (e.g., genetic, epigenetic, intra-uterine, physiological, physical and social environment) interact in the development of B?
- How is the disposition to exhibit B canalized?
- What are the different levels of organismic integration and organization at which causal/developmental processes relevant to B occur?
- Does the development of species-typical traits differ from the development of individually variable traits? (Of which type is B?)
- How do complexity of organization and specialization of function develop?
- How can intra-level and inter-level interactions be studied?

Each of the above subquestions also generates additional questions as researchers go about answering them. While this is especially true of the methodological questions, the substantive questions are also refined as research proceeds. There are of course questions that cross approaches. One could ask, for example, how genetic and environmental factors interact or how neural and other factors interact. Only the developmental systems theorists attempt to address questions about interaction, and then only in the context of developmental questions. The questions listed above, however, can be pursued independently of whatever progress (or lack of it) is occurring in the other approaches. Rather than develop methods for mutual testing, they pursue improved understandings of the causal factors they study. Given that both genetic and non-genetic factors are agreed to be involved in fixing behaviors, this is entirely appropriate. The consequence, however, is that each approach becomes self-sustaining through the modification of its questions in practice and the consequent generation of more and more refined questions. The research programs derived from an initial question are thus driven further into specificity and autonomy. Thus, the questions they actually address become more esoteric, and the meaning of the question "what is the cause of B?" is localized to the particular research program within which it is asked. The space of possible causes, then, is also localized. The assumption that there is a shared domain of investigation including common questions and a shared space of possible causes is not supported by analysis of the research approaches themselves. Common ground is not given but must be *re*constructed taking into account the sources of differentiation.

3.3 The Phenomena

If these approaches, when considered from the point of view of the *kind* of knowledge they can produce, splinter into different and non-integrable research programs, how can they persist in mutual polemics? What they do share is their way of conceptualizing behavior. Three different aspects of behavioral research interact to produce a distinct object of inquiry, authoritative knowledge of which is the prize for which the research approaches contest. These three aspects are the shared context of origin, the requirements for creating a studiable object of inquiry, and ontological presuppositions. I will discuss each of these separately and then show how they interact to produce an object of inquiry that is distinctive, but which has in many respects lost contact with the human actions and interactions that constitute everyday behavior.[6]

A shared context of origin is the point at which the different research approaches most obviously coincide. Our interest in behavior lies primarily in the domain of our moral lives and discourse—why did so and so do thus and such? What makes so and so act like that? (Why does Johnny hit his schoolmates? Why does Jimmy seek partners of his own sex for intimacy?). Here 'moral' is used broadly, as concerning the actions of which we approve or disapprove.[7] Our folk psychology has produced a system of explanation of action that coordinates with our practices of moral judgment. The sciences, on the other hand, offer the possibility of intervening in the processes they study. It is the promise of increasing the behaviors of which we approve and diminishing those of which we disapprove that channels support to behavioral research.[8] Research on aggression has made links to concerns with crime, anti-social behavior, and violence. Indeed in some instances aggression is operationalized as violent criminality evidenced by conviction rates. What warrants classifying a behavior as aggression is the direct infliction of harm on another. The framing of questions within the framework of culpability characteristic of moral discourse means, however, that only some forms of such behavior merit scientific study. Interpersonal violence is regarded as in need of explanation, while the use of force or infliction of harm that is part of the execution of organized work is not so regarded. Military aggression or other forms of state-sanctioned violence (e.g., in policing and 'corrections') are not studied, nor are the actions of other individuals employing violence on behalf of the state. Neither are the actions of decision makers in industries that damage individuals by pollution or discriminatory lending practices studied. 'Aggression' is a subset of individual infliction of harm on others; it is that set of incidents of harming or attempting to harm others performed without the sanction of the state or other corporate body.

Research on sexual orientation also has traditionally focused on the exceptional and unwanted. 'Sexual orientation' in the human context is really a code word for homosexuality, but the use of 'sexual orientation' signals the multiple functions of this research. In work on non-human animals, its study is part of understanding the range of reproductive behaviors, including mate selection, courtship, copulation, and parenting, or of understanding the physiological or genetic dimensions of these behaviors. In humans, by contrast, its study tends to be part of programs studying intersexuality and hormonal dysfunction or deviance. Until 1973, homosexuality was classified as a psychiatric disorder. In spite of growing acceptance of gay men and lesbians as evidenced in the gradual extension to them of civil protections, the rise of gay, lesbian, bisexual and transgender studies in the academy, and media tolerance, indeed exploitation, of homoeroticism, homosexuality is still reviled in many quarters. Homosexual acts are classified as sins, the forms of sex characteristic of same sex interactions (e.g., sodomy, oral sex) are illegal in many jurisdictions and often described in language that singles out homosexual performance of the acts in question for criminalization, and gay men and lesbians are frequently targets of hostility and unprovoked assault.[9]

The second aspect of these approaches is their common need to isolate phenomena that can be studied in the behavior—that is, phenomena that can reliably be identified and re-identified as of a particular type.[10] In the transition from the moral to the scientific, both the genetic and the social-environmental research approaches function as rivals for displacing moral with clinical explanations, for displacing reasons for acting with causes of behavior. Two conceptual transformations occur in this displacement. First, actions become elements in natural regularities. They must be freed from the contexts in which they can be assigned meaning so that they can be classified as belonging to types of events that are outcomes of types of initial conditions, whether these are genotypes or social/environmental conditions or a combination. Behavior or behavioral forms that stand out from the expected patterns in everyday transactions are more amenable to detachment from their context than those that are part of ritual or familiar transactions. The bank robbery, for example, stands out from the expected series of interactions constituting deposits, withdrawals, and bill payments. Thus it can be classified as something other than a standard financial transaction and entered into a different categorical system. The sequences of movement involved in the performance of standard transactions are those interactions (e.g., deposits, withdrawals) and not others (e.g., ticket purchases) because of myriad features of the context in which they are performed. They cannot be extracted from that context without losing their identity and are not susceptible to the kinds of causal explanation we typically seek for the robbery, which is assimilated to the category of crimes rather than that of financial transactions.[11] Similarly, in a dominantly heterosexual society, the choice to engage in intimate relations with a person of one's own sex stands out from the expected rituals of courtship that are woven into the fabric of everyday life.[12] Prior to becoming objects of scientific interest, behaviors are not just context dependent, but context saturated. This might be thought of as the phenotypic equivalent of polygenism: behaviors are constituted of multiple phenomena, each contributing a little bit to making the behavior what it is.

In the first phase of transformation into an object of scientific study, behaviors become event types that can participate in natural regularities.

The second phase is a recontextualization that is partly a consequence of the new venue in which the action/behavior studied is located—the clinical context—and partly of the recategorization implicit in bringing behavior within the scope of molecular and, hence, medical genetics. The behaviors studied occur in lists that also include various cancers and susceptibility to heart and circulatory problems and that extend the concept of behavior to include such conditions as Huntington and Alzheimer diseases (see note 1). The point of the clinic is to fix things, to provide cures or palliatives for unwanted conditions, or to facilitate the development of absent but desirable conditions. What holds interest and what justifies the investment of investigative resources in this context are not just any departures from the norm, but pathologies, expressed either as the presence of negative qualities or the absence of positive ones. Thus, the extraordinary behaviors that have enough distinctiveness to stand as event types are addressed with the tools for addressing pathologies, and the extraordinary is seen as the expression of pathology. Aggression is studied as antisocial personality, as childhood conduct disorder, while the commission of crime becomes criminality, and as such, the expression of a pathological condition. Same sex intimacy is the expression of an underlying condition—homosexuality—which can be understood as a function of an abnormal genotype or abnormal conditions of rearing. Studiable behavior is recontextualized not just as event type, but as pathological event type. The different research approaches compete to be the basis of therapeutic interventions.

The third aspect contributing to the construction of the object of inquiry is a shared ontological presupposition of the approaches. The behavior genetics approach, in both its population and its molecular forms, is concerned with determining factors located within the organism and treats behavior as a modification in or of the individual organism. Developmental systems approaches, even though they see the individual organism and its processes at any one time as a result of the interaction of internal and external factors in the preceding period of time, are nevertheless concerned with the individual and treat behavior as the expression of conditions interacting in and affecting the individual. Behavior is individual movement or change of state. Much of the environmental research, especially that which is in opposition to behavior genetics work, also treats behavior as a property of or modification in individuals. In these approaches, while the causal factors that are of interest are, unlike genes, external to the individual, they nevertheless work on dispositions, which are understood as inhering in individuals. Consistently with this individualist ontology, both genetic and environmental researchers have postulated an underlying dispositional property which is influenced by their preferred causal factors and which produces the behaviors at issue: impulsivity.

That individualism is an ontological presupposition or assumption can be seen by attending to alternative behavioral ontologies hinted at in some of the research carried out under the aegis of one or another of the approaches discussed. For example, the studies by Suzanne Lavigueur *et al.* (1995) and Matthew Speltz *et al.* (1995) looked at interactions in parent-child and parent-parent dyads. It is possible to see those interactions as properties of the dyads, or as relations between the members of dyadic (or triadic) complexes, rather than as properties of individuals. To see behaviors as properties of dyads or triads is to move up a level of organization: behavior is not a modification of particular individuals but of combinations of individuals. To see behaviors as relations is to expand one's ontology to include more than individuals or collections of individuals and their properties. The Lavigueur and Speltz studies are the kind of study invoked by researchers advocating the value of social- and environmentally-based research approaches for family intervention programs. While most of these programs focus on teaching parents new interaction styles in order to change the behavioral outcomes for their children, one can, by taking a relational or group property perspective, see their work as studying the effects of one kind of intervention in the dyadic (or triadic) system, rather than of intervention in individually specific, internal processes.

Another non-individualistic way of thinking about behavior is to view behavior as a property of a population. Here the focus is on the different frequencies of behaviors in different populations. Adam's anthropological work on sexual orientation, for example, looks at the varying frequencies in varieties of sexual expression in different societies. He correlates those frequencies with other population- or society-level properties like kinship structure and the division of labor. What is of interest is not individual etiologies, which can vary enormously, but which combination of population- or environment-level properties elicits which frequencies of varying behaviors in a population, regardless of what the causes of those behaviors in individuals may be (Adams, 1996). In similar spirit, Frans de Waal (1992) advocates treating aggression as a functional aspect of primate societies, rather than a feature of individuals. The ontological commitments of such an approach are similar to those in the new epidemiology of violence, where levels of violence are treated not as a function of the dispositions of individuals in a particular social context, but as a function of population- or context-level properties. Within this framework, the questions would be why one society manifests different levels of violence or aggression than another, why rates are falling in one city and not in another, etc. These are quite different questions than the research approaches discussed above are capable of answering. This is not to deny that some forms of aggressive behavior, like that associated with MAOA deficiency, or some instances of homosexual sexual preference (e.g., those provisionally associated with the Xq28 genetic marker), are strongly genetically influenced. It is instead to suggest that knowing this may be of little help in understanding society or population level questions.

Three aspects of the study of behavior in the research approaches surveyed combine in structuring a shared object of inquiry: the origin of research in the moral concerns of folk psychology, the consequences of transferring explanation from the moral and folk psychological domain of daily social interactions to the scientific clinic, and the ontological presuppositions carried into the research from its context of origin. Studiable behavior is individual performance of pathological routines. Behavior has become affliction.

As long as this model of behavior occupies center stage, debates about the causes of behavior will take the form with which we have become familiar. The course of discussion over time will continue to exhibit the oscillation between nature and nurture characteristic of debate in the last one hundred years or so. As I have shown, however, there are alternative ways of thinking about and representing human behavior. The proper characterization may depend on what we want to know in any particular context. The techniques of medical genetics that are applied in the study of behavior genetics are appropriate to the study of behavioral afflictions, such as the violent episodes associated with MAOA deficiency (or to the organic dysfunctions underlying the behavior). To suppose these methods can be used generally would require that affliction, or stereotyped individual routines, be the model of any studiable behavior. This either makes pathologues of us all or leaves most of our ordinary behavior unstudiable. But what cannot be isolated in the manner required at the individual level, may yet be so when considered at the group or population level. In moving to these levels, however, both behavior genetics and its individualist alternatives must be left behind.

University of Minnesota
Minneapolis, Minnesota, U.S.A.

ACKNOWLEDGMENTS

Research for this paper was conducted with the assistance of NSF Grant #SBR9730188. I am also grateful to the University of Minnesota for sabbatical leave and to the hospitality of the Centre for Philosophy of Natural and Social Sciences at the London School of Economics while I was conducting the research.

NOTES

1.   In a review article, Richard Rose (1995) surveys recent progress in behavior genetics. The article divides the research examined into three main categories: (1) research on cognitive ability and disability (reading disability and fragile X syndrome); (2) research on personality, lifestyles, and health habits (unspecified dimensions of personality, sexual orientation, and smoking and drinking); and (3) research on psychopathology and neurological disorders (Huntington disease, alcoholism, Alzheimer disease, schizophrenias and affective disorders, and aggression). The two areas of behavior that I discuss in this chapter are of longstanding interest to behavioral researchers. For a more focused analysis of research on aggression, see Longino (2001).
2.   This incidence is significant if one assumes a background rate in the general population of 2%. The rate of homosexuality in the general population is a matter of dispute among researchers, and the Hamer group's assumption of 2% is controversial (Fausto-Sterling and Balaban, 1993).
3.   This, at least, was the case in my sample of XX studies.
4.   While behavior geneticists do not observe this refinement in practice, it is not clear that the reaction range concept is not capable of this subtlety (see Turkheimer, Goldsmith, and Gottesman, 1995).
5.   Vitalism is, of course, the discredited view that living organisms are animated by a vital force not identifiable with any material force. To accuse any contemporary scientist of holding a vitalist view is to accuse him or her of being hopelessly retrograde.

6. An object of inquiry is a discursive object constructed by abstracting certain features of phenomena to produce a class of objects constituted by their possession of those features (see Longino, 1990, pp. 98-102).

7. 'We' here must mean any given group of persons sharing sets of standards. As these sets vary, so do the behaviors encompassed by them.

8. The 'we' here of course changes, too. The ordinary 'person on the street' who deploys folk psychological understanding and engages in moral deliberation and judgment becomes the bureaucrat speaking on behalf of the state (or on behalf of the ordinary citizen). The origin of social and behavioral sciences in bureaucratic concerns of the modern state is by now an old story.

9. Recently some gay activists have supported investigation of the biological basis of homosexual sexual orientation as a way of naturalizing, and thus normalizing, homosexuality. Such efforts assume, of course, that homophobia responds to reason.

10. This is no more than what must happen to natural objects, substances, and processes in the laboratory. Behavior is a different kind of entity. Admittedly, the scope of the term 'behavior' varies from context to context. (Compare: 'the behavior of compressed gases...' with 'langur reproductive behavior...' with 'the behavior of younger siblings...' with 'your behavior last night...') But the behavior researched in the programs under discussion is the stuff of human social life, suffused with meanings and values.

11. Of course, a crime may be committed or abetted in the course of one of those ordinary transactions, but white-collar crime is just the kind of crime that is not the object of study.

12. Binary, or presumed binary, forms of behavior, such as sex-differentiated behaviors, have a similar property: providing a pattern that is distinctive enough to be separable from its context of unique signification and integrated with (relatively) context-free behavioral categories. In the context of human research, these forms of behavior are losing their value as ideology is gradually separated from reality and as what gender differences there once were disappear.

REFERENCES

Adam B. Age, structure, and sexuality: reflections on the anthropological evidence on homosexual relations. Journal of Homosexuality 1996; 11:19-33

American Psychiatric Association, *Diagnostic and Statistical Manual of Mental Disorders*, IIIR ed. Washington, D.C.: American Psychiatric Association, 1987.

Bailey J.M. *et al.* Heritable factors influence sexual orientation in women. Archives of General Psychiatry 1993; 50:217-23

Bailey J.M., Pillard R.C. A genetic study of male sexual orientation. Archives of General Psychiatry 1991; 48:1089-96

Baumrind D. The average expectable environment is not good enough. Child Development 1993; 64:1299-317

Bierman K.L. *et al.* Characteristics of aggressive-rejected, aggressive (nonrejected) and rejected (nonaggressive) boys. Child Development 1993; 64:139-51

Bierman K.L., Smoot D.L. Linking family characteristics with poor peer relations: the mediating role of conduct problems. Journal of Abnormal Child Psychology 1991; 19:341-56

Billings P.R., Beckwith J., Alper J.S. The genetic analysis of human behavior: a new era? Social Sciences and Medicine 1992; 35:227-38

Brunner H.G. *et al.* Abnormal behavior associated with a point mutation in the structural gene for monoamine oxidase A. Science 1993; 262:578-80

Burgess R.L., Molenaar P.C.M. Commentary. Human Development 1995; 38:159-64

Byne W., Parsons B. Human sexual orientation: the biologic theories reappraised. Archives of General Psychiatry 1993; 50:228-39

de Waal, F.M. "Aggression as a Well-Integrated Part of Primate Social Relationships." In *Aggression and Peacefulness in Human and Other Primates*, J. Silverberg, P. Gray, eds. New York: Oxford University Press, 1992.

DiLalla L.F., Gottesman I. Biological and genetic contributors to violence. Psychological Bulletin 1991; 109:125-9

Eldredge, N., Grene, M., *Interactions: The Biological Context of Social Systems*. New York: Columbia University Press, 1992.

Fausto-Sterling A., Balaban E. Letter. Science 1993; 261:1257

Fishbein D.H. Biological perspectives in criminology. Criminology 1990; 28:27-72

Gottesman, I., Goldsmith, H.H. "Developmental Psychopathology of Antisocial Behavior: Inserting Genes into Its Ontogenesis and Epigenesis." In *Threats to Optimal Development*, C.A. Nelson, ed. Hillsdale, NJ: Lawrence Erlbaum and Associates, 1994.

Gottlieb G. Experiential canalization of behavioral development. Developmental Psychology 1991; 27:4-13

Gottlieb G. Some conceptual deficiencies in 'developmental' behavior genetics. Human Development 1995; 38:131-41

Griffiths P.E., Gray R.D. Developmental systems and evolutionary explanation. Journal of Philosophy 1994; 91:277-304

Haapasalo J., Tremblay R.E. Physically aggressive boys from ages 6-12: family background, parenting behavior, and prediction of delinquency. Journal of Consulting and Clinical Psychology 1994; 62:1044-52

Hall J. The mating of a fly. Science 1994; 264:1702-14

Hamer D., Hu S. A linkage between DNA markers on the X chromosome and male sexual orientation. Science 1993; 261:321-7

Herdt, G.H., Stoller, R., *Intimate Communications: Erotics and the Study of Culture*. New York: Cambridge University Press, 1990.

Latour, B., *Science in Action*. Cambridge: Harvard University Press, 1987.

Lavigueur S. *et al.* Interactional processes in families with disruptive boys: patterns of direct and indirect influence. Journal of Abnormal Child Psychology 1995; 23:359-78

Lerner R.M. Changing organism-context relations as the basic process of development: a developmental contextual perspective. Developmental Psychology 1991; 27:27-32

Lewontin, R., *Biology as Ideology*. New York: Harper Collins, 1991.

Longino, H., *Science as Social Knowledge*. Princeton: Princeton University Press, 1990.

Longino H. What do we measure when we measure behavior? The case of aggression. Studies in History and Philosophy of Science 2001; 32(4)

Luntz B.K., Widom C.S. Antisocial personal disorder in abused and neglected children grown up. American Journal of Psychiatry 1994; 151:706-35

Mason D., Frick P. The heritability of antisocial behavior. Journal of Psychopathology and Behavioral Assessment 1994; 16:301-23

McCord J. *et al.* Boys' disruptive behavior, school adjustment, and delinquency: the Montreal prevention experiment. International Journal of Behavioral Development 1994; 17:739-52

McGue, M. "Why Developmental Psychology Should Find Room for Behavioral Genetics." In *Threats to Optimal Development*, C.A. Nelson, ed. Hillsdale, NJ: Lawrence Erlbaum, 1994.

Ortner, S., Whitehead, H., eds., *Sexual Meanings: The Cultural Construction of Gender and Sexuality*. New York: Cambridge University Press, 1981.

Pillard R.C., Weinrich J.D. Evidence of familial nature of male homosexuality. Archives of General Psychiatry 1986; 43:808-12

Plomin R., Owen M.J., McGuffin P. The genetic base of complex human behaviors. Science 1994; 264:1733-9

Rivera B., Widom C.S. Childhood victimization and violent offending. Violence and Victims 1990; 5:19-30

Rose R. Genes and human behavior. Annual Review of Psychology 1995; 46:625-54

Scarr S. Developmental theories for the 1990s. Child Development 1992; 63:1-19

Scarr S. Biological and cultural diversity: the legacy of Darwin for development. Child Development 1993; 64:1333-53

Scarr S. Commentary. Human Development 1995; 38:154-8

Schaffner K.F. Genes, behavior, and developmental emergentism. Philosophy of Science 1998; 656:209-52

Speltz M.L. *et al.* Clinical referral for oppositional defiant disorder: relative significance of attachment and behavioral variables. Journal of Abnormal Child Psychology 1995; 23:487-507

Turkheimer E., Goldsmith H.H., Gottesman I.I. Some conceptual deficiencies in 'developmental' behavior genetics: comment. Human Development 1995; 38:142-53

Turkheimer E., Gottesman I. Individual differences and the canalization of human behavior. Developmental Psychology 1991; 27:18-22

Wahlsten, D., Gottlieb, G. "The Invalid Separation of Effects of Nature and Nurture: Lessons from Animal Experimentation." In *Intelligence, Heredity and Environment*, R.J. Sternber, E.L. Grigorenko, eds. Cambridge: Cambridge University Press, 1997.

Weinrich J.D. Biological research on sexual orientation: a critique of the critics. Journal of Homosexuality 1995; 28:197-213

Widom C.S. Does violence beget violence? A critical examination of the literature. Psychological Bulletin 1989; 106:3-28

Zhang S.D., Odenwald W.F. Misexpression of the white (w) gene triggers male-male courtship in *Drosophila*. Proceedings of the National Academy of Science 1995; 92:5525-9

PART THREE:
EXPLORATIONS OF ETHICAL, SOCIAL
AND LEGAL CONSEQUENCES

CHAPTER 10

LICIA CARLSON

THE MORALITY OF PRENATAL TESTING AND SELECTIVE ABORTION: CLARIFYING THE EXPRESSIVIST OBJECTION

1. INTRODUCTION

Current public and philosophical debate concerning the morality of prenatal testing and selective abortion is occurring at the intersection of a number of complex social developments. First and most obviously, the debate is occasioned by developments in medical genetics that make prenatal genetic testing possible. Second and simultaneously, the growing disability rights movement urges that disability be understood not simply as a medical fact, but as a social construction.[1] Thus, while developments in genetic technology change the way that various disabling conditions are understood and invite their "geneticization" (Lippman, 1991; Wolf, 1995), the disability rights movement advocates a shift from a strict medical (or genetic) understanding of disability.

Disability theorists urge abandonment of models that define disability as an exclusive feature of the individual and that view persons with disabilities as victims of a terrible tragedy.[2] The social model of disability challenges the personal tragedy model that equates disability with suffering, and examines how suffering that is experienced by those with disabilities may be the result not merely of their disabling condition, but the social context in which that condition is experienced. The current political and legal climate reflects a shift away from the individual, pathological, personal tragedy models of disability.[3] These different perspectives inform the debate between those who argue for a moral obligation to prevent the birth of

L.S. Parker and R.A. Ankeny (eds.), Mutating Concepts, Evolving Disciplines: Genetics, Medicine and Society, 191-213.

persons with certain disabling conditions, and those who argue that it is morally problematic to use prenatal testing and selective abortion to do so.

The debate about the morality of *selective* abortion also occurs within the context of the debate about the morality of abortion itself. How prenatal testing and selective abortion are discussed within that broader debate has implications not only for our judgments regarding abortion, but also for our attitudes toward disabilities. I return to this point at the end of the chapter, but at the outset assume both the legal right to seek abortion and the legal right to seek prenatal testing and selective abortion. My goal is to clarify considerations that bear upon the morality of seeking prenatal testing and selective abortion. In this regard, I share Laura Purdy's view that "exercising our legal rights can sometimes be morally wrong" (Purdy, 1995, p. 302).

Purdy makes this claim with respect to exercising the right to reproduce. She maintains that while women have reproductive rights, they also have a moral responsibility to prevent the suffering of future persons. Thus she presents a case *for* the use of prenatal testing and selective abortion, or refraining from conceiving altogether. In contrast, some disability theorists argue *against* such a responsibility or against the use of selective abortion to avoid the birth of disabled children. As Adrienne Asch writes, "ending pregnancies for reasons of...disability has serious moral and social consequences" that must be addressed (1995, p. 387). Thus, we can ask, as Deborah Kaplan does: "Are the social goals of those who have worked for the widespread use of prenatal screening consistent with those of the disability rights movement?" (1993, p. 605). At least on the surface, the answer appears to be "no."

It is in the context of these debates that I will address the expressivist objection to prenatal testing and selective abortion raised by disability theorists, and the rejoinders offered by philosophers like Laura Purdy and Allen Buchanan.[4] In the next section I consider three concerns targeted by the expressivist objection: possible loss of support for the disabled, the judgment that the lives of persons with disabilities are not worth living, and the sentiment that society wants "no more of your [disabled] kind." In addition, I specifically examine the way that arguments based on suffering rely upon medical models of disability and ignore the reported, lived experiences of persons with disabilities.[5] Further, I consider why prenatal testing and selective abortion must be considered as a unique case of seeking to avoid giving birth to a person with a disability, in contrast to the use of, for example, contraception or assisted reproduction technologies.

In section 3 I consider how in many cases the prenatal visibility of genotypes, in combination with the indeterminability of the severity of the condition tested for, results in the creation of "prenatal prototypes." Using Down syndrome as an example, I show how the application of prototypes prenatally commits some of the conceptual errors, and may result in ethical concerns, identified by the expressivist objection. I conclude that the concerns that ground the expressivist objection raised by disability theorists must be taken seriously by parties to the debate about the morality of prenatal testing and selective abortion, even if refutations of elements of the expressivist objection are, strictly speaking, valid.

2. THE EXPRESSIVIST OBJECTION

The expressivist objection may be characterized as the claim that some aspects of prenatal testing and selective abortion send a negative message about disability that devalues existing persons with disabilities. It must be stated at the outset that because this objection is concerned with harms to existing and possibly future disabled persons, my discussion will not focus on the fetus or a particular disabled person that does or does not come into the world. Rather, I wish to clarify precisely the nature of the "negative expression" that is captured by the expressivist objection. This will involve a discussion of the explicit and implicit content of that message and its effects, attention to how the judgment might be expressed (e.g., intentionally or unintentionally, by an individual or broader social assumptions), and a critical examination of how this objection is refuted in philosophical literature.

Consider Allen Buchanan's formulation of the expressivist objection:

> The expressivist objection, or rather family of objections, focuses on what may be called the expressive character of *decisions to use* genetic interventions to prevent or remove disabilities, and hence on the expressive character of the enterprise of developing and deploying scientific knowledge for such interventions. The claim is that the commitment to developing modes of intervention to correct, ameliorate, or prevent genetic defects expresses (and presupposes) negative, extremely damaging judgments about the value of disabled persons. (Buchanan, 1996, p. 28; emphasis mine)

Buchanan's definition suggests that the negative judgments about disability may be expressed in several ways. An individual's decision to abort a fetus that is likely to grow into a disabled child or adult might express a devaluation of persons with disabilities. The offer of prenatal testing and selective abortion, and the general structure of clinical practice may be interpreted as perpetuating negative judgments about persons with disabilities. Finally, the broader social context in which research and clinical practice occur simultaneously appears to devalue disabled persons and encourage the development of genetic testing. Lynn Gillam (1999) frames the objection in terms of discrimination, and outlines two components of the "discrimination argument": the slippery slope argument that prenatal testing and selective abortion lead to discrimination against persons with disabilities; and the conceptual version, which states that selective abortion itself implies discrimination against persons with disabilities, regardless of its concrete effects. What, then, is the nature of this discrimination? What exactly does it mean for a negative message to be "sent" or "expressed" in these various contexts?

Some argue that the negative judgment or devaluation of existing persons with disabilities is implicit in the ideology surrounding this technology (this would parallel Gillam's conceptual version of the argument). As characterized by Buchanan, "the charge is that the very conception of progress that lies at the core of the ideology of the new genetics discriminates against and devalues disabled individuals, inflicting upon them what may be the gravest injustice possible: the rejection of their very right to exist" (1996, p. 19). Here, the harm is framed in terms of injustice and the denial of rights that stems from a belief in progress. Others have made the more general claim that the ideology of the new genetics and prenatal testing is eugenic. Abby Lippman writes, "Whatever may be claimed as motives...prenatal screening cannot but be eugenic. These programs are for detecting and preventing the birth of certain babies, in fact, for 'gatekeeping'" (1991, p. 50). Angus Clarke is critical of this emphasis on prevention, and argues that "it is impossible to maintain a sincerely non-directive approach to counseling about a genetic disorder while simultaneously aiming to prevent that disorder" (1991, p. 999). According to Clarke, then, the possibility of sending a negative message is inherent in the very structure of clinical practice, and does not necessarily have to be explicitly articulated by a particular individual.

Given the prominent role that genetic counselors play in the process of prenatal screening and decisions to abort, some argue that the discourse surrounding this technology can also transmit negative messages. This suggests that counseling sessions might be vulnerable to the expressivist objection in a number of ways (Rothman, 1985; White, 1999). The counselor may present inaccurate portraits of a particular disorder, or rely solely upon a medical model of disability that presents the condition as intrinsically bad (Silvers, 1998, p. 89). Elkins and Brown substantiate this concern in their discussion of how both geneticists and genetic counselors can misrepresent the facts regarding Down syndrome:

> Even our most well-meaning geneticists seem at times to be excessively negative in their descriptive categorization of persons with Down syndrome. For example, in an effort to separate mild and serious anomalies, Down syndrome was the only nonlethal anomaly listed among serious anomalies found by genetic testing in the largest screening program in our country.... Most negative of all, however, was a published attempt to justify the cost of offering the triple screen to pregnant women of all ages.... Outdated information about institutional costs, family disruption, and loss of family productivity was used in order to justify the proposal that testing and termination for fetuses with Down syndrome would save Americans money. (Elkins and Brown, 1995, p. 16)

These misrepresentations can translate into counseling practices. For instance, while less than five percent of persons with Down syndrome are severely-to-profoundly retarded, "mothers are told that their child could be 'severely retarded' or 'normal' as if each were a 50-50 chance" (Elkins and Brown, 1995, p. 18). Some have suggested that, more generally, the complex nature of genetic information poses difficulties with regard to obtaining full disclosure and informed consent (Council, 1998, p. 17).

In addition to addressing the clinical context, the expressivist objection may focus on "what may be called the expressive character of decisions to use genetic interventions to prevent or remove disabilities" (Buchanan, 1996, p. 9). Some disability theorists argue that the negative expression is relayed in some way by a particular woman's decision to use this technology and selectively abort. Adrienne Asch suggests that at the individual level, the decision to selectively abort based on disability may express a desire for perfection or control that ultimately may not be possible:

> How is an individual woman's act of [selective] abortion detrimental to herself, possible other children, or people with disabilities in the world? The consequences for the

woman and for any other children she may bear and rear are at least as serious as those for current and future generations of people with disabilities.... The woman who decides to abort for reasons of disability may be signaling an unwillingness or incapacity to recognize that not everything in life can be controlled. (Asch, 1995, pp. 388-9)

In Asch's example, there are presumably psychologically harmful consequences for the woman herself in her desire to control what kind of child she will have. Moreover, Asch suggests that this mindset may have negative effects on her other existing or future children.

Even if a woman does not intend to send a negative message in making her decision, and no existing persons are harmed by her choice to selectively abort, Diane Paul argues that individual decisions can often result in broader, unintended consequences. She argues that, in a market system, "the commitment to personal autonomy has obscured the fact that individual reproductive decisions do have social consequences" (Paul, 1994a, p. 152). The danger of a kind of "homemade eugenics" looms over the practice of prenatal testing and selective abortion, and this suggests that the individual's decision cannot be abstracted from its broader social and economic context (Paul, 1994a, p. 152).

The individual woman's decision to selectively abort, the dynamics of genetic counseling and clinical practice, and the general ideological commitments underlying prenatal genetic testing are all potential targets for the expressivist objection. However, questions remain: what is the exact nature of the negative message that is supposedly sent? In what sense does this expression harm some or all existing persons with disabilities? In a broader context, how do these negative expressions affect general attitudes about the nature of disability and disabled lives? In an effort to address these questions, I will examine three components of the expressivist objection. The first two involve the actual content of the negative messages or judgments that prenatal testing and selective abortion allegedly express. The first negative judgment is that the lives of individuals with disabilities are not worth living (Buchanan, 1996, p. 28; Asch, 1995, p. 388), a message which often relies upon the conflation of disability and suffering. The second negative judgment is that we want "no more of your kind" (Purdy, 1995, p. 313; Asch, 1995, p. 388), a message directed at existing persons with disabilities. The third component of the expressivist objection is what Buchanan calls the "loss of support" argument, or the claim that as the number of persons with disabilities is reduced as prenatal screening and selective abortion become more prevalent, public resources for existing persons with disabilities will diminish (1996, p. 21; Gillam, 1999, pp. 164-5). In outlining these three components of the expressivist objection, I will critically discuss their treatment in philosophical arguments and ultimately argue that the expressivist objection is not readily dismissible.

2.1 Loss of Support Argument

One argument that is closely related to the expressivist objection to prenatal testing and selective abortion is the claim that "as the application of genetic science reduces the number of persons suffering from (genetically based) disabilities, public support for those who have these disabilities will dwindle" (Buchanan, 1996, p. 21). However, in addressing the question of resources, it becomes clear that the expressivist objection needs to be more broadly contextualized in philosophical discussions.

Buchanan attempts to refute the "loss of support" argument on three grounds. First, he argues that there is not enough evidence and data to support this claim. Second, loss of support is less likely in light of the growth of a vocal disability rights movement. Finally, he claims the loss of support argument overlooks interests the non-disabled may have in *not having* disabilities (Buchanan, 1996, p. 22). This argument applies more to genetic intervention than to selective abortion, thus I will not address it here.

Buchanan's refutation is not wholly successful when considered in the context of prenatal testing, where there is evidence of troubling economic and social consequences, even if the resources allocated to benefit persons with disabilities are not reduced. First, the fact that the discourse of prenatal testing is often framed in terms of cost-benefit analyses is problematic. As disability theorist Simi Linton observes, "the idea that disabled people are, in an absolute sense, an economic and social liability is rarely challenged" (1998, p. 50). This notion of persons with disabilities as liabilities can have a negative impact on the practice of prenatal testing. Paul, in her article "Is Human Genetics Disguised Eugenics?" argues that the autonomy of parents is often undermined:

> Cost-benefit considerations help explain the vast expansion in the number of women who now undergo prenatal testing. They provide a powerful inducement to test more women for more disorders at an early age. Avoiding the conception of an infant at risk for a genetic disease, or avoiding the birth of a fetus prenatally diagnosed as having one, will often be less expensive than clinical management. (Paul, 1994b, p. 73)

Michael Berube echoes this concern in his account of raising a son with Down syndrome. He argues that the danger is not in prenatal testing or selective abortion, *per se*, but is closely related to economics: "The danger for children like Jamie...lies in the creation of a society that combines eugenics with enforced fiscal austerity. In such a society, it is quite conceivable that parents who 'choose' to bear disabled children will be seen as selfish or deluded" (Berube, 1996, p. 52).

This kind of parent-blaming can be found in some philosophical arguments which maintain that it is immoral to bring a child into the world knowing that it will suffer and will drain resources.[6] Consider Purdy's argument that there is a moral obligation to avoid the birth of certain individuals that might be a drain on the system, and that resources might be spared to serve the interests of existing disabled persons:

> It *is* unreasonable, in a world of limited resources and great need to be required to allocate resources for those who didn't have to need them.... Isn't it immoral to knowingly act so as to increase the demands on these resources, resources that could otherwise be used for projects such as feeding the starving or averting environmental disaster? Isn't attempting to avoid the birth of those who are likely to require extra resources, other things being equal, on a par with other attempts to share resources more equally? But from none of this does it follow that we should reduce the concern for those who already exist; on the contrary, it is in part *their* welfare that dictates such careful use of resources. (Purdy, 1995, p. 313)

At one level, Purdy seems to successfully refute the loss of support argument when she observes that by preventing the birth of those who will incur additional costs, more resources might be available for existing disabled persons (Purdy, 1996, p. 504). While this may be true, her argument is problematic insofar as it suggests that there must be a trade-off between potential and existing disabled people with respect to resources.[7]

Furthermore, Purdy's suggestion that preventing the birth of disabled persons is somehow motivated by a concern for the welfare of existing disabled persons is questionable in light of the limited resources currently allocated to improve the social conditions of persons with disabilities. As David Wasserman correctly observes:

> the expressive significance of the sponsorship of prenatal testing by the state must be assessed in light of its responsibility for the economic and social conditions that contribute so greatly to the difficulties of raising, or being, a child with a disability. In allocating resources to enable parents to prevent the birth of children with disabilities, while failing to support facilities and services that would allow such children (and their parents) to flourish, society devalues the lives of people with disabilities less ambiguously than do the parents who avail themselves of prenatal testing. (Wasserman, 1998, pp. 282-3)

What is captured in the loss of support argument is not simply a question of resource allocation. This component of the expressivist objection exposes the tension between a genuine concern for the welfare of persons with disabilities, and an economic and ideological commitment to prevention that both suggests a devaluation of persons with disabilities and adds the moral stigma of (deliberately) giving birth to a child with a disability. Buchanan is right to point out that the growing disability rights movement may mitigate some of the negative economic effects of genetic research; however, disability theorists point to other social effects of prenatal testing and selective abortion, some evidenced in economic terms, but others evident in attitudes expressed. These attitudes cluster around two negative judgments: that persons with disabilities have lives not worth living, and that our able-bodied society wants no more of their (disabled) kind.

2.2 Lives Not Worth Living

One of the two negative judgments that Buchanan identifies as central to the expressivist objection is that "the lives of individuals with disabilities are not worth living." Many disability theorists have pointed out that our society, generally, promotes this message in many ways.[8] More specifically, the charge that disabled lives are devalued has been made against prenatal testing for abortion purposes. Asch argues that "by contrast with the improvements of the past 15 years in education and employment for disabled people, prenatal diagnosis and selective abortion communicate that disability is so terrible it warrants not being alive" (1995, p. 388). Kaplan argues that the implicit message in the practice of selective abortion based on genetic characteristics is that "it is better not to exist than to have a disability" (1993, p. 610). Lippman views the message as more explicit: "By orienting prenatal screening specifically to the detection of particular conditions, we explicitly make a social statement about the quality or the value of a fetus and on the adult it may grow up to be based solely on its genetic/chromosomal material" (quoted in Paul, 1994b, p. 68). Is this the message that is really being sent? In one sense, it seems clear that if particular conditions are tested for the purpose of selectively aborting an affected fetus, the presumption is that the birth of a child with that particular condition is undesirable. To confirm this one could examine the extent to which this sentiment is prevalent at the level of research, clinical practice, or even the individual level. However, insofar as my focus is on philosophical refutations of the expressivist objection, I will consider the ways in which philosophers present this view that certain lives are not worth living.

Some philosophers are quite explicit about the fact that, unless a certain standard of quality of life is present, it is better not to exist than to live with a particular condition.[9] Bonnie Steinbock and Ron McClamrock argue that "anyone willing to subject a child to a miserable life when this could be avoided would seem to fail to live up to a minimal ideal of parenting" (1994, p. 18). Purdy echoes this when she refers to the "argument for healthy bodies," "the claim that people are better off without disease or special limitation, and that this interest is sufficiently compelling in some cases to justify the judgment that reproducing would be wrong" (1995, p. 307). Steven Edwards observes that persons with intellectual disabilities are accorded lower moral status, and sees the justification of selective abortion in cases of intellectual disability as evidence of this (1997, p. 31). In fact, if one examines philosophical discourse about persons with disabilities, particularly the intellectually disabled, there are numerous examples of the lower moral status these individuals are granted, and the general sentiment that their lives are less worthy than "normal" lives (Carlson, 1998).

2.3 The Suffering Argument

In the context of prenatal testing, however, the primary reason that disabled lives are judged to be not worth living rests on the notion of suffering, not moral status. According to Kaplan:

> The most appealing and satisfying reason for permitting abortions based on genetic characteristics is altruism. We believe we are saving potential future children from pain and harm. Perhaps it is this justification that is most troublesome to disability rights activists. (Kaplan, 1993, p. 609)

I will examine two versions of the suffering argument that justify selective abortion in order to explain why the appeal to suffering is attractive to philosophers and problematic for disability theorists.

The first, weaker version is that the desire to prevent the birth of persons with (certain) disabilities expresses the desire to prevent suffering of future people, *not* a negative judgment about the worth of disabled persons. Buchanan's thesis is that "justice can require genetic intervention in certain circumstances" (1996, p. 26).[10] He explains that the decision to prevent this future person's existence does not necessarily express a negative judgment; rather, it may be motivated by "the desire not to bring into the world an individual with seriously limited opportunities" (Buchanan, 1996, p. 31). Thus, the woman who decides that she does not want a particular fetus after discovering it has a certain genotype or chromosomal anomaly is not necessarily expressing a value judgment concerning existing persons with that particular condition. While some, like Asch, would say that this decision may belie a desire for perfection or a misplaced anxiety regarding having a child with a disability (1995, pp. 388-9), I agree with Buchanan when he says that:

> one may wish to avoid serious strains on one's marriage, on one's ability to fulfill responsibilities to one's other children, or on scarce social resources, and yet consistently believe that the lives of many, indeed all of Down's syndrome individuals are worth living and that every child and adult with Down's syndrome has the same right to life as any other person. (Buchanan, 1996, p. 32)

Furthermore, even if that judgment were made about a particular condition, it does not follow that the negative judgment extends to *all* disabilities. Buchanan rightly states that though "some people believe that there are some disabilities so severe that they make life not worth living...this belief, whether justified or not, does not in any way imply that *all* disabilities, or even all serious disabilities, make the lives of those who have them not worth living" (1996, p. 29).

Buchanan argues against the expressivist objection by saying that not all decisions to selectively abort necessarily imply a devaluation of some or all disabling conditions and existing persons with those disabilities. Rather, these decisions signal a desire to reduce suffering or hardship. Before presenting the case made by disability theorists in response to this claim, I will outline the second version of the suffering argument.

The stronger version of the suffering argument is that there is a *moral obligation* to prevent the birth of persons with disabilities who will suffer, and fulfilling this obligation reflects the desire to prevent suffering, *not* a negative judgment about the worth of disabled persons. Though she does not discuss prenatal testing and selective abortion explicitly, Purdy makes the general argument that we have a moral obligation to prevent the birth of an individual if s/he will not attain a minimally bearable quality of life, and her argument "ultimately depends more upon the degree and inevitability of the suffering than its source" (Purdy, 1995, p. 309). Her list of persons whose suffering is both certain and severe enough includes a broad range of conditions: "I would include chronic pain, serious physical or mental limitation, and mental suffering (including the prospect of an early but not imminent death) on the list of conditions to which people shouldn't be subjected" (Purdy, 1995, p. 322). Thus, we can assume that these lives are judged to be not worth living, not on the basis of some hierarchy of value, but rather because to permit their existence would result in unbearable and inevitable suffering.

Steinbock and McClamrock make a similar claim. They argue that "the principle of parental responsibility maintains that prospective parents are morally obligated to consider the kinds of lives their offspring are likely to have, and to refrain from having children if their lives will be sufficiently awful" (Steinbock and McClamrock, 1994, p. 19). Though they do not explicitly consider selective abortion, one can assume from their position on refraining from reproducing and infanticide that terminating a pregnancy after discovering that the fetus has a genetic condition that would cause the child to suffer would be morally permissible and, in fact, required.

Both forms of the suffering argument, then, defend prenatal testing and selective abortion in the face of the expressivist objection. The claim is that these practices can prevent the suffering of future persons, and that they do not necessarily express the negative judgment that disabled lives are not worth living. The stronger version adds that there is actually a *moral obligation* to prevent such suffering. However, there are several responses to the appeals to suffering that illustrate why the expressivist objection is not entirely dismissible.

First, consider Purdy's two criteria for making a judgment about what is best for "future persons": the inevitability and degree of suffering. Though she portrays their position as opposed to her own, some disability theorists would concur that if the degree of suffering is both certain and extremely severe, selective abortion is justified. Asch, for example, argues that in cases of anencephaly, Tay Sachs, Hunter's syndrome, and certain other conditions that cause protracted physical pain or death in infancy or early childhood, selective abortion is morally permissible (1995, p. 387). However, she objects to the fact that many other conditions are lumped together, and are treated as on a par with the above cases when, in fact, neither the inevitability nor the degree of suffering are predictable: "Down syndrome, spina bifida, cystic fibrosis, or muscular dystrophy cause degrees of impairment ranging from mild to severe, the degree indeterminable at the time of prenatal diagnosis" (Asch, 1995, p. 387).

Though some philosophers do not distinguish between avoiding conception and selective abortion when discussing the morality of preventing certain births, I maintain this distinction and focus exclusively on selective abortion. Purdy writes that though some disability theorists "clearly think that aborting an existing fetus is morally worse than failing to conceive one...it seems to me that their arguments, if sound, are as telling against failing to conceive as against aborting. For this reason, and because more general questions about abortion would quickly obscure the specific question I want to consider here, my argument will focus simply on the question of what we want for future people" (1995, p. 303). Thus, in advancing the suffering argument—that "we ought to try to prevent the birth of those with a significant risk of living worse than normal lives"—Purdy does not distinguish between aborting such a child or simply refraining from reproducing if one has an identified, genetically-based increased risk of illness or disability that will cause suffering of a future child (1995, p. 302). McClamrock and Steinbock, too, shy away from the specific case of selective abortion and consider cases of infanticide and cases "where there is no child at all" when they speak generally of parental responsibility (1994, p. 18).

Insofar as determining the severity and inevitability of suffering are central to the suffering argument, their indeterminacy in many cases of prenatal testing challenges the applicability of these criteria. This suggests that the suffering argument, which in its stronger version claims that we have a moral obligation to prevent the suffering of future persons, should not be equally applied to cases of selective abortion, infanticide, or simply refraining from reproducing. (Later I will discuss more thoroughly why the case of selective abortion needs particular attention and raises unique concerns.)

Another aspect of the suffering argument that needs to be discussed is the *cause* of suffering. One of the central arguments made by disability rights activists is that much of the suffering that persons with disabilities experience is a result of societal views and discrimination, not the actual physical or mental condition itself. Purdy acknowledges the social roots of suffering for persons with disabilities. However, ultimately she wants to affirm that "disability itself" causes suffering. She writes that while arguments to improve social conditions are justifiable and urgent, "it seems to me that some of the arguments intended to further that goal can be...inadequate and counter-productive" (Purdy, 1995, p. 308). One reason for this claim is Purdy's belief that we are not moving fast enough to lessen the negative social consequences of disability (1995, p. 304). Thus, whether they suffer from the condition itself or social obstacles, Purdy assumes that persons with disabilities will inevitably face hardships that could be ameliorated by improved social conditions for the disabled (this is one variant of the personal tragedy model that relies upon the conflation of disability and suffering).

At least two responses can be made to this position on the current inevitability of the suffering of the disabled. First, despite her pessimism regarding how quickly the climate can and will change, the fact that certain types of suffering have exogenous (e.g., social) rather than endogenous causes means that they are by no means immutable or inevitable. Second, by maintaining that the cause of suffering is irrelevant to her position, Purdy is forced to make extreme claims with which I (and I think many) would take issue. In following her analysis to its logical conclusion, she is forced to admit that even for certain females or African-Americans, who "can be expected to live especially difficult lives...where we can be certain about [the degree and inevitability of the suffering], there is at least a prima facie case against reproduction in these cases too" (Purdy, 1995, p. 309).

Finally, the expressivist objection exposes the conflation of disability with suffering and disadvantage, a conceptual move, which many argue devalues the lives of persons with disabilities. Philosopher Anita Silvers explains why this equation occurs:

So inured are our historically conditioned feelings of superiority to individuals with disabilities that...[we automatically equate] disability with disadvantage. However, even though an impairment is no advantage, it does not follow that to be impaired or disabled is to be disadvantaged per se. For disadvantage is relative to context and end. (Silvers, 1995, p. 47)

Buchanan's discussion of the social construction of disability offers an example of this. Though he recognizes that disability is context dependent, and defines it as "a mismatch between the individual's abilities and the demands of a range of tasks" (Buchanan, 1996, p. 39), he assumes that disabilities and defects (physical or psychological lacks of capacities) are necessarily disadvantages: "Disabilities, as limitations on significant opportunities, are by definition undesirable, even when they do not cause suffering. They are disadvantages" (Buchanan, 1996, p. 38). Silvers responds directly to Buchanan's statement by distinguishing between intrinsic and instrumental badness (1998, pp. 89-106). She claims that though Buchanan does not fall prey to the medical model which defines disability as intrinsically bad, he does rely upon the assumption that disability is instrumentally bad and therefore necessarily a disadvantage (Silvers, 1998, pp. 103-6). For Buchanan, the very definition of disability involves the notion of disadvantage, even if the individual does not suffer, insofar as "[disabilities] interfere with opportunities for performing significant tasks and for full participation in valued social interactions" (1996, p. 38). Though there is not space to rehearse her argument here, Silvers rightly argues that while we can admit that certain tasks and modes of participation are of value and, perhaps even intrinsic value, it does not follow that the *inability* to partake in them necessarily constitutes a disadvantage, nor is it necessarily "undesirable," a term Buchanan uses to characterize disability (1998, pp. 104-5).

Generally, then, one objection made by disability theorists to the suffering argument is that by relying on the "prevention of suffering," or even Buchanan's weaker formulation of addressing the disadvantaging and undesirable nature of disability, it is assumed that living with a disability necessarily involves suffering or is undesirable in some way.

This view is often perpetuated despite claims to the contrary by persons with disabilities. Purdy admits that for many people, their disabilities are not experienced as disadvantages and do not necessarily cause tremendous suffering. However, she is quick to dismiss these accounts as either limited, or overly optimistic cases, and is ultimately "skeptical about whether the lessons learned justify the suffering they require" (Purdy, 1995, p. 311). Though there are clearly a wide range of experiences, it is imperative that these accounts be taken seriously. While Purdy charges that Adrienne Asch and Marsha Saxton denigrate the value of healthy bodies (Purdy, 1995, p. 305), I would argue that they are simply pointing out the way in which the non-disabled place disability and suffering in stark opposition to health and happiness. Silvers' explanation of the able-bodied response to disability may explain Purdy's dismissal of such claims: "Our aversion to the very idea of being disabled forestalls our understanding the disabled from their perspective" (1995, p. 37). Moreover, the fact that arguments like Asch's and Saxton's appear to be overly positive to Purdy might be attributed to the fact that persons with disabilities are often pushed into downplaying the negative aspects of their conditions. Jenny Morris explains:

> In asserting our right to exist, we have sometimes been forced into the position of maintaining that the experience of disability is totally determined by socio-economic factors and thus deny, or play down, the personal reality of disability. It is difficult to integrate this reality in a positive way into our sense of self when the disabled world has nothing but negative reactions to the physical and intellectual characteristics of disability. In this way, an assertion of our worth becomes tied up with a denial of our bodies and an attempt to 'overcome' the difficulties which are part of being disabled. (Morris, 1991, p. 70)

Thus, I do not accept Purdy's dismissal of accounts by disabled persons that challenge the equation of suffering with disability, nor do I believe that the cause of suffering is irrelevant. Furthermore, I would argue that we need to further examine the historical and psychological roots of the aversion to disability, and honestly assess the extent to which this informs suffering arguments. If we take this broader context into consideration, the force of the expressivist objection that prenatal testing and selective abortion express that certain lives are not worth living lies in the fact that disabled lives are devalued when they are presented under the "personal tragedy" model of disability, where having a disability is *necessarily* equated with a life of suffering, whatever the cause of that suffering may be.

Though the decision that an individual woman makes to selectively abort as a result of prenatal testing does not necessarily involve her own negative judgment about persons with disabilities or the value of disabled lives, the expressivist objection does expose problems with justifying this practice based on the suffering argument. The assumption that suffering and disability go hand-in-hand perpetuates a negative characterization of disabled lives. Furthermore, in making this conflation some theorists fail to closely examine the problems with determining the degree, inevitability, and causes of suffering in the context of prenatal testing and selective abortion. Finally, at a broader level, it is arguable that the very notion of prevention (an inevitable part of the goal of prenatal testing though not necessarily the primary goal[11]) relies upon the assumption that some conditions are not bearable. Silvers asks an important question with respect to this assumption: "which [source] of possible suffering is so pressing to nullify as to recommend as the means of doing so the prevention of the lives of possible sufferers?" (1998, p. 93). The expressivist objection should not be dismissed entirely because it keeps this question at the foreground, one that is particularly salient in light of the stronger version of the suffering argument that claims we have a moral obligation to prevent the suffering of future persons.

2.4 *"No More of Your Kind"*

In asking the following questions, Asch reveals another message that is a target of the expressivist objection, the charge that prenatal testing and selective abortion express to persons with disabilities society's desire for "no more of your kind." She asks:

> As a society, do we wish to send the message to all such people now living that there should be 'no more of your kind' in the future? If we use the technological solution of prenatal diagnosis to eliminate such people, what will become of our attitudes and practices toward any of those odd people who were missed by the technology and happened to be born? (Asch, 1995, p. 388)

As discussed earlier, a woman's decision to undergo prenatal testing and abort her fetus after a positive result does not necessarily express her general desire to reduce the number of persons with such a condition. She may have particular reasons for wanting to terminate the pregnancy, while still affirming the worth of persons with disabilities and not believing that all such births should be avoided. However, Paul suggests that individual decisions may have unintended eugenic consequences, and would likely agree that one consequence of individual decisions to selectively abort may be a general expression of the "no more of your kind" sentiment (1994a, p. 146).

Berube is critical of the role genetics plays in perpetuating stereotypes about disability. Although he states that "we should not assume that the technology of prenatal testing itself will impel specific human responses, as if there will be a mass extinction of fetuses with Down syndrome," he does acknowledge the danger of coercion, and warns that we must monitor the sociomedical apparatus and continuously question "whether our ability to spot fetal abnormality might not eventually become more coercive than descriptive" (Berube, 1996, p. 70).

More generally, there is little question that the development of prenatal screening for the purposes of selective abortion has the aim of preventing certain kinds of fetuses from being born. Selective abortion is only one way of avoiding the birth of fetuses with particular disorders; there are other types of intervention such as altering defective genes, or simply avoiding conception when there is a significant risk, either through contraception or using artificial insemination or embryo transplant (Buchanan, 1996, p. 30). However, all of these practices presuppose that there is a particular condition which is undesirable, and that by identifying it one may have greater possibilities to prevent its occurrence. Thus, those who maintain the expressivist objection are correct insofar as there are clearly genotypes and chromosomal abnormalities whose prevention is an underlying goal of genetic research. The question remains, however, whether this desire to prevent the incidence of particular disabling conditions expresses the judgment to existing persons with those conditions that society wants "no more of their kind." There are two responses to this question that attempt to refute the expressivist objection: the first, a version of the suffering argument, maintains that the implementation of such technology is actually an effort to reduce suffering for persons who might have such disabilities. The goal is improvement rather than the elimination of persons with particular disabilities. The second response is that prenatal testing and selective abortion express the desire to eliminate particular *characteristics*, not to devalue or advocate the non-existence of persons with disabilities. I will discuss each response in turn.

First, let us examine the claim made by Purdy and Buchanan that the underlying message is positive, rather than harmful. In a direct response to Asch's claim that prenatal testing and selective abortion express the desire for "no more of your kind," Purdy writes:

> I would dispute Asch's view that by attempting to avoid the birth of individuals with serious impairments that we either intend or in fact send such a message to the living. Wanting a world where fewer suffer implies doing what we can to alleviate the difficulties of those who now exist as well as doing what we can to relieve future people of them. (Purdy, 1995, 313-4)

Note that Purdy's claim involves two separate goals: one is to alleviate the suffering of existing people and presumably make changes that would benefit any future persons with the same condition, while the other is to alleviate difficulties of future people by failing to create them. Buchanan makes a similar argument that relies upon the distinction between relieving stigma or social causes of suffering, and eliminating biological/genetic causes of disabilities. He says that while disability rights advocates focus on the stigmatization of persons with disabilities, "a consistent commitment to equal opportunity requires that we try to reduce both the stigma and the physical causes of disabilities" (Buchanan, 1996, p. 35). The claim is that the development and use of prenatal testing reflect the motive to reduce suffering, and are not at odds with a commitment to improve the conditions of *existing* persons with disabilities (Gillam, 1999, pp. 164-5). Yet does this argument rebut the expressivist objection?

One counterargument that disability theorists might mount to the arguments of Purdy and Buchanan challenges the assumption that the individual's suffering is due to his/her physical or genetic condition. In other words, insofar as the non-disabled equate disability with suffering, "no more of your kind" means "no more of those who will suffer." At this point, we return to the earlier discussion regarding the inevitability and degree of suffering, and one can assess whether certain disabling conditions necessarily imply greater suffering. As I argued earlier, it is conceivable that when the question of suffering is taken into consideration, disabled persons can claim a harm insofar as an erroneous view of disability is perpetuated. Thus, the message "no more of your kind" where kind x = a person who will suffer, is objectionable because it relies upon the assumption that certain *kinds* (i.e., persons with disabilities) necessarily suffer. Furthermore, insofar as the cause of the suffering is presumed to be genetic or physical (another claim that disability rights advocates wish to challenge), it is assumed that "future people" will be relieved by eliminating the defect. However, in the case of prenatal testing and selective abortion, that necessarily means eliminating the fetus or future person (who thus far is defined solely by its genetic characteristics). Because it does not distinguish between cases of avoiding conception and selective abortion, Purdy's claim that we are simply concerned with "alleviating suffering" for future people is not applicable to cases where eliminating suffering (if its cause is assumed to be genetic) necessarily involves eliminating the future person (the fetus).[12]

The second argument against the expressivist objection contends that disability theorists fail to separate qualities from persons. In addressing the claim that prenatal testing and selective abortion send the message "no more of your kind," Purdy and Buchanan both argue that the expressivist objection is guilty of conflating qualities with persons; in other words, they argue that it is important to distinguish between devaluing certain characteristics and devaluing the persons who have them (Purdy, 1995, p. 305; Buchanan, 1996, p. 34). Buchanan gives the example of hypertension: "we seek to prevent hypertension without devaluing people with hypertension" (1996, p. 34). Moreover, "it is important to keep in mind that to value some characteristic is not necessarily to look with *contempt* on those who lack it" (Purdy, 1995, p. 305). Once we "think through more carefully any leap from qualities to persons" (Purdy, 1995, p. 305), the desire to eliminate specific characteristics (particularly those that would inevitably cause suffering) should not be viewed as sending a negative message to existing disabled persons. Rather, it is an expression of our desire to be rid of attributes, not disabled persons generally.

There is no question that this conceptual distinction can be made, and that we in fact distinguish between persons and traits all the time. However, at an experiential level, some persons with disabilities do not view their condition as separable, and claim that their disabilities are integral to their sense of self (Asch and Fine, 1988; Morris, 1991; Silvers, 1998). Morris writes, "we reject the meanings that the non-disabled world attaches to disability but we do not reject the differences which are such an important part of our identities" (1991, p. 17). This view that disabilities are not detachable from selves (Edwards, 1997, p. 40) suggests that we may not be able to treat certain disabilities as we do hypertension, and contradicts Purdy's statement that "it should be possible to mentally separate my existence from the existence of my disability" (1995, p. 314). It is an assumption on the part of able-bodied persons that persons with disabilities can or wish to make this separation, and that they would rather not have a disability. Instead, Silvers argues that the issue of justice for persons with disabilities should not be "couched in expectations that they be cured. How can laying claim to the right to be altered so as not to be an impaired individual be central to any person's self-respecting self-identity as an individual with a disability?" (Silvers, 1998, pp. 137-8). Silvers' view is by no means representative of all persons with disabilities (Tucker, 1998). However, challenging the assumption that one can or should separate qualities (in this case, disabling conditions) from persons has important implications for the expressivist objection.

Edwards argues that persons who are intellectually disabled are accorded lower moral status because they fail to conform to the "individualist model" of the self, where the legitimate self is independent and autonomous (1997, p. 36). However, disability theorists like Morris and Silvers have proposed an alternate view of the disabled self. Edwards argues that, if we accept that disabilities are not detachable from selves, "one will not be able to undertake the kind of separation that is required in order to claim that terminations of pregnancy on grounds of disability in the fetus imply nothing about the moral status of more developed humans who have disabilities" (1997, p. 32). For Edwards, then, the expressivist objection holds once we acknowledge that for existing persons with disabilities, their impairments are not separable from their "selves" as a whole.

This might suggest, however, that the validity of the expressivist objection rests entirely upon the subjective experiences of existing persons with disabilities. In refuting the "discrimination argument" (a version of the expressivist objection), Gillam argues that while a person with a disability might find certain assumptions deeply offensive, "feeling offended is not equivalent to being discriminated against" (1999, p.169). However, it is too simplistic to view the expressivist objection as only addressing a "personal insult" or offense perceived by persons with disabilities. In addition to raising concerns regarding the conflation of suffering and disability, the expressivist objection also challenges the assumption that disabilities are detachable from selves. This view of disability is problematic insofar as it ignores the lived experiences of many persons with disabilities. Moreover, when we consider mental or intellectual disabilities, the possibility of detachment seems even more unlikely. The commitment to the individualist/detachable disability model may explain why persons with disabilities that diminish autonomy and independence are no longer considered "fully human" or legitimate selves.[13] A striking example of this can be found in some animal rights literature. Peter Singer, for example, uses the "profoundly retarded" to make the case for better treatment of animals, and writes that "when we consider members of our own species who lack the characteristics of normal humans, we can no longer say that their lives are always to be preferred to those of other animals" (1995, p. 21).

3. PRENATAL PROTOTYPES

Philosophers like Buchanan and Purdy dismiss the expressivist objection that prenatal testing and selective abortion send the message "no more of your kind" based on the distinction between traits and persons. However, given the complex relationship between disabling conditions and self-identity described by persons with disabilities, the conceptual separation between qualities and persons is inadequate to justify dismissing the expressivist objection entirely. Furthermore, this distinction between qualities and persons cannot be made at the prenatal level. Even granting that medical genetics seeks to ·eliminate certain characteristics or abnormalities, and not to devalue the existence of *particular existing individuals*, the separation between qualities and persons is impossible when the only means of eliminating the trait is abortion. The very characterization of future persons *prenatally* for the purpose of selective abortion, then, rests upon this leap from genetic qualities to future persons, a fact that is not addressed by either Purdy or Buchanan. To elaborate upon this point, I will briefly consider a feature of classification that I believe is relevant to the expressivist objection: the generation of prototypes.

In his discussion of categorization, George Lakoff points to the fact that in some categories, certain members become representative of the whole category (1987, p. 41).[14] My interest in how prototypes are generated grows out of my study of mental retardation as a classification. There are both concrete examples of how at various points in history certain sub-classes of persons labeled "mentally retarded" became representative of the group as a whole,[15] and examples of prototypes in philosophical literature about mental retardation. Often in philosophical discussions, the severe case becomes representative of all "mentally retarded" individuals, and the construction of this prototypical case often rests upon stereotypical assumptions or over-simplifications about the nature of mental retardation. The "mentally retarded" as a group are also treated as the prototypical "marginal case" in moral theory, again in ways that fail to take into account the social, political, and internal complexities of this category (Carlson, 1998, pp. 120-59).

Similarly, prototypes can appear both in the practices surrounding prenatal testing and selective abortion, and in the literature debating the morality of these practices. The etiologic paradox of this new screening technology is that the genotype or chromosomal anomaly is visible prenatally, yet its phenotypic manifestation remains invisible until the child is born or years later (depending on the condition). I maintain that this indeterminacy creates the possibility of what I call prenatal prototypes: cases which are applied prenatally but are taken as representative of an entire class of future persons. The dynamics of how these prototypes are created and function in the clinical setting, and in the broader social fabric surrounding prenatal testing and selective abortion, are worthy of examination. For the purposes of this chapter, however, I will focus on how philosophical arguments regarding the morality of prenatal testing employ prenatal prototypes that often take the following forms: the morally unproblematic case and the severe case.

Fetuses diagnosed with a genotype or chromosomal abnormality that could cause a disability are often considered the "easy" cases when it comes to abortion; it is only when we consider the possibility of sex-selection or choosing other non-disability/disease related features (e.g., length of limb, eye color) that things get complicated. In their discussion of the ethics of sex-selection, Wertz and Fletcher argue that one of the strongest reasons to oppose prenatal testing for the purposes of selecting the sex of the child is that "it undermines the major moral reason that justifies prenatal diagnosis and selective abortion—the prevention of serious and untreatable genetic disease. Gender is not a disease" (Wertz and Fletcher, 1989, p. 24). The assumption is that in the case of diseases that are "serious and untreatable," prenatal testing is entirely morally justified. Yet, given that the severity of some conditions are unknown, what counts as a "serious disease," or *potentially* serious condition? Asch points to the relative acceptability among feminists of aborting fetuses with disabilities, as opposed to the vehement objection to sex-selective abortion: "I know of no feminist who countenances abortion for sex-selection. An overwhelming number, along with at least 80 percent of the nation, condone abortion for fetal 'deformities,' 'defects' or 'abnormalities'" (Asch, 1995, p. 387). It is indisputable that in some instances abortion would prevent a life of suffering. However, as discussed earlier, there is a tendency in moral arguments about selective abortion to conflate disability with profound suffering and pain and, in doing so, to apply prototypical cases of the future disabled person prenatally.

Another prototype is created when the "severe" case becomes representative of the entire range of a particular condition. Despite the fact that the severity is currently indeterminable prenatally, Down syndrome is often grouped together with other "severe" afflictions. Jeffrey Botkin argues that "the standard of disclosure for prenatal information should be designed to prevent harms to parents" (1995, p. 37). He says parents might reasonably expect significant harm in conditions that are fatal in childhood, cause chronic illness or repeated hospitalization, conditions that will not allow the child to achieve adult independence, and those that "are of such severity that there are constant demands on the parents for time, effort, and financial resources" (Botkin, 1995, pp. 37-8). He presents Down syndrome as the "prime example" of a condition where parents suffer, not because their child is suffering, but because of the support, time, and effort required (Botkin, 1995, p. 37). Despite the fact that the severity is unknown at the time of diagnosis, that seventy-five to ninety percent of persons with Down syndrome are capable of living independently of their families and are employable as adults, and that less than five percent are severely-to-profoundly mentally retarded (Elkins and Brown, 1995, p. 18), the prototypical future person with Down syndrome is presented as the severe case who is a drain and burden on his or her parents (echoing the personal tragedy model of disability to which many disability theorists object).[16]

How is this discussion of prototypes relevant to the expressivist objection? First, there is a danger that prototypes will be created (either in the clinical setting or in philosophical debates regarding prenatal testing) by relying upon stereotypes of persons with a particular condition and emphasizing the personal tragedy model of disability. This falls under the scope of the expressivist objection insofar as it promotes negative and perhaps erroneous images of persons with disabilities, which in turn may inform decisions regarding the nature and worth of future persons. Taking the most severe case as representative of a condition that has a wide range of phenotypic manifestation is one example of this.

An examination of how prototypes are constructed and applied prenatally points to the unique complexity of prenatal testing and selective abortion. Purdy counters the expressivist objection by claiming that disability theorists wrongly jump from qualities to persons. First, insofar as this distinction cannot be maintained prenatally, it is important to consider this leap in the specific context of prenatal testing and selective abortion; it is not enough to speak generally about "loving future persons." Furthermore, the creation of a prenatal prototype is far more complex than a simple jump from genotype x to phenotype/future person y. There is a host of social and normative factors that are involved in constructing the image of this future person, a process that should be closely examined in the context of clinical practice as well as in moral debates regarding prenatal testing and selective abortion and in arguments about what we want for future persons.

4. CONCLUSION

In making a series of clarifications and distinctions with respect to the expressivist objection, I have suggested that, at the very least, the expressivist objection be reconsidered and placed under greater scrutiny. First, there are valid concerns about the concrete practices surrounding prenatal testing and selective abortion, insofar as the clinical setting and broader research agendas are susceptible to expressions of devaluation regarding persons with disabilities. At the level of philosophical discourse, I have demonstrated how, in refuting the expressivist objection, some theorists perpetuate the very assumptions about persons with disabilities that this objection is meant to address. This occurs in a number of ways: by focusing on the prevention of suffering as the main reason for avoiding the birth of some future people, and in doing so equating disability with suffering; by perpetuating a personal tragedy model of disability; by dismissing or ignoring the personal accounts of persons with disabilities; and by failing to take into account the specificity of prenatal testing and selective abortion. Though I have not reconciled the two positions, the distinctions I have made are an attempt to clarify elements of the philosophical debate between those who believe there is a moral obligation to prevent the birth of certain kinds of individuals, and those who argue that prevention through prenatal testing and selective abortion is morally problematic. Thus, this chapter is a call both to broaden the context in which the expressivist objection is discussed by taking seriously claims made by persons with disabilities, and to narrow the focus and deal with prenatal testing and selective abortion as a particular case that raises a unique set of questions regarding the definition and treatment of future and existing persons.

Seattle University
Seattle, Washington, U.S.A.

NOTES

1.   I will not use the language of social construction in this chapter though it is popular among disability theorists. For a discussion of the merits and limitations of social constructivist discourse in the context of disability, see Carlson, 1998.
2.   For some examples of this literature, see Browne, Conners, and Stern, 1985; Davis, 1997; Fine and Asch, 1988; Lane, 1995; Mitchell and Snyder, 1997; Morris, 1991; Silvers, 1995; Oliver, 1990; Silvers, Wasserman, and Mahowald, 1998; Wendell, 1989, 1996. This list is by no means exhaustive.
3.   I agree with Anita Silvers (1995, 1998) that the 1990 Americans with Disabilities Act is legislative evidence of these changes, though some argue that the ADA instantiates the pathological view of disability, particularly with respect to genetics (Wolf, 1995).
4.   Evidence of the vitality of this debate may be found in a number of significant publications on the expressivist objection that have appeared since I formulated this chapter's argument, most notably a collection, *Prenatal Testing and Disability Rights*, edited by Erik Parens and Adrienne Asch (2000), and a Special Supplement of the Hastings Center Report (Special Supplement, 1999).
5.   As a non-disabled philosopher I do not pretend to speak on behalf of persons with disabilities. However, I think it is crucial to include the disability rights perspective in philosophical discussions concerning prenatal testing, and this chapter is an attempt to take such perspectives seriously.

6. Glenn McGee (1997) presents an alternate argument with respect to parental responsibility, outlining five parental sins that genetic testing makes possible.

7. I thank Lisa Parker for articulating this problem with Purdy's analysis.

8. Discussions have included the devaluation of disabled lives in ethical debates (e.g., reproductive technology, euthanasia, sterilization), literature, film, public policy, discrimination, and a host of societal attitudes and practices. See note 2.

9. I will not address the details and merits of arguments concerning whether one could prefer non-existence to existence.

10 Selective abortion following prenatal testing is only one of four possible kinds of interventions he discusses, and the only one that is vulnerable to the expressivist objection (Buchanan, 1996, p. 31).

11. For a discussion of whether prevention is/ought to be the aim of clinical genetics, see Clarke, 1991.

12. Brock's (1995) distinction between harms that result from same person choices, and harms that affect same number choices (choices that affect which child will exist) is relevant here.

13. Even Purdy, in her discussion of jumping from qualities to persons, makes an exception for cases of "mental disability." The use of the 'profoundly mentally retarded' in animal rights literature reflects this view of them as non-persons (Carlson, 1998, pp. 141-65).

14. Prototype effects were first defined by Eleanor Rosch, who conducted experiments and found that subjects judged certain members more representative of a category than others. For example, a robin was more representative of the category BIRD than a duck; desk chair more representative of CHAIR than rocking chair. Contrary to the classical view of categories, there are often "asymmetries among category members and asymmetric structures within categories" (Lakoff, 1987, p. 41). She also found asymmetry in generalization, where "new information about a representative category member is more likely to be generalized to non-representative members than the reverse" (Lakoff, 1987, p. 41).

15. In "Mindful Subjects," I trace the generation of prototype effects in the history of mental retardation as a classification. In the mid-nineteenth century when institutions for "idiots" were first developed, the idea of cure and treatment gained prominence, and the prototypical case was educable or at least "trainable." However, as the hereditarian explanation of feeblemindedness became dominant around the turn of the century, the prototypical feebleminded individual was one who came from a long line of "defectives" and who threatened to perpetuate his or her "bad stock." If one examines the history closely, it becomes evident that prototypes dictated the treatment for all individuals, regardless of the severity of their condition. Thus education gave way to institutionalization, and this was generalized to the entire category regardless of the severity of particular cases. For an excellent history of mental retardation, see Trent, 1994.

16. Mary White addresses the difficulties of generating lists like Botkin's (1999, p. 15). For an example of an appeal to the personal tragedy model presenting Down syndrome as a prototypical case in an argument for pre-embryo selection, see Persson, 1999.

REFERENCES

Asch, A. "Can Aborting Imperfect Children be Immoral?" In *Ethical Issues in Modern Medicine*, J. Arras, B. Steinbock, eds. Mountain View, CA: Mayfield Publishing Co., 1995.

Asch, A., Fine, M. "Shared Dreams: A Left Perspective on Disability Rights and Reproductive Rights." In *Women and Disabilities*, A. Asch, M. Fine, eds. Philadelphia: University Press, 1988.

Berube, M., *Life As We Know It: A Father, and An Exceptional Child*, 4th ed. New York: Pantheon Books, 1996.

Botkin J. Fetal privacy and confidentiality. Hastings Center Report 1995; 25(5):32-9

Brock D. The non-identity problem and genetic harms. Bioethics 1995; 9(3/4):269-75

Browne, S.E., Conners, D., Stern, N., eds., *With the Power of Each Breath: A Disabled Women's Anthology*. Pittsburgh: Cleis press, 1985.

Buchanan A. Choosing who will be disabled: genetic intervention and the morality of inclusion. Social Philosophy and Policy 1996; 13(2):18-46

Carlson, A.L., Mindful Subjects: Classification and Cognitive Disability. Ph.D. diss., University of Toronto, 1998.

Clarke A. Is non-directive counseling possible? Lancet 1991; 338(8773):998-1001

The Council on Ethical and Judicial Affairs, American Medical Association. Multiplex genetic testing. Hastings Center Report 1998; 28(4):15-21

Davis, L., ed., *The Disability Studies Reader*. New York: Routledge,1997.

Edwards S. The moral status of intellectually disabled individuals. Journal of Medicine and Philosophy 1997; 22(1):29-42

Elkins T., Brown D. Ethical concerns and future direction in maternal screening for Down syndrome. Women's Health Issues 1995; 5(1):15-20

Gillam L. Prenatal diagnosis and discrimination against the disabled. Journal of Medical Ethics 1999; 25(2):163-71

Kaplan D. Prenatal screening and its impact on persons with disabilities. Clinical Obstetrics and Gynecology 1993; 36(3):605-12

Lakoff, G., *Women, Fire and Dangerous Things: What Categories Reveal About the Mind*. Chicago: University of Chicago Press, 1987.

Lane H. Constructions of deafness. Disability and Society 1995; 10(2):171-89

Linton, S., *Claiming Disability: Knowledge and Identity*. New York: New York University Press, 1998.

Lippman A. Prenatal genetic testing and screening: constructing needs and reinforcing inequities. American Journal of Law and Medicine 1991; 17:15-50

Mitchell, D., Snyder, S., eds., *The Body and Physical Differences*. Ann Arbor: University of Michigan Press, 1997.

McGee G. Parenting in an era of genetics. Hastings Center Report 1997; 27(2):16-22

Morris, J., *Pride and Prejudice: Transforming Attitudes to Disability*. Philadelphia: New Society Publishers, 1991.

Oliver, M., *The Politics of Disablement*. London: Macmilan, 1990.

Parens, E., Asch, A., eds., *Prenatal Testing and Disability Rights*. Georgetown: Georgetown University Press, 2000.

Paul, D. "Eugenic Anxieties, Social Realities, and Political Choices." In *Are Genes Us? The Social Consequences of the New Genetics*, C.F. Cranor, ed. New Brunswick, NJ: Rutgers University Press, 1994a.

Paul, D. "Is Human Genetics Disguised Eugenics?" In *Genes and Human Self-Knowledge: Historical and Philosophical Reflections on Modern Genetics*, R. F. Weir, S.C. Lawrence, E. Fales, eds. Iowa City: University of Iowa Press, 1994b.

Persson I. Equality and selection for existence. Journal of Medical Ethics 1999; 25(2):130-6

Purdy, L. "Loving Future People." In *Reproduction, Ethics, and the Law*: Feminist Perspectives, J. Callahan, ed. Bloomington, IN: Indiana University Press, 1995.

Purdy L. What can progress in reproductive technology mean for women? Journal of Medicine and Philosophy 1996; 21(5):499-514

Rothman, B., *The Tentative Pregnancy: How Amniocentesis Changes the Experience of Motherhood*. New York: Knopf, 1985.

Silvers A. Reconciling equality to difference: caring (f)or justice for people with disabilities. Hypatia 1995; 10:1

Silvers, A. "Formal Justice." In *Disability, Difference, Discrimination: Perspectives on Justice in Bioethics and Public Policy*, A. Silvers, D. Wasserman, M. Mahowald, eds. Lanham, MD: Rowman & Littlefield Publishers, Inc., 1998.

Silvers, A., Wasserman, D., Mahowald, M., *Disability, Difference, Discrimination Perspectives on Justice in Bioethics and Public Policy*. Lanham, MD: Rowman & Littlefield Publishers, Inc., 1998.

Singer, P., *Animal Liberation*. London: Pimlico, 1995.

Special Supplement. "The Disability Rights Critique of Prenatal Genetic Testing: Reflections and Recommendations." Hastings Center Report 1999; 29(5)

Steinbock B., McClamrock R. When is birth unfair to the child? Hastings Center Report 1994; 24(6):15-21

Trent, J.W., *Inventing the Feeble Mind*. Berkeley: University of California Press, 1994.

Tucker B.P. Deaf culture, cochlear implants, and elective disability. Hastings Center Report 1998; 28(4):6-14

Wasserman, D. "Distributive Justice." In *Disability, Difference, Discrimination: Perspectives on Justice in Bioethics and Public Policy,*, A. Silvers, D. Wasserman, M. Mahowald, eds. New York: Rowman & Littlefield Publishers, Inc., 1998.

Wendell S. Toward a feminist theory of disability. Hypatia 1989; 4(2):104-25.

Wendell, S., *The Rejected Body: Feminist Philosophical Perspectives on Disability*. New York: Routledge, 1996.

Wertz D., Fletcher J. Fatal knowledge? Prenatal diagnosis and sex selection. Hastings Center Report 1989; 19(3):21-7

White M.T. Making responsible decisions: an interpretive ethic for genetic decision making. Hastings Center Report 1999; 29(1):14-21

Wolf S.M. Beyond genetic discrimination: toward the broader harm of geneticism. Journal of Law and Medical Ethics 1995; 23(4):345-53.

CHAPTER 11

ANITA SILVERS

MELIORISM AT THE MILLENNIUM: POSITIVE MOLECULAR EUGENICS AND THE PROMISE OF PROGRESS WITHOUT EXCESS

1. MEDICINE'S MELIORIST HERITAGE

1.1 The Promise

In a volume of elegant essays entitled *Democracy and DNA: American Dreams and Medical Progress*, the belletrist physician Gerald Weissmann identifies the application of genomics to molecular medicine as the most lasting and powerful expression of the century-old social movement called meliorism (Weissmann, 1996). This social movement was inspired by the conviction that the quality of human life can be improved through our own intelligent efforts. From a melioristic standpoint, raising people with less than species-typical capabilities to normal and raising normally functioning people to new heights of capability are equally enhancing. As a correlative benefit, human suffering occasioned by current lacks in human understanding and achievement could abate under such a program. Yet, enhancing humans' capabilities is more directly a melioristic goal than relieving their suffering.

In our new century, genomics seems an enormously promising instrument for making people's lives better. As Weissmann points out, scientists and laypeople alike evince melioristic confidence in the beneficial applications of genomics. There is enormous public support for the prospect of liberating unfortunate individuals for whom, until now, inheriting an atypically limiting biological endowment has meant a life of biologically destined disadvantage. Raising some of those worst off to a life of normal opportunity seems an admirably democratic end.

L.S. Parker and R.A. Ankeny (eds.), Mutating Concepts, Evolving Disciplines: Genetics, Medicine and Society, 215-234.

Nevertheless, the melioristic outlook that invites public support of the technology of engineering human genes strikes some thoughtful observers as troubling. In this chapter, I will explore why certain melioristic medical applications of genomics have been thought excessive rather than progressive. In particular, I will assess policy proposals to constrain or ban them. Contrary to the proponents of these proposals, I will argue that permitting such applications can be progressive rather than excessive.

1.2 The Threat

The history of the melioristic programs initiated a century ago is a source of worry. Meliorism commits us to employing reason and its products to transform how we live so that human well-being is enhanced. To further this end, meliorists cultivated hygienics and eugenics, conceiving of these as complementary policies for promoting physical well-being through broad, publicly supported applications of advancing medical knowledge. Indeed, the novelist George Eliot, who named the movement, considered medicine the paradigm of meliorism because it puts intellectual achievement at the service of people by directly improving their lives.

Eliot was right. Medicine clearly is meliorist in temperament. Meliorist convictions are an important source of medicine's energy, but they also introduce an ominous potential for transgressive medical policies that violate moral boundaries. Meliorism has been the eugenics movement's driver it from nineteenth-century inception, and melioristic objectives suffused the rationales, and nourished the excesses, of that familiar list of horribles linked to eugenics.

These are the programs carried out in the name of enhancing the level of collective human performance by removing underachieving performers—namely, negative eugenics programs like the involuntary sterilization or termination of putatively warped and burdensome kinds of people, the prohibition of interracial marriages, and the banning of immigration from parts of the world designated by the 1924 U.S. Immigration Restriction Act as having "biologically inferior" indigenous populations (Duster, 1990, p. 13). Enjoining immigration from Eastern Europe because feeble-mindedness was supposed to be endemic among Jews (Duster, 1990, p. 13), extirpating the reproductive capability of individuals imagined to carry inheritable dispositions to perform poorly or behave dangerously (Burleigh, 1994; Duster, 1990, pp. 29-31; Gallagher, 1990; Miller, 1996; Proctor, 1988), and eliminating fetuses or neonates deemed so defective as to make their existence onerous are just a few of the schemes devised to further negative programs of eugenics (Pernick, 1996).

As a consequence of this history, eugenics now is identified as a genocidal practice conducted by dominant or strong classes and aimed at eliminating people who are targeted as belonging to inferior or weak classes (Wertz, 1999a). Weak classes consist of people whose biological properties are taken as proxies for such social traits as disruptiveness, dependence, and vulnerability. In contrast, strong classes consist of people whose biological properties are taken as proxies for such social traits as productiveness, self-sufficiency, and power.

Meliorism argued for reducing or erasing the incidence of these devalued social traits in order to improve the quality of community life. Equating socially devalued traits with biological properties found in only a minority of the population, eugenics theory strategized that an effective route for doing so was to contain the biological transmittal of undesirable properties by constraining or eliminating their carriers. With excessive faith in the doctrine that biology is destiny, eugenics thus urged the adoption of medical interventions to alter the course of biological inheritance. These included diagnosing individuals as socially undesirable on the basis of physiological properties and preventing such individuals from reproducing by confining, sterilizing, or euthanizing them.

2. EUGENICS: THE NATURE OF ITS EVIL

2.1 Negative Eugenics

Eugenics programs are now so notorious they have become symbolic of how science invites us to overstep the boundaries of proper human intervention. In this regard, it has become virtually automatic to scrutinize new proposals for altering human genetic material to guard against any further venture into eugenics. Scarcely any contemporary bioethical conversation on the general subject of the benignity or peril of inserting genes into human somatic cells or germ cells omits referencing the Nazi practice of euthanizing "defective" Germans or the U.S. practice of sterilizing "defective" Americans, notoriously endorsed by the Supreme Court in Buck v. Bell (Colker, 1995, pp. 3-6). To illustrate how widespread and palpable is fear of the deleterious consequences of human intervention in biological inheritance, we might note the Council of Europe's call for "explicit recognition...of the right to a genetic inheritance which has not been interfered with" (Council of Europe Parliamentary Assembly, 1982).

Why the practice of eugenics has been so egregiously disposed is something of a puzzle though. For initially, eugenics was meant to put the science of human inheritance at the service of enhancing human lives. Was it simply unfortunate historical confluences of malignant cultural and social factors that diverted melioristic intentions so as to make the road to the eugenic hells which history documents so broad and inviting? Or is eugenics' characteristic method—attempting to elevate the general level of human performance through mechanisms suggested by or derived from the science of human biological inheritance—inescapably dangerous? That is, did eugenics programs of the past accidentally precipitate evils that could be avoided in a less biased society? Or are eugenics programs inherently evil?

2.2 Positive Eugenics

Is this deplorable history of *negative* eugenics, which is aimed at *preventing* the existence of people less capable than people usually are, proof for an adverse assessment of a different eugenics strategy? *Positive* eugenics aims at *promoting* the existence of people more capable than people usually are. Originally, positive eugenics programs focused on inducing the healthiest people to increase the number of their children.[1] When we shift from the level of phenotype to the level of genotype—that is, to the domain of molecular medicine—positive eugenics programs alter individuals by making inheritable modifications to modify or elevate certain aspects of biological performance.

In light of the historical record of oppressive practices carried out in the name of eugenics—such as the sterilizing or euthanizing of people with disabilities—is the very idea that eugenics can be positive an oxymoron? Some argue that all eugenic practices display the morally problematic aspects of negative eugenics. Witness debates about the problematic termination of pregnancies where pre-natal testing identifies genetic deficits in the fetus. This practice is heavily criticized by many disability advocates and their allies, who say that the practice of terminating fetuses in virtue of their predicted disability is meant to rid the population of the disabled and therefore expresses the same evil as sterilizing or euthanizing people with disabilities (Kaplan, 1989).

Some go further and insist that genetically altering individuals so as to surmount inherited biological flaws is tantamount to rejecting the lives of people with disabilities because doing so alters the inherited traits that make them who they are—that is, identify them as the progeny of ancestors who carried genes for these traits (CAHGE, 1998). (People who are "deaf of deaf"—that is, whose social status in Deaf society is elevated because they are third or fourth generation deaf—constitute only one of several groups that sometimes press this claim.)

This thought is expanded into the charge that whatever public health initiatives aim at, adjusting the molecular conditions occasioning these traits cannot help but promote an oppressively exacting standard of biological function. Such a standard, it further is claimed, is impossible for anyone with a corporeal or cognitive impairment to meet. Consequently, such a standard threatens people whose genetic heritage incorporates a heightened potential for their being corporeally or cognitively impaired (CAHGE, 1998).

Responding to such a broad-brushed identification of genetic alteration with negative eugenics involves ascertaining whether eugenics is only coincidentally transgressive or is inherently so. In what follows, I will begin by considering what is traditionally thought to be the sole justification for medical intervention, namely, that the goal of medicine is to restore or maintain health, where health is equated with species-typical modes and levels of human biological functioning. But the aptness of this equation cannot be assumed.

Following upon this analysis, we will see how molecular medicine's techniques and possibilities may alter the strategies that traditionally have been equated with eugenics. In respect to eugenics, history shows that eugenic interventions have been executed in the name of raising collective humanity to a standard of biological functioning that rises above the level currently taken as commonplace. Eugenics' strategy has been, at core, homogenizing. Raising the level of collective performance has been pursued either by reducing the percentage of low-performing members by eliminating them or the progeny who would inherit their deficits, or by promoting increases in the progeny of certain high-performing individuals in the population.

Molecular medicine appears to offer eugenics an alternative strategy to the eliminative program cited above. Molecular medicine promises to elevate individuals' performance by raising certain aspects of their biological functioning above the currently commonplace level without necessarily homogenizing them. Nevertheless, promoting any such enhancement is condemned by people who associate it with eugenics.

To the contrary, I will argue, the association of enhancement at the molecular level with eugenics is apt, but categorically condemning enhancement is not so. Far from eliminating low-functioning members of the human collective, applying genetic medicine so as to compensate for an inadequate biological process by inducing elevated performance sometimes promises both to increase these individuals' presence in the population and enlarge their opportunities for social participation. Thus, the use of molecular medicine to achieve the goals of positive eugenics has the potential to shatter the divisive conceptualization of strong and weak classes that informs and drives negative eugenics.[2]

Very broadly, molecular medicine could, if practiced beneficially, contravene a central assumption of the conceptual framework within which the bifurcation of humans into strong and weak classes flourishes. By extending the lives of individuals with certain genetic conditions into the reproductive years, or by altering their germ cells, molecular medicine may change the distribution of the biological properties traditionally thought to mark weak classes. By changing this distributive process as well as its outcomes, molecular medicine could challenge a doctrine central to negative eugenics' thesis about the immutability of the differences between strong and weak classes. Whether it does so will depend on whether its practice is informed by negative eugenics, which is pursued at the molecular level to search for individuals to keep out of the population, or by positive eugenics, which at the molecular level seeks ways of enhancing individuals' participation in the population.

3. SHOULD NATURE SET BOUNDARIES FOR MOLECULAR MEDICINE?

3.1 A Naturalistic View

Bioethicist Daniel Callahan speaks for the many people troubled by meliorist excess when he warns that medicine must not be "in the business of promoting the...pursuit of general human happiness and well-being.... [M]edicine should limit its domain to promoting and preserving human health" (1992, p. 52). Further, Callahan cautions, we must respect nature's domain. Thus, biology should not be displaced by human agency.

So, for example, we should not employ medicine to improve humans' longevity but should, instead, accept the average life span as a natural fact (Callahan, 1988). From this perspective, eugenics programs, whether negative or positive, dangerously breach the boundaries of proper medical practice because they pursue a goal not dictated by, or at least not limited by and to, the natural facts of human biology. That is, because eugenics programs are dedicated to enhancing human performance as well as to maintaining the incidence of health, they cannot help but overstep the domain of medicine.

Initially, Callahan's definition of medicine appears to solve the problem of identifying the boundaries of medicine. It does so by adopting a self-limiting definition of appropriate medical intervention, one that serves as a stipulative safeguard against programmatic excess. For if medicine is understood narrowly as aimed only at the maintenance of health, medical intervention should be initiated only to restore or maintain individuals at the common level of biological performance, but never to further elevate it. When health is unthreatened or has been restored, there is no impetus for continued intervention.

In contrast, interventions with the broader goal of advancing human beings' happiness and well-being—or, for that matter, elevating the general level of human performance—do not similarly appear to limit themselves. Where enhancement of the level of biological performance, rather than its restoration or maintenance, motivates intervention, the process appears to lack any element of self-containment and self-regulation. For in respect to human capabilities, there seems to be endless room for improvement. Melioristic programs thus seem inherently disposed to such open-endedness as to make excess a constant peril. Viewing enhancement as potentially excessive in turn suggests that positive eugenics is much too risky to be acceptable moral policy, advisable public policy, and responsible medical policy.

But this argument begs the question. For it equates health with the individual's original state—prior to deterioration, disease, or injury—and assumes that (re)establishing the patient in this natural condition is the unquestionable *desideratum*. However, it is odd to think that maintaining a patient's original, albeit dysfunctional, state could be an aim of medicine. What should be the aim of medicine in respect to individuals who are naturally endowed with potentially dysfunctional biological traits? Traditionally, in such circumstances, medical intervention seeks to improve on the patient's original condition in order to make it approximate that typical of humans.

Callahan (1988, 1995) does not agree, for he opposes expending medical resources to restore capabilities that deteriorate through aging. It is, of course, a short slide from insisting that sub-optimal functioning associated with aging is a natural boundary medicine should not transgress to contending that other inherited limitations also are so because they are natural developments of constitutents of the patient's original state. That there is no bright line between naturally induced kinds of sub-optimal functioning that deserve intervention and those that do not indicates that applications of his distinction may be of no practical help.

Further, if intervening to alter a potentially dysfunctional trait, why not aim to include the patient in that group of individuals who function at a high-normal rather than low-normal level? In any single case, it is difficult to imagine a good basis for opposition to such a practice. After all, it privileges the individual patient only slightly by shifting the prognosis in her favor to make it more likely that her condition will be repaired in a way that results in her performing slightly above, rather than slightly below, average performance. On balance, it seems only fair that any imprecision in projecting the outcome of a medical procedure be designed to the patient's advantage rather than her detriment. For instance, in intervening to extend an older patient's life, it seems wiser to make the intervention as effective as possible rather than to craft it so that the patient will live, but not so long as to exceed the average expectancy of life.

Notice that what is typical of any species cannot help but improve if more and more members of the group display higher and higher concentrations of biological properties that enhance their functionality. Consequently, taken in the aggregate, the reasonable and humane practice of aiming to restore patients to high-normal rather than low-normal performance eventually would raise the common level of performance. Technically, such a practice would be melioristic, regardless of whether pursued by encouraging optimally functional individuals to reproduce (the method of traditional positive eugenics) or altering genes to make more people function optimally (the method of molecular positive eugenics). With more and more citizens advanced to somewhat better than average health, more and more will perform better, so that the average level of performance in the population is raised.

3.2 The Moral Distinctiveness of Genetic Technology

In an influential analysis of the moral and social dimensions of "biologists' newly gained ability to manipulate...the material that is responsible for the different forms of life," published in 1982, the President's Commission for the Study of Ethical Problems in Medicine and Biomedical and Behavioral Research assessed "interventions aimed at enhancing 'normal' people." These were convicted of being problematic because "the difficulty of drawing a line suggests the danger of drifting toward attempts to 'perfect' human beings once the door of 'enhancement' is opened" (President's Commission, 1982, pp. 1, 3). But the President's Commission was not much more precise than this about the nature of the problem. In expanding on the reported concerns, the document offers several very different illustrations that introduce quite distinct locuses of concern.

Some of the prospectively disturbing consequences mentioned in the report are clearly peripheral to the character of the technology. For instance, the prospect of policies mandating medical intervention into pregnancies to repair fetuses predisposed to certain inherited genetic conditions is cited with deserved concern

(President's Commission, 1982, p. 66). But as the report goes on to say, there already are circumstances where a pregnant woman's right to refuse treatment is over-ridden in view of the medical condition of the fetus she carries. It is clear, then, that genetic technology is not the source of what is problematic about this kind of procedure. Rather, genetic technology is simply a contingent instrument; like earlier medical technologies such as the treatments for certain highly infectious diseases, genetic technology has the potential to be used as a vehicle for restoring health at the expense of a person's democratic right to self-determination or her right to privacy. As such, even when it is used to promote the existence of certain kinds of people— for instance, through altering fetuses so they become neonates who are not compromised by genetic diseases—this concern offers no reason to construe molecular medical technology as inherently dangerous.

In quite a different vein, the report goes on to claim that unlike previous medical methods, genetic technology poses a direct challenge to a traditional tenet of democratic political morality. For as humans gain the ability to alter inherited health conditions, it is speculated, we unsettle the conceptual foundations of social justice theory. Genetic engineering does so because it transforms the "natural lottery" into just another kind of "social lottery" (President's Commission, 1982, p. 67). That is, as our rapidly advancing knowledge of genetics offers new scope for intervening in humans' inheritance of biological traits, the indifferent natural forces that govern biological inheritance no longer will determine who is to be disadvantaged by malfunctioning biological processes, and who is to benefit from effectively functioning biological processes. Instead, biological advantage will be procurable by those already socially privileged but will be beyond the reach of those who are not. Thus, which individuals are disadvantaged by their biology could become as much a matter of social choice as who is to be marginalized politically, culturally, or economically. Suffering from a disadvantageous biological inheritance then would be as much a product of biased social arrangements, and thereby as unfair, as being oppressed because one has inherited an inferior social rank.

4. SHOULD EQUALITY SET BOUNDARIES FOR MOLECULAR MEDICINE?

4.1 Biological Privileging

Traditional democratic political morality is suspicious of social processes that advantage some individuals over others. Restoring ill or injured individuals to the level of performance with which they were originally endowed appears neither to advantage nor to disadvantage them beyond the competitive positions they originally acquired through the natural lottery. Even if treatment sometimes restores patients to a privileged rather than an average position—for instance, restores to a baseball player his ability to throw accurately at such great speed that his services earn him millions of dollars—the social mechanisms that prompt developing and allocating the treatment do not entrench unearned advantage by similarly privileging the physical prowess of the patient's heirs. (The literature on genetic enhancement does not find inheriting acquired financial advantage as problematic as inheriting acquired genetic advantage, despite their being equally unearned.)

In contrast, artificially improving some people's biological performance above the level apportioned to them in the natural lottery that originally determined their genetic inheritance has seemed to offer the beneficiaries of such technology a socially contrived rather than a natural advantage. Heretofore, in a democratically organized competitive society in which vigor, industry, and talent theoretically constitute the paramount determinants of success, individuals' natural biological and moral endowments have been regarded as relatively impervious to the influence of social position and thereby as offering an antidote to artificially induced social privilege. (In the last half of the twentieth century, we began to attribute the competitive disadvantages associated with certain biological differences to biased social arrangements, but this enlightened perspective does not reach much beyond differences in pigmentation and secondary sex characteristics.) But gene transfer technology purportedly makes humans' most fundamental biological characteristics so malleable that biological superiority could be realigned so as to be a product of, rather than a constraint upon, the privilege social rank or power bestows.

If this is the case, positive molecular eugenics programs appear to imperil democratic values by magnifying the competitive advantage social perquisite confers. Fears of an increasingly tilted playing field thus have suggested prohibiting gene transfer protocols that enhance rather than merely restore individuals' biological performance. In response, I will argue that restoration is itself too rigidly narrow and oppressive a standard for distinguishing between morally acceptable and transgressive applications of gene transfer technology. I then will extend this line of thought by applying it to positive molecular eugenics, understood as a program that improves people by altering them genetically. I will show that positive molecular eugenics programs inherently benefit the kinds of individuals for whom the dynamic between inherited biological performance and acquired social status heretofore has been prejudicial. From a perspective informed by the considerations I will raise, positive molecular eugenics inherently furthers rather than fragments democratic aims.

4.2 Germ-line Impact

In rare cases it now is possible to alter human performance through gene transfer that improves on an individual's somatic inheritance. In the best known of these protocols, gene transfer techniques have restored missing DNA segments to a few individuals born with severe combined immune deficiency (SCID), adding a protein producing mechanism lacking in these individuals' genetic inheritance. Well-publicized and heavily capitalized efforts are underway to develop many more such protocols, and the gene therapy industry has not yet emerged from infancy.

Thomas Murray proposes that "the most important question in the debate over the ethics of gene therapy is whether gene therapy is ethically distinctive from other forms of medical therapy...[and that] the ethically distinctive element of gene therapy is only characteristic of germ-line manipulation" (1991, p. 484). Although, as of this writing, the National Institutes of Health endorses only therapeutic protocols that alter somatic cells (Department of Health and Human Services, 1986) and the pharmaceutical industry has directed its efforts at somatic-cell engineering, there is no reason in principle why germ-line engineering might not occur, whether intentionally or by accident.

Accidents like this can happen. For example, an early effort to repair defective eye color in fruitflies by inserting the gene for a missing enzyme inadvertently created a heritable repair (U.S. Congress Office of Technology Assessment, 1984, pp. 17-8). Furthermore, the success of somatic therapies well may have outcomes that make subsequent germ-line intervention attractive to cost-conscious policymakers. For, as LeRoy Walters (1986) observes, successful somatic-cell therapies eventually may permit many more individuals with dysfunctional biological conditions to live to an age at which they reproduce.

However, even in the absence of effective therapy for these specific conditions, medical advances in areas such as the mechanical assistance of breathing and the control of respiratory infections already have reduced early (pre-reproductive) mortality in individuals with such conditions as cystic fibrosis and spinal muscular atrophy, and presumably will continue to do so. Thus, whether prompted by the advent of successful somatic-cell therapies or by improvements in non-molecular therapies, such an eventuality will necessitate either greater health care resource expenditures to intervene in the cases of the increasingly large number of individuals inheriting these conditions, or else the introduction of germ-line alteration that can transmit therapeutic benefits to succeeding generations.

It sometimes is argued that the most benign approach consistent with such considerations of efficiency would be to eliminate genetically defective embryos, and to provide sperm or egg donation for parents for whom the probability of defective embryos is too high (Arras and Steinbock, 1995, p. 428). Yet parents may quite reasonably prefer transmitting their own advantageous characteristics to their offspring rather than risking the child's inheriting a less gratifying aggregate of the traits of one or more donors, as long as they can avoid transmitting dispositions to defectiveness to their offspring. Moreover, while germ-line alterations pose uncertain risk (as well as uncertain benefit) to future children, it is becoming clear that egg donation and egg reception procedures also carry risks. Among other problems, we have not had the opportunity for longitudinal study of the effects on women of the various interventions that implement these procedures, especially when no resulting pregnancy is brought to term. Thus, we cannot assume that germ-line alteration automatically puts people at more peril, or puts more people at peril, than the use of reproductive technologies as an alternative for preventing the transmission of genetic defects by individuals who want to have their own (in some sense) biological children.

Consequently, the thrust of my discussion applies equally to somatic and to germ-line alterations. Contrary to Murray's proposal, if gene transfer technology is ethically distinctive from other forms of medical intervention, the implications which confer its special moral status hold for both kinds of applications. One is not inherently more or less morally problematic than the other. Further, if there is a sense in which positive eugenics is not an oxymoron, it will be equally intelligible in respect to somatic and germ-line intervention.

4.3 Genetic Therapy and Genetic Improvement

The public, through its policymakers, has found cause for concern at the prospect of positive eugenics programs employing the techniques of molecular genetic technology. The President's Commission (1982) urged that any opening of the "door of 'enhancement'" could also initiate attempts to "'perfect' human beings." Applications that enhance are problematic, the report warns, because it is difficult to draw a firm and objective line that bars their being used for nontherapeutic purposes. This caution constrained subsequent research policy.

It is unclear, however, that the 1982 Report's distinction between therapy and enhancement plausibly separates troubling from innocuous uses of molecular genetic technology. Previously, I showed that meliorist proposals to apply genetic engineering have been rejected as morally troubling and politically threatening because they appeared to disconnect us from fundamental principles of democratic morality. Accordingly, we should recognize that in this public conversation about which forms of genetic intervention are distinctively transgressive, political and moral considerations take precedence over medical ones. In this regard, the question becomes whether interventions that aim so broadly as to raise aspects of humans' biological performance to more than ordinary proficiency inherently oppress members of groups purported to be biologically deficient. If this is so, the line between troubling and innocuous interventions at the level of molecular genetics is as the President's Commission supposed it to be. To the contrary, I shall argue, it is instead the Commission's too narrow delineation of medicine's aims that threatens to be oppressive.

4.4 The Standard of Normality

The Commission defines the permissible medical applications of genetic technology as those that restore the patient to familiar functional levels and modes rather than enhance the patient by facilitating elevated performance. The most thoroughly argued defense of this conviction that it is democratic to use medicine to promote normal functioning is found in the influential work of Norman Daniels (Satz and Silvers, 2000). Daniels notices that our rapidly growing technological proficiency permits us to recast the disadvantages occasioned by illness or injury. To the extent that these now are avoidable through preventative or curative medical care, people who suffer them now may be considered victims of remediable biases—thoughtless or corrupt healthcare resource distribution—rather than victims of inescapable misfortune. Daniels' view suggests that individuals disadvantaged by the outcomes of an illness or injury need not be suffering but for lack of funding for research, or for inadequate allocation of preventative, reparative, or compensatory health care (Daniels, 1985, 1987).

By focusing medical intervention on treatment that restores or raises diseased or injured people to common or normal functioning, Daniels intends to specify what kind of intervention will improve people's lives and elevate human happiness generally while avoiding excess. The idea is to designate the level of health care citizens deserve without diagnosing all disadvantageous performance as pathological and thereby medicalizing it. For Daniels (1987), melioristic medical programs are sufficiently constrained in their service to suffering people as long as they address deficits in typical or "normal" functioning and do not attempt to elevate us, individually or collectively, above the species-typical functioning common and familiar to us. The argument for this level of constraint consists not in an appeal to natural boundaries, similar to Callahan's appeal, but rather in a warning against exaggerating social inequalities by enabling already privileged individuals to acquire greater than normal capabilities.

In the absence of a firm and fair standard for the correction or "setting right" of functioning, however, Daniels' line is too abstract to be practical and too arbitrary to be fair or advisable. To imagine that we can separate interventions that restore from interventions that improve, and give permission and priority only to the former, is to assume that there is a morally non-controversial, politically non-privileging standard for determining when an intervention is therapeutic rather than aggrandizing. The standard usually proposed is so-called normal healthiness, a notion that is applied in some confusion to physical or mental conditions, also to the modes and levels of the performances affected by these conditions, and as well to the functionality such performances effect. How fair a standard for authorizing intervention and assessing its success is normal healthiness? As I now go on to show, the standard Daniels and others agree on as defining therapeutic goals is neither firm nor fair.

Daniels equates being normal—that is, functioning at normal levels and performing in normal modes—with being typical of one's group or species. Being typical is supposed to be the definitive functionally effective condition. The functional organization typical of the species is imagined to be the one best suited to meet our biological goals, so that to fall away from this standard is to be definitively disadvantaged in meeting these goals (Daniels, 1987). Moreover, facilitating typical modes and levels of performance, rather than singular ones, is presumed to be the invariably reasonable standard for arranging our social environment. These or similar valorizing assumptions about the advantage of being biologically typical or normal must be accepted if we are to justify the scale on which restoration eclipses improvement and becomes the sole permissible objective of medical intervention.

These assumptions are not perspicacious, however. First, we surely are unprepared to abandon all attempts to apply medical technology to improve on nature. Vaccination programs are a familiar example of compensatory medical strategy that enhances our physical performance beyond the level that is native to our species. Although now considered to be unexceptional and to provide the protection deserved by fragile populations like the elderly and the very young, early in their history vaccination protocols were widely denounced as dangerous attempts to elevate individuals above species-typical states. Undoubtedly, the vaccination process may have undemocratic results. Freeing some but not all people from the consequences of an illness clearly makes those protected from widespread illness more competitive.

Further, it now is unexceptional to suppose that some gene therapeutic protocols may raise the performance of one function above the common level to compensate for the adverse impact of other performances. For instance, one approved gene transfer protocol enhances the low-density lipoprotein receptor to above normal to achieve an acceptably low level of cholesterol in the context of a genetic condition that expresses as hypercholesterolemia. At the time this protocol was proposed, it did not seem more threatening, or less natural, than the SCID protocol that aimed at restoring a typical genetic sequence in patients whose biological inheritance omitted it (RAC, 1997).

Second, the assumptions leave us confused about how, for policy purposes, to categorize performances that are elevated above the common or normal level but are not the result of unusual medical intervention. For instance, it is not common for people to be cognitively competent when they surpass a hundred years of age; yet, a few people are so. Is it abnormal for them to be so? It clearly is atypical, but should we therefore eschew trying to develop artificial means of giving everyone the somatic or germ-line genetic configurations that up to now have naturally facilitated a few fortunate people's being so?

Success in extending the durability of people's cognitive competence could not help but provoke some amount of social disruption, for very many of our current arrangements are predicated on tying cognitive incompetence to advancing age. Of course, it might be argued that raising people above the level of functioning currently typical of the species is unfair because it redistributes more opportunity to whoever is enhanced. But many advances in compensatory medical technology have had a similar impact, as individuals whose physical or cognitive conditions formerly displaced them from social participation improve their capacity to function and consequently seek to reform practices they perceive to be barriers to their further flourishing. So, for example, the development during the last half-century of mechanical devices that effectively compensate for compromised capacities to mobilize led to widespread reform of the built environment to permit wheelchair users to participate equitably in civic and commercial activities.

Third, because normal performance is defined with reference to familiar modes and levels of performance, what is thought of as normal often is artificially skewed by patterns of social domination that favor some kinds of performances over others and consequently ensure that those will continue to be the most common and thereby the seemingly normal ones. For instance, the domination of individuals for whom text is the most efficient conveyer of information has led to social arrangements that presume the ability to read text is normal. This bias disadvantages whoever is far better at grasping pictorial communication than at comprehending the linear activities of reading and writing texts. Consequently, energy and expenditures are applied to altering people in the latter group so they can perform as normal readers, at a sometimes enormous and painful cost. Such costs might not be extracted if normalcy were less excessively prized.

4.5 Enhancing Strengths, Compensating for Weaknesses

We should neither build such domination into the policy that governs the development of medical interventions, nor permit policy to further the conflation of artificial social disadvantage with natural functional disadvantage. Unfortunately, the 1982 Report fails to distinguish, with sufficient precision for policy purposes, among the different kinds of objectives that procedures for elevating performances above the common or normal level may advance. Therefore, the Report does not come to grips with how enhancement may avert, and even may rectify, artificially imposed social disadvantage.

Of course, enhancements might be allocated so as to disadvantage certain classes of people, but appropriately applied enhancements can further fairness. While improving oxygenation mechanisms above normal would unfairly advantage the rich athlete, it might equalize opportunity for the poor athlete who cannot afford to train in Kenya, or for the African-American athlete with sickle disease who must play part of the season far above sea level. Treatment and enhancement are two approaches to the same goal of reducing disadvantage. Each has appropriate applications, as well as ineffective or dangerous ones, which policy should discourage. One strategy is not automatically more suspect than the other.

In this regard, it is morally troubling when the conceptual framework we adopt to guide therapeutic policy assigns priority to restoring individuals to the predominant way of functioning instead of promoting their successful, if idiosyncratic, functioning. Should anyone doubt that this is so, I invite them to consider the last sixty years of treatment for children with cerebral palsy, polio, congenital limb loss, and similar orthopedic conditions, in which interventions to make them appear more normal or perform in normal modes were systematically permitted to compromise their functionality. An illuminating illustration is the imposition of dysfunctional prosthetic arms on children with thalidomide-shortened upper extremities who were better served by enhancing the dexterity of their feet and toes so as to perform manual functions. Yet in an effort to make them depend on their prosthetics, their feet were placed in restraining bindings (Baughn et al., 2000).

To illustrate this point further, a penetrating look at how we practice risk assessment reveals how over-stressing normality, and its appearances, harms some kinds of people while privileging other kinds. The fact that a subject is already ill or otherwise in deficit or impaired or not normal often has been thought to justify placing them at higher risk in regard to pharmaceutical testing. But policy should not distinguish invidiously between the degree of risk acceptable for individuals who function typically and those who function in singular or anomalous fashion (RAC, 1997).

Finally, fourth, eliminating underlying pathological conditions to restore patients to their previous or to the common state is not equally the preeminent strategy in all medical domains (Silvers, 2002). To illustrate the broad application of compensatory interventions, corrective lenses, cochlear implants, and genetically boosting the low-density lipoprotein receptor above normal range all improve functioning in instances of impairment without repairing the underlying defect and without restoring performance to the common, typical, or normal mode. (The last named exemplifies a genetic intervention where compensation is more feasible than repair.) In compensatory strategies, effective functioning well may be achieved without restoration of normal performance modes, nor should our practices assume that anomaly compromises success.

Thus, for example, individuals who experienced successful post-polio rehabilitation often developed an exquisite sense of balance to compensate for one of the disease's sequellae, namely, the marked disparity of strength between the right and left sides, and upper and lower parts, of the body. Similarly, non-oral deaf communicators often surpass most others in their ability to express themselves in body language (Corker, 1998, p. 48). Our growing medical success in maintaining the lives of individuals whose physical or cognitive capabilities have been compromised to enormous degrees by accident or illness brings a concomitant increase in the importance of developing imaginative compensatory strategies for functioning and flourishing.

Although therapeutic intervention to remedy disease or injury once was the tactic of choice, now interventions to manage susceptibility to illness, to manage chronic illness, or to manage the results of illness or injury have become important elements of medical practice. For example, this change in outlook is documented and emphasized by the World Health Organization in the beta revision of the International Categorization of Impairments, Disabilities, and Handicaps (1998). The new emphasis is guided by a social model in which dysfunction is understood to emerge from a mismatch between the individual's mode and level of biological performance and the demands of the environment.

Under the revised classification, three different strategies are evoked as important for rehabilitative intervention (World Health Organization, 1998). These are (a) restoring the individual's biological condition to permit typical performance, (b) compensating for anomalous conditions by improving the functionality of alternative modes of performance, and (c) reforming the environment so that it is no longer hostile to a functionally unsuccessful individual's anomalous performance. Thus, restoration of a dysfunctional individual to the common mode and level of functioning, and enhancement of some aspect of an individual's nonimpaired functioning to compensate for a deficit in another aspect of functioning, are seen to be useful alternates for attaining the same goal of reducing functional disadvantage.

To suppose that anomalous physical or cognitive performance must be functionally inferior to species-typical levels and modes of performance is to make two mistakes about humans' biological functioning. It is, first, to assume that human biological organization is functionally rigid, when it is instead gloriously adaptive precisely because it is so flexible. Second, it is to assume that human social organization also is functionally rigid, when, instead, it can be gloriously progressive in expanding the opportunities of different kinds of people precisely because it is subject to moral and political reform. Moreover, any individual's successful functioning is the product of a dynamic between the individual's biological capabilities and the degree to which the environment favors or is inimical to her strengths. As environments evolve, the attributes that have been most important in making people functional may be superseded by other kinds of strengths.

So far, I have argued that there is no reason to assign reparatory interventions—gene transfer applications that restore individuals to species-typical biological condition—priority over compensatory interventions—gene transfer applications that improve the level of one biological performance mode to compensate for dysfunctional deficit in another biological performance mode (see Silvers, 1998a, for further examples). Already of importance at the level of rehabilitative medicine, compensatory interventions appear to be similarly apt at the level of molecular medicine. Further, there is no more reason to fear, as the disability advocates cited earlier appear to do, that engineering genes to secure compensatory performance at the level of molecular medicine is any more threatening to an individual's identity—to her continued existence as the person she takes herself to be—than the engineering of more effective prosthetics and orthotics at the level of rehabilitative medicine (CAHGE, 1998).

5. RECLAIMING MELIORISM FOR MEDICINE

5.1 Normalizing is not Equalizing

In this section, I extend the argument by proposing that the preeminence afforded normalizing interventions in medicine is not just unwarranted but is morally troubling as well. Because normal functioning is defined with reference to the typical or common modes and levels of achievement, what is thought of as normal performance often is artificially skewed by patterns of unfair social domination that favor some ways of performing a function over other equally effective ones, and consequently entrench the socially dominant ones. Public policy that governs genetic intervention should not further privilege, and so further fix and fortify, common or dominant modes of functioning by favoring medical practice that privileges the modes of functioning typical of our species over anomalous but comparably efficient modes.

Ironically, identifying the dominant class's fashion of functioning as a biologically superior mode is the core of negative eugenics. For example, Herbert Spencer, the social Darwinian sociologist whose writing promoted eugenics, wrote the following about a group he labeled as "weak" because its members' "defective" biology prevented their functioning in the fashion of the dominant group. Enhancing the members of this group by rescuing them from their "natural" subservience would result in "a puny, enfeebled and sickly race," he warned in his repeated explanations of how the biology of the female human made her inferior to human males (quoted in Miles, 1988, p. 187).

As historian Rosalind Miles explains in *A Women's History of the World*, "Women's imputed physical and mental frailty thus became the grounds for refusing her any civil or legal rights, indeed any change from the 'state of nature' in which she dwelt" (1988, p. 187). Of course, Spencer and his colleagues did not see the inferiorities of women's biology as being a reason to prevent them from participating in reproduction. So women, as a class, were not primary targets of negative eugenics. Not eugenics but domestics was invoked to contain their influence.

Women's emancipatory movements have freed women from being labeled as a weak group. In doing so, feminist theory has revalued women's ways of functioning, introducing diversity by acknowledging alternative modes. Enhancement protocols for intervention at the molecular level offer a stimulus for an analogous revaluing because they can introduce and promote nonspecies-typical biological mechanisms that enable successful human functioning. Thus they counter the oppressive view that the typical mode of functioning is the sole successful or propitious one. They also undercut the pernicious idea of assigning humans to either a strong or a weak class, categorizing them as functional successes because they function typically or as functional failures because they function anomalously.

5.2 Reconnecting Science to Democracy

As I have envisioned it here, positive molecular eugenics programmatically challenges the propensity of negative eugenics to divide humans into so-called strong and weak groups, and to eliminate or decrease the numbers of the latter. Instead, positive molecular eugenics applies genetic technology to expand upon existing biological strengths where doing so compensates for biological deficits. In so doing, enhancement protocols challenge precisely that drive toward homogenization that has made negative eugenics so enormously troubling by inviting the suppression of supposedly weak people. Far from sharing the inherently morally troubling character of negative eugenics, positive molecular eugenics thus promises to safeguard some of negative eugenics' potential victims.

This is not to make light of the challenges posed by placing genetic technology in the service of human diversity. To illustrate, let us consider the effects of a gene transfer process designed to increase production of IGF-1, a protein that induces muscle repair and growth. During the first three decades of a person's life, vigorous physical exertion stimulates IGF-1 production. In subsequent years, IGF-1 production decreases, with resulting loss of muscle strength from atrophy and injury. Introduction of genes whose expression results in more IGF-1 would not only increase muscle bulk in young adults, but retain mass and strength without decline as the patients age. (These results have been observed in aging mice which, when young, received IGF-1 genes in a process developed by H. Lee Sweeney at the University of Pennsylvania; see Swift and Yaeger, 2001.)

Let us suppose that, for people with certain conditions, the intervention successfully reverses muscle deterioration. The pathological results of muscular dystrophy would be slowed. The incidence of hip fractures in elderly people would decrease. In some patients, enhancing the power and resilience of some muscles above the commonplace might compensate for other, untreatable limitations. So, for example, spinal-cord injured individuals might benefit from gene-enhanced shoulder, arm, and wrist muscles to enhance healing of the upper-extremity damage incurred by years of rolling in a manual wheelchair.

For others in the population, recovery from injuries to slow healing tissues, such as anterior cruciate ligaments and knee cartilage, would be improved. As well, increased muscle mass could help protect against the stress fractures experienced by people who run for exercise. Some patients might remain in deficit in respect to muscle strength and mass, but for others muscular capability would be elevated above that of the average person, and vigorous muscle strength and resilience would persist long past their youth.

Of course, therapies for injured untalented amateurs cannot be denied to injured competitive athletes. Suppose, for instance, muscles crucial to the speed of throwing a baseball, or the distance of throwing a football, are treated? It is unlikely to be possible to control IGF-1 production to the fine degree needed to ensure no muscular improvement beyond recovery to a pre-injury state and thereby to prevent enhancement. In such cases, gene therapy could enhance patients' physical prowess above normal. Yet it seems unfair to disqualify individuals from participating in competition merely because, by seeking relief from injuries, their physical state has become better than ever, and better than other people's.

To extend Thomas Murray's well-known concerns about a similar kind of intervention, the IGF-1 gene transfer procedure has the potential to become a genetic version of the use of anabolic steroids for muscle enhancement. According to Murray, such interventions are ethically suspect in competitive sports, whether they are achieved through traditional medical technology or through gene transfer (1991, p. 485).[3]

Yet there are many situations in which muscular enhancement would not be ethically suspect. For instance, the same elevation of performance that might permit an athlete who is genetically muscularly enhanced to triumph in a game might give an otherwise biologically typical individual the strength to save another person's life, or an otherwise biologically compromised individual the energy to be productive in the service of others. Thus, it is the value of the function individuals will achieve, not the biological normality of how they achieve it that should be the main criterion for researching and distributing genetic interventions.

Noticing this suggests a final important distinction—a difference of emphasis and aim—between negative and positive molecular eugenics. Whereas negative eugenics emphasizes the competitive dimension of social interaction and focuses on eliminating individuals who would suffer because of their poor prognosis for success in competitive endeavors, positive eugenics emphasizes the cooperative dimension and focuses on constructive alterations that permit individuals to function and flourish together. By promoting compensatory strategies through a program of positive molecular eugenics, genetic technology reclaims medicine's melioristic tradition by reconnecting the aims of science and democracy.

San Francisco State University
San Francisco, California, U.S.A.

NOTES

1.   When Francis Galton introduced the term 'eugenics' in 1883, the program he pursued was one of 'positive eugenics,' a practice of encouraging people whose traits are valuable to have more children than people whose traits are not. However, after the turn of the century, 'negative eugenics,' a program of preventing people viewed as inferior from having children, gained momentum and eventually became the approach identified with the eugenics movement (Wertz, 1999b). My use of the expressions 'positive' and 'negative' eugenics is different from their original meanings because, I believe, any program incorporating a distinction between strong and weak classes will devolve into negative eugenics. Nevertheless, my terminology preserves the earlier notion that positive eugenics programmatically focuses on promoting the existence of certain sorts of people, thereby increasing their numbers, while negative eugenics programmatically focuses on deterring the existence of certain sorts of people, thereby decreasing their numbers.
2.   See Silvers, 1998b for a discussion of the framework that must constrain genetic counseling if it is to achieve this end.
3.   Although readers may rightly wonder whether principles of democratic distribution of advantage and handicap are best explicated in terms of sports and other competitive games, these analogies are favored by some bioethicists to illustrate their claims about genetics and social value. (see also Buchanan *et al.*, 2000). Readers may notice, as well, that the rules of different sports competitions are not uniform in regard to banning pharmaceutical and other enhancements. The sports example is Murray's, however, not mine. I introduce it only to dispose of it.

REFERENCES

Arras, J., Steinbock, B., *Ethical Issues in Modern Medicine*. Mountain View, CA: Mayfield Publishing Company, 1995.
Baughn B., Degener T., Wolbring G. Email correspondence on file with the author, 1/11/2000, 1/12/2000, 1/18/2000.
Buchanan, A., Brock, A., Daniels, N., Wikler, D., *From Chance to Choice: Genetics & Justice*. New York: Cambridge University Press, 2000.
Burleigh, M., Death and Deliverance: "Euthanasia" in Germany c. 1900-1945. Cambridge: Cambridge University Press, 1994.
Callahan D. Aging and the ends of medicine. Annals of the New York Academy of Science 1988; 530(June 15):125-33
Callahan D. When self-determination runs amok. The Hastings Center Report 1992; 22(2):52-5
Callahan, D., *Setting Limits in an Aging Society*. Washington, D.C.: Georgetown University Press, 1995.
Colker, R., *The Law of Disability Discrimination*. Cincinnati: Anderson Publishing Co., 1995.
Corker, M., *Deaf or Disabled or Deafness Disabled?* Buckingham, U.K.: Open University Press, 1998.

Council of Europe Parliamentary Assembly. Recommendation 934. 23rd Ordinary Session. Strasbourg: 1982.

Daniels, N., *Just Health Care*. Cambridge: Cambridge University Press, 1985.

Daniels, N. "Justice and Health Care." In *Health Care Ethics: An Introduction*, D. Van DeVeer, T. Regan, eds. Philadelphia: Temple University Press, 1987.

Department of Health and Human Services. National Institutes of Health points to consider in the design and submission of human somatic-cell gene therapy protocols. Recombinant DNA Technical Bulletin 1986; 9:221-42

Duster, T., *Backdoor to Eugenics*. New York: Routledge, 1990.

Gallagher, H.G., *By Trust Betrayed: Patients, Physicians, and the License to Kill in the Third Reich*. New York: H. Holt, 1990.

Kaplan, D. "Disability Rights Perspectives on Reproductive Techniques and Public Policy." In *Reproductive Laws for the 1990s*, S. Cohen, N. Taub, eds. Totowa, NJ: Humana Press, 1989.

Miles, R., *A Women's History of the World*. London: Michael Joseph, 1988.

Miller, M., *Terminating the "Socially Inadequate": The American Eugenicists and the German Race Hygienists, California to Cold Spring Harbor, Long Island to Germany*. New York: Malamud-Rose, 1996.

Murray T. Ethical issues in human genome research. FASEB 1991; 5(1):55-60

Pernick, M., *The Black Stork: Eugenics and the Death of "Defective" Babies in American Medicine and Motion Pictures Since 1915*. New York: Oxford University Press, 1996.

President's Commission for the Study of Ethical Problems in Medicine and Biomedical and Behavioral Research. Splicing Life: The Social and Ethical Issues of Genetic Engineering with Human Beings. Washington, D.C.: U.S. Government Printing Office, 1982.

Proctor, R., *Racial Hygiene: Medicine under the Nazis*. Cambridge: Harvard University Press, 1988.

Recombinant DNA Advisory Committee (RAC). Discussion regarding the use of normal subjects in human gene transfer clinical trials; 1997 Minutes. March 6-7; Maryland: National Institutes of Health, 1997.

Satz, A., Silvers, A. "Disability and Biotechnology." In *The Encyclopedia of Biotechnology: Ethical, Legal, and Policy Issues*, M. Mehlman, T. Murray, eds. New York: John Wiley and Sons, 2000.

Silvers, A. "A Fatal Attraction to Normalizing: Treating Disabilities as Deviations from 'Species-Typical' Functioning." In *Enhancing Human Capacities: Conceptual Complexities and Ethical Implications*, E. Parens, ed. Washington, D.C.: Georgetown University Press, 1998a.

Silvers, A. "On Not Iterating Women's Disabilities: A Crossover Perspective on Genetic Dilemmas." In *Embodying Bioethics: Feminist Advances*, A. Donchin, L. Purdy, eds. Lanham, MD: Rowman & Littlefield, 1998b.

Silvers, A. "Bedside Justice: Personalizing Judgment, Preserving Impartiality." In *Health Care and Social Justice*, R. Rhodes, M. Battin, A. Silvers, eds. Oxford: Oxford University Press, 2002.

Swift E.M., Yaeger D. Unnatural selection: genetic engineering is about to produce a new breed of athlete who will obliterate the limits of human performance. Sports Illustrated 2001; May

The Campaign against Human Genetic Engineering (CAHGE). Email distributed to disability-related listservs. Copy on file with the author, 1998.

U.S. Congress Office of Technology Assessment. Human Gene Therapy: Background Paper. Washington, D.C.: Office of Technology Assessment, 1984.

Walters L. The ethics of human gene therapy. Nature 1986; 320:225-7

Weissman, G., *Democracy and DNA: American Dreams and Medical Progress*. New York: Hill and Wang, 1996.

Wertz, D. Eugenics: Definitions. The Gene Letter 3, 2 1999a; February. Available at: http://www.geneletter.org/0299/eugenicsdefinitions.htm.

Wertz, D. Eugenics:1883-1970. The Gene Letter 3, 2 1999b; February. Available at: http://www.geneletter.org/0299/Eugenics1883-1970.htm.

World Health Organization. International Classification of Impairments, Disabilities, and Handicaps. 1980 Geneva: World Health Organization 1998. Versions of ICIDH-2, Available at: http://www.who.int/msa/mnh/ems/icidh/icidhtrg/sld033.htm.

CHAPTER 12

DAVID WASSERMAN

PERSONAL IDENTITY AND THE MORAL APPRAISAL OF PRENATAL THERAPY

1. INTRODUCTION

The medical promise of genetic therapy on gametes, zygotes, and early fetuses (which I will refer to hereafter as 'prenatal therapy') lies in the plasticity of the entities on which it is performed. Genetic changes in one cell, or a few cells, may result in major phenotypic changes in the child that emerges, changes that would be impossible or impractical to introduce at a later stage. But the very plasticity that makes prenatal therapy medically attractive makes it morally problematic. Genetic changes made in a zygote or early fetus seem more likely to alter its identity, and the identity of the resulting person, than similar genetic changes made (if that is even possible) in a late fetus or child.

The possibility that genetic therapy will alter fetal and personal identity complicates the appraisal of "successful" therapy that achieves its medical objectives, since it may replace the fetus it was intended to repair. It also complicates the appraisal of unsuccessful therapy that fails to achieve its medical objectives or causes harmful side effects, since its consequences may be borne by a creature who would not have otherwise existed. In section 2, I will outline the metaphysical and moral issues that underlie the appraisal of prenatal therapy and identify two standard sets of views on those issues. In section 3, I will examine the critical roles that have been assigned to the genome in giving the fetus a moral claim to prenatal genetic therapy and in maintaining the identity of the fetus subject to therapy. I will question whether the genome can fulfill either of these roles, and suggest that identity-preservation is the more problematic. In section 4, I will look at unsuccessful therapy, whose moral appraisal appears to depend on questions of fetal and personal identity that, as I have argued in section 3, are beset with uncertainty. I

L.S. Parker and R.A. Ankeny (eds.), Mutating Concepts, Evolving Disciplines: Genetics, Medicine and Society, 235-264.
© *2002 Kluwer Academic Publishers. Printed in the Netherlands.*

will consider various attempts to identify a complaint or a harm in cases where identity may be altered by harmful genetic changes, and offer my own suggestion about the moral assessment of actions that cause harmful conditions while affecting personal identity.

2. AN OUTLINE OF METAPHYSICAL AND MORAL ISSUES

The first issue raised by prenatal genetic therapy is whether the zygote, fetus, and person should be seen as distinct entities or as phases of the same human organism. If I say that I was once in the womb, am I speaking literally? Many people are inclined by the language they use and the identification it conveys to see themselves as the same entities that gestated in their mothers' wombs. On this view, the terms 'embryo,' 'fetus,' 'child,' and even 'person' refer to phases of the same entity, not to distinct entities. I will call philosophers who defend the view that we are essentially human beings, not persons, 'humanists.'

This intuitively appealing view is reinforced by arguments against the alternative position, that each of us is essentially a person, conjoined with but not identical to the human organism in the womb. Those arguments, elaborated most recently by Eric Olson (1997, 1999), emphasize the awkwardness of defining the relationship between a person and its organism as distinct entities. Do I—a person—act when my body moves? Which of us performs basic bodily functions, has sensory experiences, feels, thinks, and decides? Those who defend the view that we are essentially persons, to whom I will refer as 'personalists,' [1] dismiss these problems as merely rhetorical, and argue that a coherent and plausible association between the two entities can be described, either in terms of constitution or some other intimate relationship (e.g., Baker, 1999).

There is disagreement, among both humanists and personalists, about when human beings come into existence (see Persson, 1995). Some insist that the single-celled zygote cannot be a human being despite its genetic identity with the fetus that develops, because it ceases to exist on replication (Stone, 1987, p. 819), or because it separates into embryonic and placental parts; others hold that the mere possibility of twinning prevents the zygote from being a single human being (Kuhse and Singer, 1990).

Defenders of conception as the origin of human beings insist that the zygote does not cease with replication, but becomes a single unit of undifferentiated cells; that its contribution to the placenta does not preclude its humanity, any more than a caterpillar's contribution to the cocoon precludes its identity with the emerging butterfly; and that the threat of twinning is simply a threat that one human being will be destroyed and replaced by two others, a loss unmourned because it is unknown at the time and quickly overlaid with happier developments (Oderberg, 1997).

The issue of when the human organism begins is relevant to the assessment of prenatal therapy because much of that therapy may be performed during the ambiguous two-week period between conception and cell-differentiation. Many philosophers believe that changes in the material or components of an entity not yet in existence are more likely to affect identity than similar changes occurring after the entity has come into existence—a claim I will explore shortly. For simplicity's sake, however, since this is really an internecine dispute among humanists, I will assume that conception is the point of origin for the human organism, and that the normal processes of cell-differentiation, growth, and development bring no change in identity.

The second issue is how robust the identity of the zygote, fetus, or person is against the changes in structure and functioning wrought by prenatal therapy. Does any substantial structural or functional change result in a different zygote, or must the change be quite comprehensive? Are changes in *genetic* structure especially likely to alter identity?

Different entities composed of the same matter will have different identity conditions. Because of its protean nature, a lump of clay may survive its transformation into indefinitely many different statues, but the statues it successively constitutes will not be identical to each other. If persons are not the same entities as the human organisms that constitute them, they too will have different identity conditions, and personal identity may be affected by changes that do not affect the organism's identity. This is a familiar theme in recent discussions of death, where it is often claimed that the human organism, but not the person, can survive in a permanent vegetative state (McMahan, 1995). In the prenatal context, the roughly analogous claim is that genetic alterations which preserve the identity of the zygote or early fetus may result in a different person coming into existence. Obviously, this is a prospect that only personalists can entertain.

The possibility that prenatal therapy may affect personal but not fetal identity is reinforced by a widely-shared intuition about the "necessity of origin": roughly, that what determines the identity of an entity is a continuous history back to a point of origin (Brennan, 1988, pp. 166-9; Noonan, 1983). Once in existence, an entity may undergo substantial changes without losing its identity—changes that might have resulted in a different entity coming into being if they occurred in its constituent parts before its point of origin.[2]

Thus, a given change in the DNA of an early fetus might preserve its identity, while a change or changes in the DNA of its gametes with the same net effect on fetal DNA would have resulted in a different fetus coming into being. Similarly, if the person comes into being after the fetus, there may be genetic modifications that would preserve her identity if they occurred after she came into being, but would have resulted in a different person coming into being if they had occurred in the early fetus (McMahan, 1998, pp. 212, 245, n. 4), whether or not they altered fetal identity. Although this view about the significance of origins appears to influence much of the debate on fetal and personal identity, it is not often articulated in that debate, and it is not universally accepted.[3]

The third issue concerns the moral status of the zygote and fetus, a status relevant to their claims both *for* therapeutic intervention and *against* replacement. If the fetus and the resulting person are the same entity, does that give the former the interests or rights of the latter? Why should the early fetus have such rights, since, even if it is the same entity as the resulting child or adult, it has none of the properties that are inherently worthy of respect, such as self-consciousness and deliberative agency?

Some philosophers regard the simple humanity of the fetus as the essential property it shares with the resulting person, and believe that property suffices to secure it many of the same rights as the child or adult.[4] But for humanists, who find an appeal to the fetus' species membership inadequate to explain its moral status,[5] the challenge has been to define a relationship between fetus and adult that confers on the former some of the rights or status enjoyed by the latter. It cannot be potentiality in the sense of mere possibility, as that would confer such rights or status on any cell that could be made into an adult human through cloning or other technology. Nor can it be potentiality as a high, or minimum, degree of probability, because the fetus' moral status does not appear to depend on its statistical odds of developing into an adult human, e.g., it is not affected by the rate of spontaneous abortion.[6]

Several philosophers have argued that the relationship between the fetus and adult human must be understood in terms of a concept of strong, inner, or active potentiality—a concept that cannot be reduced to possibility or probability (Annis, 1984; Wreen, 1986; Stone, 1987). Jim Stone, whose account I will focus on in the rest of this section, understands the fetus' potential as the realization of its nature as a self-conscious, social being; its rights are based on its interest in achieving those "conscious goods like self-awareness" which it is in its nature to actualize (1987, p. 281).

Personalists generally deny that we acquire significant rights or interests until we actually acquire such properties as self-consciousness and rational agency—properties that are inessential for humanists, essential for personalists. This position, however, is vulnerable to claims that it must help itself to some notion of potentiality to explain the rights and interests ascribed to very young children, who lack self-consciousness and deliberative agency. Few personalists are willing to follow Michael Tooley (1983) in denying, on this basis, that infanticide is a serious moral wrong.

Humanists and personalists, then, differ on these first three issues. For humanists, you and I are essentially human beings and not essentially persons; we come into being at conception (or some analogously transformative event), and our identities, established at that point, are comparatively robust against the changes that prenatal genetic therapy will bring about. As fetuses, you and I had moral claims to therapy based on our potential to become, through an identity-preserving developmental process, self-conscious, rational agents. Because the interests or rights of a human do not depend in any wholesale way on its developmental stage, prenatal therapy has the same general moral warrant as medical treatment for neonates, children, or adults. At the same time, that therapy risks two kinds of morally significant failure. It can leave the existing human organism no better off, worse off, or dead; or it can cause identity-altering changes. The latter, as well as the former, involves a substantial loss, because a human being has ceased to exist, even if the total number of human beings has not changed. For the humanist, the fetus' genetic constitution may be relevant not only in determining if it has that potential, and therefore a moral claim to therapy, but also in determining if the therapy will alter its identity.

Personalists hold that you and I are essentially persons; that we are not identical with the human organisms we are embodied in, which come into existence at conception or implantation; and that we come into existence with the onset of consciousness, or of self-consciousness, which is usually thought to occur no earlier than birth (and often thought to occur significantly later). Therefore, you and I are neither created at conception nor present in the womb. Prenatal therapy and, *a fortiori*, preconception therapy, may make it the case that a different person arises from the zygotes or gametes, so that it may be that "we" would not have existed had such interventions occurred. The moral warrant for such interventions is weaker than for the medical treatment of children and adults, since fetuses are not identical with children or adults and lack their moral standing. A parent may have a duty to prevent harm from occurring to a person who does not yet exist, but whom she intends to bring into existence, and that duty may compel her to employ prenatal therapy. But she has no duty to employ that therapy to preserve a fetus with a genetic defect rather than to abort and start over.[7]

The fourth issue concerns the relevance of personal identity to the moral appraisal of unsuccessful therapy that fails to achieve its objectives or causes harmful side-effects. If such therapy does not affect personal identity, it can be assessed the same way as any other pre- or post-natal intervention that leaves the person worse off than she might have been. But moral assessment is more complicated if that therapy affects identity, bringing into existence a person who would not have existed without it. The effect of unsuccessful therapy on identity presents a problem in moral appraisal for personalists as well as humanists.

It is easiest to isolate this issue by looking at preconception therapy, which does not risk the loss and replacement of an existing human being, but may alter the identity of any fetus or person created from the gametes on which it is performed. If, for example, the careless action of a doctor in the treatment of a gamete causes harm that could have been avoided, but thereby causes a person to come into being who could not have existed without suffering that harm, there is a wrong in search of a victim.[8] The possibility that slight differences in the extent of genetic modification will cause different people to come into existence, but that it will be difficult or impossible to ascertain if identity has been affected, raises problems for any moral theory that judges actions and states of affairs by their effect on specific people. It suggests that doctor's or parent's action can only be appraised impersonally, for making the world a worse place. But that form of appraisal is no more congenial to personalists than humanists.[9]

3. WHEN THERAPY "SUCCEEDS":
ASSESSING THE THREAT TO FETAL IDENTITY

To make the first three issues, concerning identity and potentiality, less abstract, consider an extreme case of prenatal genetic "therapy," then a number of more moderate (if equally fanciful) ones. The doctor has just completed a battery of genetic tests on the zygote, and gravely informs the mother that "it" has a number of fairly serious genetic defects spread around its genome. Furthermore, the only feasible way to eliminate them is to discard the nucleus and substitute one from a bank at the hospital, an expensive procedure with high odds of failure and a small risk to the mother. Almost everyone would agree that what the doctor is proposing is the termination of the existing zygote and its replacement by another.

Humanists and personalists might both object to this procedure, but for different reasons. For humanists, the procedure would be at the very least *prima facie* wrong because it involved the killing of a very young human being who would, despite his genetic defects, be likely to have a life worth living—a wrong hardly offset by the creation of another human being. Indeed, it might be seen as an especially objectionable killing, because it involved the sacrifice of one entity to create another; the defective zygote, gutted to provide a home for the replacement nucleus, would arguably be killed as a means. For personalists, the objection would be less categorical. While there would be no morally significant loss in causing a zygote to go out of existence, or a different person to come into being, it might be pointless to make a new zygote from the "raw material" of the old: why go through all the trouble just to preserve some remnants of the original organism?

But perhaps the situation is not so grim. The doctor informs the mother that the defects are all on four chromosomes, 6, 12, 21, and Y, which he can replace. However, the hospital is out of Y chromosomes. Would she mind a girl? Or perhaps the substitution is less radical, involving only a small sector of one chromosome, but the replacement sector comes from a monkey or a pig. Here, our intuitions about zygote identity may begin to diverge. Some would argue that such wholesale chromosomal replacement would alter the identity of the zygote, especially the switch from a Y to an X chromosome. Others would doubt that this specific set of modifications would alter identity, but would concede that identity would be affected at some point, e.g., if more than a third of the chromosomes were replaced. These examples lack biological verisimilitude, but there is no reason to think they greatly exaggerate the uncertainty and disagreement that may arise over the effect of some prenatal interventions on fetal and personal identity.

The uncertainty about fetal identity poses a severe problem for the humanist. Unlike the uncertainty about personal identity revealed by adult fission cases, it does not have its source in the persistence of something else that matters to us almost as much as identity. We may have reason to care about psychological continuity and the survival of our "continuers"—those who share our memories, intentions, values, and projects—even if they are not identical with us, or if we cannot determine or decide whether they are. But there is no counterpart to psychological continuity in the context of fetal identity. The fetus replaced by genetic modifications leaves nothing but a residue of unaltered genetic and cellular material to its successor. For someone who cares about the loss of the original fetus, it is small consolation that most of its DNA is retained in a different being. Perhaps a parent would like to retain as much of her genetic contribution as possible, but that would reflect a concern about the relationship between her and her offspring that could usually be satisfied more directly and completely by replacing the defective zygote with another "original," bearing a full complement of her genes.

3.1 Genetic Criteria for Potentiality and Identity

The difficulties that humanists face in assessing prenatal therapy arise in part because they have looked to the genome as the source of identity as well as potentiality, making prenatal therapy both more risky and more urgent than it is for the personalist. There is, however, strong intuitive appeal in genetic criteria for both potentiality and identity, an appeal not limited to those who regard the fetus and person as identical. However formidable the difficulties in basing moral status on potentiality (see Tooley, 1983), we appear to rely on some notion of potentiality in the moral appraisal of post- as well as prenatal developments. Jeff McMahan maintains that we rely on a baseline of "native potentiality" in assessing harm to developing infants and children, a baseline that requires, or at least invites, "a firm distinction between genuinely native potential—i.e., that potential that is grounded in the physical constitution of the individual—and a broader notion of potential that includes all a being could become, compatible with preserving its identity, by being externally augmented" (1996, p. 22). The genome appears to supply the desired grounding in the individual's physical constitution.

The appeal of a genetic criterion of identity is similarly broad and strong. There are several reasons to regard a living and developing entity as having quite different identity conditions than an inanimate object, whose survival arguably depends on continuities in location, size, shape, and other structural and functional features. The intuition that the genome has a special or unique role in determining identity is reflected in our reaction to the case of nuclear replacement. I suspect we would have the same reaction if it were only the invisible DNA, not the prominent nucleus, that was replaced wholesale.[10] We would be far more inclined to regard the identity of the zygote as altered by the replacement of its nuclear DNA than by the replacement of the same amount of molecular material elsewhere in its cells. Although it is only recently that we have assigned a critical role in development to nuclear DNA, we have a more general intuition that the identity of a developing organism depends upon whatever internal features guide its development.

This intuition is hardly limited to humanists. Noham Zohar, who regards the fetus and person as distinct entities, holds that:

> If there is any sense at all in regarding the embryo's existence as the beginning (or: origin) of a particular human life, then it surely comes from the fact that once the genotype has been specified, personal identity has been significantly determined. Genetic make-up—the 'blueprint' of a person's organization and functional

capabilities—is the aspect of personal identity relevant to embryonic identity.... This
perspective, therefore, provides a frame of reference for evaluating any particular
change. (1991, pp. 283, 285)

Humanists may regard personal identity as more robust than personalists like Zohar,
because they regard the person as having an earlier origin. But while they may be
less inclined to see prenatal therapy as threatening personal identity, they see more
of a moral problem when it does.

Much of the criticism of the humanist position on the fetus has concerned its
attempt to ground the potentiality of the fetus in its genetic constitution. I will
suggest, however, that the problem of assessing the potentiality of a being with an
unaltered genome is more tractable than the problem of assessing the identity of a
being with an altered genome, although the humanist response to both problems may
involve recourse to the same misleading metaphor of a genetic blueprint.

3.2 Inner Potentiality and External Interference

The notion of strong or inner potentiality has frequently been criticized for assuming
that there is a single path of normal development (see Fisher, 1994).[11] According to
Stone, the fetus' potential resides in the unique "developmental path...determined by
[the fetus'] genetic consitution." It will follow that path, and actualize its nature,
unless it is diverted by "environmental interference or genetic change" (Stone, 1987,
p. 821, n. 11). Critics maintain that the distinction between genetic determination
and environmental interference cannot be sustained in post-Darwinian biology.
Thus, Eliot Sober argues:

> According to the Natural State Model, there is one path of foetal development which
> counts as the realization of the organism's natural state, while other developmental
> results are consequences of unnatural interferences.... Or more modestly, the
> requirement might be that there is some restricted range of phenotypes which count as
> natural. But when one looks to genetic theory for a conception of the relation between
> genotype and phenotype, one finds no such distinction between natural states and states
> that are the result of interference. (Sober, 1980, p. 374)

On this view, phrases like "the unique developmental path determined by a
creature's genetic code" lack a clear reference, unless relativized to a particular
cellular, somatic, physical, and social environment.

The extent to which modern biology does not discriminate among developmental
paths is, however, a matter of dispute among scientists and philosophers of science.
Christopher Boorse, for example, maintains that biology can restrict itself to
standard environments in defining normal development and functioning for an
organism, and he ridicules Sober's assumption that all developmental outcomes are
equal:

> It is obviously false that there is no interesting or fundamental biological distinction
> among outcomes of development.... If one pours gasoline on a corn plant or child and
> ignites it, the organism dies, but its only contribution is to be vulnerable to 1000°
> temperatures. In burning, it is expressing only that part of its nature, which it shares
> with all organisms, indeed most substances, on earth. (Boorse, 1998, p. 111)

Even if Sober overstates the neutrality of modern biology among developmental paths, it is still clear that biology does not recognize what Stone's account assumes: a unique path determined by the organism's genetic constitution. Widely divergent paths may be possible within a range of environments that excludes factors inimical to survival and reproduction.

There are, however, two ways in which Stone could modify his account of strong potentiality to accommodate Sober's objection. First, he might hold on to the notion of a unique developmental path but require a unique result only within a small subset of physically possible environments. He might, for example, argue that biological theory provided a basis for restricting normal or standard environments to a narrow range relatively congenial to the development of human cognitive capacities.[12] Alternatively, he might treat potentiality as a moral, not a biological concept, and restrict the range of environments to those that were considered relevant on normative grounds. On that basis, he might, for example, exclude not only environments generally inimical to human life, such as Boorse's inferno, but also those sufficiently rare in human experience to have played little or no role in shaping our expectations about the possibilities of human development.

As long as Stone insists that normal development has a unique path, however, he is vulnerable to claims about the plasticity of embryonic development within any range of environments held to be standard by biological or moral theory. Given the uncertainty about developmental variation, the more plausible humanist response might be to forgo the assumption of a unique path. What strong potentiality would require is not the same result in every standard environment, but merely the achievement of self-consciousness and deliberative agency in *some* standard environment. The frailty of a fetus that can only acquire those attributes in a very narrow range of environments does not preclude its potentiality. Nor does it matter if the fetus would achieve widely-varying levels of cognitive development in different environments, as long as it would achieve some minimum level in at least one.[13]

The difficulty with this notion of potentiality is not that it is too permissive, but that it is too restrictive. It would deny potentiality to any human organism that could not achieve adult consciousness in at least some standard environment, even if it needed only a simple intervention to do so. Thus, an infant with PKU would lack the potential for adult consciousness if all standard human diets included phenolallenines, however easy it was to deliberately exclude them from the infant's diet. An otherwise normally developing fetus with a heart defect that was invariably fatal in the third trimester would lack the potential for adult consciousness even if that defect could be fully cured with a simple intrauterine injection.

It would be tempting to claim potentiality for such fetuses by including routine dietary and even medical interventions in the range of standard conditions. If nursing and nurturing are included, why not restricted diets and intrauterine injections? The problem is not that the latter interventions are of recent origin—the humanist might concede that the moral status of the fetus depended on the state of medical technology, or else claim that the fetus could have strong potentiality even if her potential could not be realized with current medical technology. The problem is rather in distinguishing interventions that help the fetus realize her potential from those that exceed or enlarge her potential.

The humanist might avoid this problem altogether by refusing to consider any recent technology in assessing potentiality, restricting herself, for example, to the care available in the environments in which human beings evolved—a restriction that would exclude both the PKU infant and the fetus with a bad heart. Or she might allow only interventions, like heart repair, that secured the development of "support systems," but not interventions that secured the development of the central nervous system, like dietary restrictions for PKU. If, however, the humanist wanted to treat both the fetus with a bad heart and the PKU infant as having strong potentiality, she would have to distinguish interventions which merely facilitated or corrected fetal development from those which redirected it. In making that distinction, she would not have to assume that the fetus' genome determined a unique developmental path. But she would have to come up with a basis for deciding whether an intervention imposed a new path of development on the fetus, different from the one it had already taken or from any that it might have taken.

A variety of considerations might affect this judgment. The later or less specific the intervention, the less likely it would be seen as superceding. The fetus with a bad heart and the PKU infant have already acquired much of the neural structure underlying adult consciousness; the modest interventions they need seem to facilitate the completion of a course of development already far advanced.[14] And many early interventions, such as altering material diet or lowering amniotic pH, would be too non-specific to be regarded as superimposing a distinct course of development on the fetus.

Admittedly, these are rough, intuitive judgments. The humanist, however, may not need a precise answer to the question of whether a fetus has the potential for adult consciousness—it may have more or less, depending on the range of environments in which it would acquire adult consciousness or the types of interventions needed to secure that outcome.[15] But even with a plausible account of potentiality, which would determine the strength of the fetus' claim to genetic therapy, the humanist must still find some basis for determining whether such therapy would threaten fetal identity. This, I will argue, is the more formidable challenge.

3.3 Genetic Modification and Fetal Identity

There are at least two reasons why determining whether a genetic modification would alter the identity of a fetus poses even greater difficulties for the humanist than determining whether a somatic or environmental intervention would fulfill rather than enlarge its potential. The first, suggested above, is that potentiality is more plausibly treated as a matter of degree than identity (as opposed to similarity or survival; see Brennan, 1988). The second is that, in the case of genetic modifications to a zygote or early fetus, it is difficult to make even a rough intuitive judgment about the extent to which the original physical constitution of that creature is the exclusive or predominant cause of the attributes it acquires.

The latter problem becomes apparent in considering Stone's account of genetic modification and fetal identity. In establishing criteria for identity preservation, Stone again has recourse to the notion of a developmental path determined by the fetus' own genetic constitution:

> The survival of the embryo requires only that the original genetic code is the primary determinant of either the developmental path the embryo follows, or where genetic expression is interfered with, the path he would have followed had he been left alone. (Stone, 1994, p. 290)

Applying this notion of 'primary determination' to prenatal therapy, Stone suggests that the removal of the third copy of chromosome 21 to prevent Down syndrome would preserve the embryo's identity, since it merely prevents "interference" with the complete genetic code found in the standard complement of chromosomes. Similarly, it would not alter the embryo's identity to fill in a "slight genetic gap [that] prevents the development of an organ, the features of which are already determined by the genetic code," since "plugging the gap merely enables the developmental process to proceed as determined" (Stone, 1994, pp. 290-1).

The case of removing interference appears to be the most straightforward, since that intervention merely permits "the complete expression of the creature's genetic code," which remains the primary determinant of its developmental path. But the notions of 'complete expression' and 'interference' are ambiguous and problematic. Understood to mean the expression of each gene, the complete expression of a creature's genetic code would be impossible, since the expression of some genes suppresses or interferes with the expression of others. Alternatively, understood to mean the elimination of *any* genetic suppression or interference, the complete expression of a creature's genetic code would seem to require the maximum possible output of the proteins each of its genes coded for—at least those proteins that did not suppress or inhibit the production of other proteins. But that, I suspect, would be likely to yield an amorphous mass of malignant cells. Some interference is critical to normal development, which involves complex sequences of epistasis and activation; other interference results in disease and death; and still other interference may, for all we know, stifle the development of superior functioning.

It would certainly be useful to distinguish these kinds of interference genetically, but even if we could, the question would remain of whether eliminating any of them would preserve fetal identity. If an eliminative intervention resulted in a creature of a different sort—a blob, butterfly, or kangaroo—it would clearly alter identity. There might be some temptation to conclude from this that interventions that pushed the creature away from species norms were less likely to preserve identity than interventions that pushed the creature toward those norms. But even if this conclusion could be defended, it would not help Stone: there would be no reason to assume that the original genetic code of the fetus played a greater role in determining its developmental path in cases where an eliminative intervention brought the fetus toward, rather than away from, species norms.

Stone's confidence that eliminating a trisomy would preserve fetal identity, despite the pervasive changes in cognition and agency it would bring, appears to have a different source: the fact that the extra chromosome merely repeats the genetic code found on another chromosome. The organism's design can be clearly discerned because the extra chromosome is just a redundancy, an additional copy of a page in the blueprint. More broadly, the blueprint metaphor suggests that we can consult the genetic code to distinguish genetic interference that is part of the design from interference that is not. Because a blueprint "expresses"—to use that word in a different sense—the designer's intentions, it may be possible to correct or interpolate a great many details.

The work required of the blueprint metaphor is even more apparent in Stone's claim that identity would not be altered by filling in a "slight genetic gap [that] prevents the development of an organ, the features of which are already determined by the genetic code." This suggests a notion of 'preformationism' that most modern geneticists reject—the notion that the genome contains what Ken Schaffner calls "'traituncli'—little copies of the traits genes determine" (1998, p. 233). If the genome does not actually contain such copies, it is not clear how we can tell if the fetus' unaltered genetic code has "already determined" the features of an organ that, *ex hypothesi*, it cannot produce without genetic modification. The assumption that a "slight gap" in the genome corresponds with a slight gap in the developing organism assumes an isomorphism not found in molecular genetics.

Without recourse to a blueprint, the humanist could place varying weight on different dimensions of genetic modification that appear to affect identity—the number of genes altered, the number of alterations made, the number of cells ultimately bearing those alterations, the magnitude of phenotypic changes attributable to those alterations, and the "directedness" of those alterations—their creative or authorial character. The first three are straightforward to assess, the latter two less so, but it is not only the difficulties in assessment which should concern the humanist. It is, rather, the obvious ways in which these dimensions can come apart, and the lack of consistent intuitions about "primary determination by the unaltered genome" in cases where they do.

There is, for example, no reason to expect a close correlation between the number of genes altered and the magnitude of the phenotypic change. Small, localized changes in the genome may have pervasive phenotypic effects (pleitropy), while extensive changes in the genome may be necessary to achieve slight phenotypic changes (polygeny). I suspect that we have no clear or consistent intuitions about which dimension matters more to identity when they diverge.

Or consider the contrast between "eliminative" interventions with pervasive effects and more "constructive" interventions with minor or localized effects, e.g., the contrast between the removal of a complete set of third chromosomes, an intervention with comprehensive effects but little re-engineering, and the intricate modification of several chromosomes required to create an otherwise missing liver or ventricle. The former merely removes interferences; the latter creates an organ from scratch, but either or both may appear to alter identity. (Genetic changes that affected brain development would appear especially likely to alter fetal and personal identity, though their assessment would be complicated by the environmental contingencies and stochastic elements involved in brain growth and organization.)[16]

3.4 Genetic Criteria for Potentiality but Not Identity

The humanist could try to minimize the uncertainties about fetal identity that arose in the context of prenatal therapy by treating the zygote or fetus as akin to an ordinary physical object for identity purposes. The recourse to criteria unrelated to the fetus' moral status as a potential person would appear to make fetal identity robust against extensive genetic modifications, which would, for the most part, have little immediate effect on the appearance of the organism, and would affect its size, shape, structure, or function only gradually. The humanist could claim that the genetic features necessary for the emergence of self-consciousness and deliberative agency were not necessary for the fetus' identity, even if they were the source of its moral status. A genetic modification that destroyed the fetus' potential to become self-conscious would cause a great loss, akin to the loss resulting from a change that destroyed a caterpillar's potential to become a butterfly (assuming, controversially, that it had such potential). The loss of potential, however, need not mean the loss of identity—it would be the same fetus or caterpillar, divested of those features that gave its life, respectively, moral or aesthetic value.

But even if the identity conditions for ordinary inanimate objects offered more manageable standards for assessing the effect of prenatal therapy (and there is much debate about the clarity and adequacy of continuity accounts; see Brennan, 1988), their adoption would conflict with a strong intuition that the organism's identity is primarily determined by its genetic code. Perhaps that intuition should be rejected as biologically naive. But the humanist may be especially reluctant to part with it, since it links the identity of the fetus to its potential—its source of moral value.

Adopting criteria for fetal identity similar to those for fetal potentiality maintains a close link between identity and value. (Genetic identity criteria also maintain a link between the conditions of identity-preservation for an organism and its conditions of origin—the creation of a new genome in sexual reproduction or the initiation of a new developmental process by asexual reproduction.) Somewhat analogously, the appeal of psychological criteria of personal identity derives in part from the conviction that it is self-consciousness and deliberative agency that give people their moral standing.

It may, however, be unreasonable to demand that identity and value go together. As Olson suggests, "the motivation behind this demand may be the idea that any acceptable account of our identity must explain the moral significance of our identity." Although Olson regards this as an attractive thought, he does not think it is a compelling one: "I think it is fair to say that no account of our identity has yet been proposed that guarantees...the coincidence of what is important in our identity with the actual conditions of our identity" (1997, p. 165).

Accepting some divergence would allow the humanist to develop a genetically-based account of potentiality without worrying about identity. In those cases where prenatal genetic therapy seems most clearly warranted, e.g., where a Huntington disease or cystic fibrosis mutation threatens health or longevity but does not preclude the acquisition of adult consciousness, the kinds of genetic interventions likely to prevent the disorder are unlikely to bring about radical discontinuities in the organism's growth and development. To avoid worries about identity, however, a humanist might be wise to deny strong potentiality to any fetus that could not achieve adult consciousness without genetic modifications, however minor. For to find potentiality in the case of some genetic modifications but not others would be to raise questions about the range and type of modifications that fell within the purview of the fetus' potentiality, questions parallel to, and possibly as vexing as, those raised about the range and type of modifications that would be consistent with its identity.[17]

4. WHEN THERAPY "FAILS": HOW SHOULD WE ASSESS GENETIC INTERVENTIONS THAT CAUSE A DIFFERENT, LESS HEALTHY PERSON TO EXIST THAN WOULD HAVE EXISTED OTHERWISE?

If the success of (radical) prenatal therapy poses a problem only for the humanist, the failure of such therapy poses a problem for the personalist as well, since the effect on identity seems relevant in appraising the harm suffered by the person who actually comes into existence. This difficulty becomes clearer if we look at therapy on gametes, which may be even more likely (because it predates origin) than therapy on a zygote or fetus to alter the identity of the zygote, fetus, or person, but which does not risk the complicating loss of a human being. Consider a doctor who treats an egg with a genetic defect, which will then be fertilized with an already-isolated sperm. The radiation he uses to treat the egg may cause a mutation which is more harmful than the defect being treated, but which will also alter the identity of the resulting zygote. If any person who develops from that zygote will have a life worth living despite that mutation, then the therapy cannot be appraised in conventional person-affecting terms. The harm of non-treatment will be borne by one person, who would have existed without the therapy, while the harm of treatment—the mutation—will be borne by another, who would not have existed, and had a life worth living, without the therapy.

To simplify the doctor's options, assume that there are only two levels of radiation. At level 1, there is some chance of correcting the defect, and no chance of causing the more harmful, identity-altering mutation. At level 2, there is a slightly better chance of correcting the defect, but also a significant chance of causing the mutation.

Were there no threat to identity, the appropriate practice, based on the priority of avoiding harm, would clearly be to stay at level 1: going to level 2 creates a substantial risk of greater harm for a small reduction in the risk of a lesser harm. But the identity-altering character of the harmful mutation complicates the analysis. The greater chance of correcting the defect gives the doctor a weak person-affecting reason to go to level 2—to give the person (the person, that is, who would develop after level 1 treatment) a better chance of a cure than she would have at level 1. Moreover, the significant risk of harm at level 2 does not give the doctor a countervailing person-affecting reason for staying at level 1, because the person who bears the harmful mutation will not be the same person who would have come into existence with level 1 radiation; she will owe her worthwhile existence to the mutation.[18] So if the doctor, in his single-minded zeal to cure the defect, goes to level 2, he does not appear to subject himself to reproach from anyone except possibly the parents, even if the odds of a cure are only marginally greater than at level 1, and the odds of the more harmful mutation are extremely high. (If he goes to level 2 and the harmful mutation does not occur, the identity of the resulting person will not have been altered, but the doctor will have done everything possible to eliminate her defect, whether or not he succeeds.)

This perverse result raises difficulties for personalists as well as humanists, since the putative victim—the creature with the harmful mutation—*is* a person, who appears to owe her existence to that mutation. Moreover, it is a result that is at least as plausible on a personalist account of identity, which holds that the identity of the person may be altered even if the identity of the zygote or fetus is not.

Cases like this raise the now familiar specter of the "non-identity" problem—the problem of how to frame the moral complaint against conduct, like the use of the higher radiation level, that risks or causes harm to a future person while making it the case that that person, rather than someone else (or no one), comes into existence. If that harm were inflicted on someone who would have existed otherwise, she would have a personal grievance. It still seems wrong to inflict that harm on someone who would *not* have existed otherwise, but it is more difficult to express that wrong as a personal grievance. Because no one is made worse off by the action that causes harm, it may appear that the only objection to that action is that it leads to a poorer outcome, assessed impersonally.

Derek Parfit (1984) introduced "non-identity" cases to argue that the underlying basis for moral appraisal is impersonal. His examples are designed to suggest that our evaluation of alternative states of the world with the same number of people depends entirely on how well or badly off people are in those states, not on whether the same people do better or worse in some states than others. Thus, Parfit invites us to consider two medical testing programs: Pregnancy Testing detects a maternal condition that will cause a disability in any existing fetus, a defect that can be treated only prenatally; Preconception Testing detects a maternal condition that lasts two months, during which any child conceived will have a similar, but untreatable, disability, an outcome that can be avoided merely by postponing pregnancy for those two months. As a result of the former program 1,000 children are born normal rather than disabled; as a result of the latter, 1,000 normal children are born rather than 1,000 disabled ones. Parfit suggests that there is no moral reason to favor Pregnancy over Preconception Testing if there are funds for only one program, even though the consequences of the former are person-affecting, the consequences of the latter impersonal. For Parfit, our indifference between the two programs suggests that an impersonal standard of appraisal governs both. Applying such a standard to the radiation case, we would clearly favor the lower level because it has a better expected outcome than the higher level, even though no one would be worse off (or at greater risk of harm) under the latter than they would be under the former.

In the literature on the non-identity problem two approaches have emerged to resist the sway of impersonal morality. One, which I will call containment, tries to set terms for the peaceful co-existence of impersonal and person-affecting considerations. It concedes that the former account for our indifference between Parfit's two medical programs and for our preference for lower over higher radiation. But it tries to avoid the array of perverse implications associated with a general utilitarianism—in particular, that people do wrong by failing to have children who would significantly increase net utility. One containment strategy is to adopt a negative utilitarianism, which aggregates harm but not benefit, and would not impose a duty to create happy children (Brock, 1995; Wolf, 1997). Another strategy is to limit the positive utilitarian imperative to maximize happiness to "same number" choices about which people to add to the world.

A second approach, which I will call assimilation, tries to identify a person-affecting objection in identity-affecting cases. Thus, it could be claimed that if the use of the higher radiation level caused a harmful mutation, it would violate the rights of the person who owed her worthwhile existence to that mutation. Or it could be claimed that the doctor who uses the higher level of radiation is responsible for the harmful mutation but not for the worthwhile existence caused by the mutation. A variant of this approach, which I favor, claims that the doctor who uses the higher level of radiation should be treated as if he risked harm to a person who could have existed otherwise, as long as he did not act from a desire to create a person whom he believes could only exist with the mutation. (A third general approach is to deny that "contingent future people" can have any objection to the acts necessary for their existence. But this claim, made by David Heyd (1992, pp. 21-64) strikes me as utterly implausible—it denies a person-affecting objection even for the deliberate creation of a child with the worst conceivable genetic disorders, a child who would be almost universally regarded as better off dead. I will not discuss this approach further.)

4.1 Containment

Negative utilitarianism appeals to a commonsense, non-comparative notion of harm: pain, suffering, and frustration count as harms whether or not the person they affect would have been better off without them. The advantage of requiring the minimization of harm rather than the maximization of utility is that the former imposes no duty to procreate early and often, or, more generally, to engage in the relentless pursuit of utility (Brock, 1995). Its disadvantages, however, have the same source. Since it considers only harm, it cannot take account of the pleasure, happiness, and fulfillment in people's lives in assessing the morality of creating them. It appears to condemn procreation altogether, since even the happiest, healthiest child experiences some pain, suffering, and frustration. (Interestingly, some "assimilationists," who favor a person-affecting account of non-identity cases, have been prepared to argue just this: that the benefits of a life worth living cannot compensate, or fully compensate, for the harms it contains; see section 4.2.) If we try to amend negative utilitarianism to forbid only the infliction of gratuitous or uncompensated suffering, then it would not condemn procreation, but it would also not condemn the creation of a disabled child whose life is barely worth living instead of a disabled child with an abundantly worthwhile life, as long as their disparity in well-being was attributable to differential happiness rather than differential suffering.

A second containment strategy is to circumscribe positive utilitarian duties rather than recognize only negative ones—specifically, to limit the scope of impersonal appraisal to choices among the same number of future people. If we are creating a fixed number of new people, we should pick the happiest people possible, even if we have no general duty to make the world as happy as possible. The challenge for this strategy is to explain how we can have a conditional duty to select the happiest person *if* we are bringing a person into existence, without having an unconditional duty to create happy people or otherwise maximize happiness. Jeff McMahan has reviewed various attempts to provide such an explanation, notably Parfit's own, and found them wanting (1998, pp. 234-41); I will not repeat his arguments here.[19]

4.2 Assimilation

Given the difficulties of restricting an impersonal account so that it does not impose onerous duties of procreation or yield other perverse results, the more promising approach to the non-identity problem may be found in accounts that try to tease out a person-affecting complaint in identity-affecting cases. When the agent, through indifference or excessive zeal, causes a person to come into existence in a harmful state, he appears to wrong that person, even if she could not have existed in any other condition. The containment approach cannot account for this appearance, except as the persistence of illusion.

Some assimilationist accounts begin, like negative utilitarianism, with a non-comparative notion of harming, under which a person can be harmed by actions that do not leave her worse off than she would have otherwise been. Others adopt a comparative notion of harming, but insist that a person can be wronged even if she is not harmed. In either case, they take the position that the agent can violate someone's rights, or wrong her, by actions that do not leave her worse off overall. To take one of James Woodward's (1986) more striking examples, the Nazis violated Viktor Frankl's right not to be abducted, imprisoned, and enslaved—a violation in no way justified or mitigated by the spiritual transformation that left Frankl with a better life than he would have had if he had not been sent to a death camp. Or, to take a more mundane example from Joel Feinberg (1982), a taxi-rider's right to competent service is violated, and the violation is in no way mitigated or excused, if the driver's lack of familiarity with the area causes the passenger to miss a plane that later goes down in mid-flight. It is also clear that the agent can violate the rights of someone who does not yet exist; Feinberg gives the example of a person who secretly places a bomb in an elementary school, set to go off years later. The bomber surely violates the rights of the children killed or injured in the blast, even if he acted before any of them was conceived. If a person can have her rights violated by acts that predate her existence and by acts that do not leave her worse off overall, then perhaps the putative victim in non-identity cases can, at least in some of those cases, claim a rights-violation.

The challenge is to identify the right violated. It is not obvious that a child has the right not to be born with a disease or susceptibility that leaves her with a life worth living. Most of us do not regard parents who pass on genetic diseases or disease susceptibilities to their children, rather than screen or adopt, as violating the rights of those children, so long as the children have lives worth living and they receive appropriate treatment for their conditions. We do not regard those parents as excused for violating their children's rights by their strong desires or limited options (as we might, say, if they failed to feed their children adequately because of their own poverty); we do not think there is any rights-violation to excuse. But how to distinguish such parents from the overzealous doctor, whose intervention also brings into existence a child who could not have existed except in a harmful condition?[20] My own account, which I will offer shortly, emphasizes the reasons for creating that child rather than some other. First, however, it is instructive to consider a more radical response: that a child is in fact wronged, by her parents or her doctor, merely by being placed in harm's way without her consent, even if the harms she faces are not the extraordinary ones associated with genetic diseases, but the routine ills to which all flesh is heir.

Several assimilationist accounts argue for an asymmetry between harms and benefits, in order to claim that the creation of a normally healthy child constitutes a presumptive wrong or rights-violation. Thus, David Benatar challenges the assumption that a life with a favorable ratio of pleasure to pain is better, all things considered, than non-existence. That assumption, he argues, ignores a critical asymmetry: while the absence of pleasure is not bad, the absence of pain is good. No matter how much pleasure a life contains, or how favorable a ratio of pleasure to pain, it is not bad to fail to produce that pleasure; no matter how little pain a life contains, it is good to avoid that pain. "Because there is nothing bad about never coming into existence, but there is something bad about coming into existence, all things considered non-existence is preferable" (Benatar, 1997, pp. 348-9). Seana Shiffrin (1999) argues that we are only justified in causing harm to a person without her consent to prevent a greater harm, not to confer a benefit, however great. A surgeon can operate on an unconscious trauma victim to prevent her death or permanent injury, but not to improve on her pre-trauma health or longevity. Since it is no harm not to come into existence, and since consent to being brought into existence is impossible, we are not justified in creating a new life, with its inevitable tribulations, no matter how great its anticipated benefits.

Benatar and Shiffrin both treat the creation of any new person as a presumptive wrong, and Shiffrin's account, at least, would give any child, however happy, a grievance against her parents for creating her. This strikes me as a *reductio ad absurdum* of the claimed asymmetries. To invert Abraham Lincoln's famous remark about slavery, if it is not right to create people we expect to lead worthwhile lives, what is right? It is more plausible to infer from the presumptive rightness of creating people with worthwhile lives that even if we are not in general justified in causing harm to a person without consent to bestow any other benefit, we *are* justified in causing harm to a person to bestow the foundational benefit of existence. But if we conclude that the mere inevitability of pain and suffering does not make procreation wrongful, we need to explain why some people appear to have a grievance for some of the harms caused by the very acts responsible for their worthwhile lives.

Matthew Hanser (1990) argues that an agent who causes harm that is necessary for a person's worthwhile existence may wrong her, in that he may be responsible for her harm but not her existence. If a future generation is wiped out by a massive leak of nuclear waste, another of Parfit's (1984) examples, our risky energy policy would make us responsible for the deaths of the people in that generation, but not for their worthwhile lives, despite the fact that none of them would have existed if we had adopted a safer policy. What puts our moral balance sheet in the red is a basic asymmetry, not between preventing harm and conferring benefits, as Benatar and Shiffrin claim, but between being responsible for causing harms and for conferring benefits. Although we could have foreseen, but did not desire, either the life-threatening or the identity-altering effects of our energy policy, foresight alone makes us responsible for the former but not the latter.

Hanser suggests, however, that the attribution of responsibility for harm and benefit may be different for "intrinsically reproductive acts": "A woman's act in becoming pregnant *does* make her responsible for the resulting child's existence," since "her act is one whose purpose it is to bring a human being into existence." But "it is not clear that act makes her responsible for the child's foreseen defects," since the child's existence but not its defects are a direct result of her procreative acts (1990, pp. 68-9). Even if the mother is responsible for the harm as well as the benefit, her moral balance sheet is still in the black, as long as the benefit outweighs the harm. Thus, parents rarely face the reproach, or the burden of justification, that they routinely confront on Shiffrin's or Benatar's account.

Hanser is, I believe, on the right track. But his account is overly permissive in one respect and overly restrictive in another. As to the former, Hanser would hold that a couple who gave birth to a child with Tay Sachs disease only because they were too busy to obtain prenatal testing would be responsible for whatever benefit there was for the child in his brief life, but not for the painful, debilitating disease that ended it. As to the latter, it is not clear why an act's "intrinsically reproductive" nature, rather than the intentions with which it is performed, should affect the attribution of responsibility.[21] I will argue that the character of the agent's intentions with respect to the identity of the person he creates determines whether he can be regarded as having wronged that person by bringing her into existence in an unavoidably harmful state. And I will claim that the relevant intentions can be sought in a variety of acts beyond the species-typical ways of making babies.

4.3 Discriminatory Intent and Reproductive Acts

Consider a variation on a real case. A few years ago, Dr. Jacobson, director of a northern Virginia fertility clinic, was convicted of using his own sperm to fertilize the eggs of women who thought they were receiving sperm from their husbands or anonymous donors. Jacobson claimed that he had used his own sperm to assure the health of the resulting children (a nicely impersonal pretext), but the consensus was that he had yielded to the Dawkinsian urge to replicate his own genes as often as possible.

We can alter the case to make Jacobson both more poignant, and in one respect, more blameworthy: he is childless, and homozygous for a dominant genetic condition that causes a debilitating neurological disorder. He badly wants a child with a genetic link to himself, even knowing that any child with that link will have the same neurological disorder that afflicts him. He fertilizes a "spare" egg from one of his clients, and hires a surrogate to bear the child, who has, as he expected, a life well worth living despite her disorder. This Dr. Jacobson has surely committed an offense against his client, stealing her reproductive material and making her, without consent, the genetic mother of a child with a serious disorder. But in contrast to the overzealous doctor, whose use of the higher radiation level brings a child into existence with a harmful mutation, Dr. Jacobson does not even appear to wrong the child he helps create, and that child does not appear to have a grievance against him. In impersonal terms, the doctor's actions appear even worse than those of the overzealous doctor. But the fact that his actions are impersonally worse does not contribute to the appearance of a person-affecting wrong; if anything, they suggest his profound, if excessive, partiality toward the child he creates.

What appears to deny Jacobson's child a grievance is not merely his knowledge that she could only exist with the genetic defect he transmitted. The overzealous doctor may have known that the harmful mutation he risked was necessary for the existence of the child born with that mutation. But that child still seems aggrieved in a way that Jacobson's child is not. What denies Jacobson's child a grievance is rather his intention to create a child who (he believes) can only exist in a harmful state. The overzealous doctor has no such intention: he simply wants to cure the defect, and he does not care whether in doing so, he causes one person to come into existence rather than another.[22] Dr. Jacobson's reason for wanting a child of his own may not be a good one, and may even be regarded as a bad one by people who regard Dawkinsian urges as atavistic, but it gives him a clear preference for a child created with his own defective sperm rather than a child created with someone else's healthy sperm. His intention to create such a child seems to preclude any complaint from the resulting child about the genetic harm he inflicts.

We can confer at least the appearance of a grievance on Dr. Jacobson's child by assuming that both Jacobson and the unwitting egg donor are *heterozygous* for the genetic susceptibility gene, and that Jacobson negligently fails to screen his sperm for that gene and thereby assure that he contributes a healthy gamete. As a result, a child is born with that susceptibility. I think most people would have a stronger inclination here than in the homozygous case to see the child as aggrieved, although he could not have existed if the doctor had dutifully screened his sperm. Even those who insist that the appearance of victimization is an illusion in all non-identity cases would, I suspect, find it a more powerful illusion in the case of a negligent failure to screen than in the knowing selection of a gamete with the disease susceptibility. While Dr. Jacobson does have reason to want a child with his genes, he presumably has no reason to want a child with the susceptibility gene (although we could imagine cases where, however perversely, he does, e.g., where it was a genetic heirloom from a beloved forebear). We could draw the same contrast in a single case by making Dr. Jacobson homozygous for one disorder and heterozygous for the other. In that case, the child he created would have a grievance with respect to the latter disorder but not the former, even though she could not have existed without either condition.

But is there any moral basis for this felt difference? After all, in both the homo- and heterozygous cases, the child could not have enjoyed a worthwhile existence without the genetic susceptibility—a child lacking that susceptibility would not have been *her*. Those who doubt that the child has a grievance in either case (or for either condition) might claim that Dr. Jacobson's desire for his own child merely makes his contribution to the child's worthwhile life salient. But Jacobson's contribution to the child's life is no less salient in the heterozygous than in the homogenous case, so something else must lie behind our different reaction to the two cases.

Those who believe, on the other hand, that the child has a grievance in both cases might claim that the grievance is simply mitigated or excused in the homozygous case by Jacobson's intense longing for a child of his own. That longing cannot mitigate or excuse his creation of a diseased child in the heterozygous case, however, since he could have had a healthy child of his own by screening his sperm.

I do not think the psychological burden on Dr. Jacobson can account for the differing appraisal of the two cases. His longing for a child of his own gives him a reason, not an excuse, for creating such a child. The difference in apparent victimization persists if we reduce the burden in the homozygous case and increase it in the heterozygous case. The child created by the homozygous Dr. Jacobson does not appear to have any more of a complaint (though others may) if the doctor would have been almost as happy to adopt, but finally decided to try for a child of his own. The child created by the heterozygous Dr. Jacobson does not appear to have any less of a complaint (though others may) if the screening that would have assured a healthy gamete was expensive and painful. Even if his longing were not so intense, the homozygous Dr. Jacobson would still have wanted to create a child he believed could only exist in a harmful state; even if he were faced with great expense and discomfort, the heterozygous Dr. Jacobson would *not* have wanted to create a child he believed could only exist in such a state.

Hanser's account lacks the resources to distinguish the homozygous and heterozygous cases. In both, Dr. Jacobson engages in "intrinsically reproductive acts" intended to bring a new person into existence; in both, he is responsible, in one familiar sense of 'responsibility,' for the worthwhile existence of the child. In both, he is responsible for its being a child with a disease susceptibility that is conceived, even if he is not responsible for the susceptibility itself (to echo a distinction Hanser makes; see Hanser, 1990, p. 69). In the homozygous case, he knowingly makes that the case; in the heterozygous case, he does so negligently. If anything, we hold people more responsible for harm they cause intentionally than harm they cause negligently. The challenge is to explain why only the homozygous Dr. Jacobson can invoke the child's worthwhile existence to justify the harm she unavoidably suffers, when he is no more responsible in a conventional sense for her existence than the heterogenous Dr. Jacobson, and at least as responsible for the harm.

One basis for distinguishing the two cases lies in the connection between the harmful condition and the identity of the child Dr. Jacobson wants to bring into existence. Because the homozygous Dr. Jacobson sees the genetic disorder, and its attendant pain, as necessary to the life of the child he wants to create, while the heterozygous Dr. Jacobson does not, the former but not the latter can be said to cause harm to the child for its own sake. The homogenous Dr. Jacobson would be similar in this respect to a pregnant woman who declined to abort a fetus with a genetic defect. Because of her attachment to the child who would develop from that fetus (whether or not the fetus and child were the same entity), she could be said to expose that child to harm for its own sake.

Clearly, however, Dr. Jacobson cannot have the kind of particularized attachment to an unconceived child that parents often have to existing children, or even the less particularized attachment they sometimes have to fetuses, and to the people they will develop into. Benatar asserts that "children cannot be brought into existence for their own sakes" (1997, p. 351). Robert Adams suggests that only God has that capacity: "Knowing more about the future than we do, perhaps he can rightly love in advance things that we can love only in retrospect" (1979, p. 61). Loving a future child who can only exist in a harmful state, God is not obliged to create a happier child instead. Jacobson, in contrast, knows nothing about the child he brings into being except his pedigree, and he can hardly possess the kind of love that Adams' God has for the people he has not yet created.

And yet, however abstract or threadbare his attachment, Dr. Jacobson does desire to give life to a child whom he believes can only enjoy it in a harmful state. Although his motivation may be narcissistic, it is also, in a minimal sense, other-regarding. This may seem like a thin reed on which to rest his immunity from reproach for the harm he causes. But for creatures like us, capable of loving others only partially and conditionally, it may be enough.[23'] It is difficult to comprehend how we, unlike Adams' God, could ever regard an unconceived person as an end-in-herself.[24] But perhaps we can have sufficient attachment to a future person to enjoy, to a limited extent, the divine prerogative of bringing that person into existence, rather than a better or happier person.

It is this possibility that Laura Purdy overlooks when she suggests that a woman who fails to abort a fetus that will develop into a child with significant health problems is not sufficiently sensitive to the pain and frustration that are likely to result from those problems: "the thought that I might bring into existence a child with serious physical or mental problems when I could, by doing something different, bring forth one without them, is utterly incomprehensible to me" (1996, p. 58). For the pregnant woman, at least one who adopts what William Ruddick (2000) calls the posture of "maternalism" toward her fetus, those problems are attached to a creature she is already attached to, and she has a morally significant reason, born of that attachment, for preferring the existence of the child who will develop from that fetus to the existence of some other child she might create. In declining to abort, the woman acts for the sake of a child whom she views largely in prospect.

We can attribute the same person-affecting motivation to a pregnant woman who declines extensive prenatal therapy because of its threat to the identity of the fetus she is carrying, and to the child that will develop out of the fetus without such intervention.[25] She may reasonably fear that that child will not come into being if the fetus is subject to extensive therapy, and she may reasonably dread that loss, even if it does not involve the death of a person. If the mother declines therapy in order to protect the identity of the person who will emerge from the untreated fetus, she can be regarded as having put that person at risk of harm for his or her own sake. While she may exaggerate the threat to identity, or underestimate the risk or magnitude of harm without therapy, she has in principle a justification for declining therapy that would preclude any grievance on the part of the child who is born without its benefits.

There are limits to the motivation that will count as minimally other-regarding or person-affecting. We would not regard a person as having the requisite motivation if he deliberately created a diseased child only to get the large sum of money offered for lending it to a research laboratory, nor would we attribute that motivation to a doctor who deliberately substituted a defective embryo for a healthy one to avenge himself on a couple who damaged his reputation. Both have a desire to bring a specific kind of child into existence. But there must also be some desire to benefit the child by bringing him into existence, even if that is not the only, or the predominant, motivation.[26]

The underlying intuition I have been attempting to tease out can be expressed in the following way: if the agent is motivated by a desire to give the benefit of existence to a child whom he believes he can only bring into existence in a harmful state, we regard him as engaging in the harm-producing action for the sake of the child he brings into existence and permit him to invoke the child's worthwhile life to justify the infliction of that harm. We treat an agent who has no preference for the existence of a child with a harmful condition over that of a child without that condition *as if* he gratuitously harmed the former by creating him.[27] In canceling Preconception Testing, the state does not act from any preference for the 1,000 children who will be born with a disease susceptibility over the healthier children who would have been created otherwise. For this reason, we treat the cancellation of

Preconception Testing as no different than the cancellation of Pregnancy Testing. This is also why the overzealous doctor gains no immunity by altering the identity of the fetus.

For those committed to a person-affecting account of the non-identity problem, however, it may not be enough to treat the agent indifferent to the identity of the person he harms *as if* he harmed someone who could have existed otherwise, since he did not in fact harm someone who could have existed otherwise. A motivational account may seem too agent-centered, failing to explain why the person harmed has a complaint despite owing her worthwhile life to the agent's conduct. There are two ways of teasing out a rights-violation that attempt to capture the special standing of the putative victim to complain. One recognizes a "first level" right against being exposed to harm, whose violation is only justified when the agent exposes the victim to harm for the right reason; the other recognizes a "second level" right against being exposed to harm except for the right reason.

The first approach would give the child (per Shiffrin, 1999) a presumptive grievance if she was born in a harmful state, but would also give the child's progenitor (contra Shiffrin) a possible justification for placing her in that condition, if she had a worthwhile life that she could only enjoy in that condition. This justification, however, would be available to the progenitor only if he was motivated by a desire to bring into existence someone he believed could only exist in a harmful state.[28]

Alternatively, we could recognize a person's right not to be harmed except for his own benefit (and for other reasons that are not relevant here), a right which is violated if the victim's benefit—including his worthwhile existence—is not (among) the agent's reasons for causing him harm. This would be a second-order right in the sense that it would give a person a claim against an action done in the absence of a particular intention or motivation. Some philosophers, such as Jonathan Bennett, would resist the recognition of such a right on the ground that morality, as a guide for deliberating agents, concerns acts, not intentions or motives (Bennett, 1981). The formal issue of whether the reference to the agent's motive occurs in the formulation of the right or of the conditions justifying its violation is less important than the moral issue raised by both formulations: whether our claims against each other depend not only on what we do to each other, but on the intentional attitude with which we do it.

University of Maryland
College Park, Maryland, U.S.A.

ACKNOWLEGEMENTS

I have benefited greatly from comments on earlier drafts from Alan Strudler, Jeff McMahan, Dave Wendler, Robert Wachbroit, and Lisa Parker.

NOTES

1. Persson (1995) uses this term to refer to roughly the same configuration of views; he opposes it to animalism--humanism with a biological face.

2. This appears to be entailed by some accounts of the necessity of origin; e.g., Noonan (1983). The significance of origin is suggested by the familiar example of a statue made of clay. Imagine that the sculptor is making a human figure; the statue comes into being when it receives a sufficient degree of articulation to count as a distinct artifact. After that point, the statue can undergo substantial changes without a loss of identity. It would remain the same statue even if vandals, or the sculptor herself, cut off its limbs or even its head (recall the Little Mermaid of Copenhagen). But if the sculptor had decided early in her work to produce a statue without limbs, or had found herself without enough material to fashion limbs, then the limbless statue she produced, even if identical in appearance to the dismembered one, would arguably not have been the same statue she would have produced if she had endowed the figure with limbs. (While changes at the earlier stage might alter the identity of the statue, however, they would not alter the identity of the clay mass from which the statue was made.)

3. Thus, Persson (1995) implicitly rejects the significance of origin in proposing what he calls the "origination" view: that only one person can arise from one fetus, even if the person and fetus are not identical. In arguing against the view that different people can arise from a single fetus, he assumes that a difference in identity would have to be attributable to different psychological characteristics. He argues that no psychological characteristic could be essential to a person, since any such characteristic could be altered without altering personal identity—a baby severely brain-damaged by a drop would lose its cognitive potential, not its identity. But even if Persson is correct about the identity of the baby before and after it is dropped, he can extend his conclusion to changes occurring before a person comes into existence only by assuming that origin does not matter, so that similar brain damage inflicted on a fetus would not have resulted in a different person coming into existence (see Eliot, 1995).

4. This appears to be Oderberg's (1997) view.

5. Although those who accept species membership as the ground for the moral claims of fetus or zygote would seem to be paradigmatic "humanists," I will reserve that term for those who ground the moral status of the fetus or zygote on some kind of potentiality, since that is a more widely-held position, which attempts to capture a deeply entrenched intuition about potentiality.

6. Recently, however, Harman (2000) has argued that an early fetus enjoys some moral status if and only if it will *in fact* become a person. I find this claim highly implausible. It seems to imply, for example, that not only the mother but any third party with the means to induce abortion can determine at t2 whether the fetus was a morally significant being at t1. Moreover, it is unclear on Harman's view that the unsuccessful attempt to abort the fetus would be a wrong to it, unless the attempt caused some non-lethal harm. Attempting to kill a creature is typically regarded as wrong in part because the success of the attempt would gravely wrong that creature. But this would not be true of the attempted killing of a fetus, since the success of that attempt would make it the case that the fetus lacked the moral status to be gravely wronged. It is hard to see how an early fetus could be regarded as having any significant moral status prior to becoming a person if it was not wronged by the gratuitous or malicious attempt to kill it.

7. These conclusions also appear to follow from Harmon's (2002) position that an early fetus has moral status if and only if it will become a person. The fetus would have no moral claim against abortion, or for treatment of a fatal condition, but it would have a claim for treatment of a non-fatal condition if it were going to be brought to term. And if that treatment altered fetal identity, the fetus initially treated would go out of existence before it became a person, so there would be no morally significant loss.

8. Some philosophers would regard the phrase 'harms necessary for existence' as an oxymoron, since they hold that harms can only be understood counterfactually—in comparison to how things might have otherwise gone for the person (see Feinberg, 1982). Other philosophers accept a non-comparative notion of harm as pain, suffering, misery, and frustration, but would still deny that someone who causes a necessary harm could be said to harm or do harm—they regard "harming" as counterfactual. I will nevertheless use the terms 'harm' and 'harming' with respect to such necessary evils. I think those terms have a fairly clear non-comparative sense; someone who doubts this can substitute 'limitation' or 'defect.' I will also speak of causing a person to exist "in an unavoidably harmful state," or "with an unavoidably harmful condition"—in a state of infirmity or misery that she could not have avoided.

9. Humanism and personalism represent two standard constellations of views on the issues I have outlined. There are, of course, a variety of intermediate positions. Jeff McMahan (1995), for example, argues that "we" are embodied minds, who come into being with the neural structure necessary to support conscious activity—early in the third trimester. We are not essentially human organisms on McMahan's view, because we do not exist for the first several months of that organism's existence. But we are also not essentially persons, because we exist well before we are, or are capable of being, conscious or self-conscious.

Moreover, the opposition between humanism and personalism is by no means complete. Both hold that a creature's moral claims are not based on its mere membership in a biological species, but on its actual or potential exercise of certain mental or psychological capacities, such as self-consciousness and deliberation. Both thus deny that a human organism with no potential to acquire such traits would enjoy most of the rights of normal adult humans. Finally, humanists and personalists agree on the fourth issue, about the relative strength of person-affecting and impersonal concerns. For both, it is worse if a "successful" medical intervention causes one *person* to cease to exist and another to come into existence than if it preserves personal identity, even if the net gain in happiness or health is the same. The humanist, but not the personalist, would say the same thing about the loss and replacement of a human being, even if he were not (yet) a person.

10. And we would certainly be more inclined to treat the replacement of the nucleus as identity-altering than the replacement of some equally salient cell-part, like the mitochondrion. We might regard nuclear replacement as analogous to a brain transplant, but this suggests that we would endow the genome with the same sorts of executive features, which would be a mistake.

11. Fisher (1994) offers an extended critique of Stone, to which I am much indebted, that stresses his reliance on a single genetically-determined path of development. But Fisher mistakenly imputes to Stone the view that any modification of the zygote's DNA alters its identity.

12. In a footnote, Stone suggests that a "principled way to distinguish the "complete expression" of the human genome from cases of 'interference'" can be found in evolution: the path that counts as the complete expression will be the one selected by evolutionary selection "which in turn evolved because it leads to conscious goods" (1994, p 287, n.13). But this conceives of evolution as a far more coherent, purposeful process than it now appears to be: many genetic features have been "fixed" by accidental conjunction with other features that were the objects of selection pressures, and many others by their contribution to genetic replication in environments very different than the ones we find ourselves in.

13. This broad, coarse-grained notion of potentiality does not require any enumeration or ordering of human environments, which may be necessary to assess heritability—the quantitative measure of genetic and environmental contributions to a trait.

14. Jeff McMahan suggests that if "eye transplants were a routine procedure, we would have no trouble thinking of people born blind because of defective eyes as having the potential for sight" (1996, 22). If this is so, it is because the transplant recipients are already wired, and sculpted for sight; I doubt we would regard people as having the potential for sight if they required extensive neural reconstruction or implants, no matter how routine the procedure became. McMahan thinks our intuitive sense of potentiality requires the retention of existing "neural hardware," but I believe that the intuition rests on assumptions about "software"—a person requiring a large amount of brain tissue to see might still be regarded as having the potential for sight if the missing parts played only a passive or facilitating role in vision, e.g., as (mere) medium for neural impulses. What seems to matter is the extent to which the structure or plan for vision is already in place; the extent to which the intervention merely supplies the material rather then the formal causes of vision; the extent to which the direction of the visual processes resides in the surgeon or the patient.

15. In a somewhat similar vein, Tooley suggests that the potentiality of an entity to acquire certain properties could be treated as more or less "passive" depending on how many of the "positive causal factors" sufficient to bring about those properties it lacked (1983, 167-168). Tooley doubts, however, that there is a non-arbitrary way of defining a potential person in these terms; he also doubts that there is a good account of why it would be a serious moral wrong to kill a potential person.

16. A rough analogy to plastic surgery suggests that these dimensions, or their somatic counterparts, have uncertain or debatable weight even in the case of already-developed phenotypes. When does the surgeon give you a "new face" rather than improve your old one? Does it depend on how much of your original facial tissue is preserved or replaced in the surgery? Does it depend on the extent to which your old contours guide the reconstruction? Would it matter whether, in the absence of such "internal" cues, the

surgeon was guided by standard human anatomy or by his own idiosyncratic vision of human beauty? Would your old face be less likely to survive in a novel than a standard reconstruction? The uncertainty is that much greater when a genetic "surgeon" brings about changes to a phenotype that has not yet emerged.

17. Stone must confront these questions, because he believes that a creature may have the strong potential for conscious adulthood even if it requires minor genetic alterations to achieve that outcome. How extensive or substantial can genetic alterations be and still fulfill rather than enlarge the fetus' potential? Perhaps a genetic modification that merely achieved the same end result as a later somatic intervention, e.g., correcting a circulatory disorder, could be seen as fulfilling rather than enlarging the fetus' potential. But if the genetic modification required to effect such a result were quite extensive, it might not be seen as identity-preserving, and therefore as beyond the potential of *this* fetus.

18. A countervailing person-affecting reason might be found in the interest of the person who would come into being with level 1 radiation in not being replaced by another person. But that person does not yet exist, even in embryonic form (if she could be said to exist in that form). If her interest in having herself rather than someone else come into existence supplies the doctor with a person-affecting reason for making that the case, then he will have an indefinite number of such reasons, corresponding to the number of alternative persons who could be created by his intervention.

19. McMahan proposes an alternative containment strategy, which he calls the "Encompassing Approach" (1998, p. 243). He asserts that person-affecting and impersonal considerations are "distinct and non-additive," and that the latter may in some contexts have almost as much weight as the former. McMahan suggests that this approach can explain the lack of felt difference in the moral urgency of two medical testing programs described by Parfit. The cancellation of Pregnancy Testing *is* marginally worse than the cancellation of Preconception Testing because it involves an adverse person-affecting consequence: children who could have been healthy are born diseased. But it is *only* marginally worse, since the impersonal consequence of canceling of Preconception Testing—the creation of diseased rather than healthy children—is nearly as bad as the person-affecting consequence of canceling Pregnancy Testing. The impersonal considerations in the latter case do not get added to its person-affecting considerations, which would make Pregnancy Testing *far* worse than Preconception Testing. Rather, the person-affecting considerations absorb or subsume the impersonal ones when they are both present, leaving Pregnancy Testing only marginally worse.

But the Encompassing Approach is silent on how impersonal considerations weigh against person-affecting ones when they do not concern the same narrowly defined effect, e.g., "being born with a disease susceptibility." In regarding the impersonal consequences of canceling Preconception Testing as offering almost as strong a reason against cancellation as the person-affecting consequences of canceling Pregnancy Testing, McMahan seems to allow that impersonal considerations can readily outweigh person-affecting ones, as, for example, in a case where the disease susceptibility prevented by Preconception Testing is slightly worse than that prevented by Pregnancy Testing. Moreover, it provides no explanation of why impersonal considerations should weigh against the cancellation of Preconception Testing but not for the creation of happy children, unless those considerations accord a strong priority to avoiding harmful conditions.

20. Melinda Roberts (1995) proposes a right whose violation would confer a grievance on many people harmed by the very acts that bring them into existence. She argues that a procreative agent has a duty to do as well as possible for the people he creates, and may violate that obligation in causing harm to the persons he brings into existence. Even if that person *would* not have existed without the harm-inflicting act, she will have a complaint if she *could* have existed without that harm. Roberts understands "could" in terms of "practical possibility": more than logical possibility, but consistent with vanishingly small odds. Thus, Roberts holds that it is practically but not logically impossible for a child who has a chromosomal abnormality with complete penetrance and no known treatment to avoid the harm associated with that abnormality, while it is not practically impossible for an egg to be fertilized by a given sperm, regardless of the number of sperm released. If the harm-producing act did function like a gamete lottery, then even if the child arising from a particular combination would not have existed if the lottery had not been conducted at that time, it could have existed even if the lottery were conducted at that time under less harmful auspices. For example parents who produce a child after signing a contract to sell him into slavery (Kavka's slave-child case) could have produced the same child without signing the contract. The *ex-ante* odds of producing that particular child are equally and infinitesimally small with or without the

contract, so even in terms of expected outcome, the parents who sign the contract would not have done the best they could for that child.

That is clearly not the case, however, for the overzealous doctor. Although an indefinite number of children could have been produced by the high level of radiation the doctor employs, it is "practically impossible" (in Robert's sense of not being feasible under current science and technology) that the child who is in fact produced could have come into existence without those harmful effects. Thus, however slight the odds of producing a particular child with the high level of radiation employed, the doctor did as well as possible for that child, since that child *could not* have existed, and had a worthwhile life, with a lower, "safer" level of radiation.

21. It may be that by "intrinsically reproductive acts," Hanser means acts with the purpose of bringing a new person into existence, but that is not clear.

22. I will refer interchangeably to the "intention," "desire," "reason," or "motive" to create a child or kind of child whom the agent believes can only exist in a harmful state. I will leave it to others to correct my usage; I do not think that the plausibility of my account depends on such matters of terminology, important as they are in other contexts.

23. Adams himself maintains that we have a duty not to create children "notably deficient...in mental or physical capacity," because those children will have less of a capacity "to enter into the purposes that God has for human beings as such" (1972, p. 331). But he does not think that we wrong such a child if it could not have existed without such a deficiency (p. 327).

24. The proposal I have sketched has affinities with one other account of the non-identity problem which tries to accommodate the conviction that the child harmed by the actions that bring him into existence may have a grievance against his progenitor. Gregory Kavka (1982) suggests a modified categorical imperative, which condemns the creation of a person as a means only. This imperative makes it wrong to create a child with a serious disease or susceptibility, or even a perfectly healthy child, solely for cash, even if it would have a life worth living. It would also explain why we mitigate our condemnation of the homozygous Dr. Jacobson—while he may create a child for narcissistic reasons, he is not creating that child as a means only.

It might seem that the heterozygous Dr. Jacobson would not violate Kavka's imperative either, since he does not create his child as a means only. Kavka, however, intends his imperative to apply where the creation of a child in a harmful state is a side-effect rather than a means. The heterozygous Dr. Jacobson would violate it by creating a child in an unavoidably harmful state as a result of his failure to screen his sperm.

This application of the imperative to side-effects, however, extends it to a very different kind of moral offense, the heart of which is insufficient regard, not exploitation. Kavka would maintain that the heterozygous Dr. Jacobson exploits the child he creates in "drawing against" his existence to justify actions he carries out "for his own reasons." But it is difficult to regard this justification, however disingenuous, as exploitative, especially since his only reason for failing to screen his sperm may be his impatience to give life to a biological heir. The objection to the heterozygous Dr. Jacobson is rather that he cannot offer even a minimally adequate justification, since his reasons did not require the creation of a child with a serious disease or susceptibility. Kavka's account cannot capture this objection.

Moreover, Kavka's imperative seems too demanding. He applies it to condemn parents who have a second child to provide a kidney for the first. But it is not clear, to me at least, that those parents do anything wrong, if they expect to love and nurture the second child and believe he can have a life well worth living with one kidney. To the extent Kavka's imperative requires concern for an unconceived child as an end-in-itself, it condemns the birth of many, perhaps most children.

25. This attachment may be further attenuated if any one of an indefinite number of possible people may develop out of the fetus in the course of a normal or expected pregnancy, as Elliot (1993) suggests.

26. It may also be that the desire to benefit a child by bringing her into existence can only be found in the desire to create a child with a certain relationship to his progenitor, not (merely) in the desire to create a child with certain physical or mental traits. A child with a disease susceptibility created because his progenitor valued the red hair or tall stature associated with that condition would seem to have much the same grievance as a child created to get money from a research laboratory; a child would not have the same grievance if he were conceived with the preserved ovum of a beloved deceased spouse, an ovum known to carry that disease susceptibility. The desire to have a child with or of a beloved other may sometimes be hard to distinguish from the desire to have a child who possesses some of the traits

associated with that other. But the difficulty in making this distinction merely corresponds to the difficulty in making the more familiar distinction between loving another person and valuing his traits.

27. This suggestion was anticipated in a general way by Feinberg's (1982) analysis of the non-identity problem. Although it is often read as a negative utilitarian account, it has the germ of a person-affecting one. Feinberg concedes, perhaps unnecessarily, that a negligent doctor would not violate the rights of the child unavoidably born in a harmful state, since he does not leave her worse off than she could have been nor adversely affect her interests. His only wrong is impersonal, "wantonly introducing a certain evil into the world." But Feinberg also suggests that the child may have special reason to resent the doctor, because the harmful condition she is in is the result of the "indifference to the possibility of human suffering" the doctor displayed by his willingness "to bring people with harmful impairments into existence for no morally respectable reason." The question is whether Feinberg would regard the mere desire to create a child who could not exist without a harmful impairment as a "morally respectable reason" that would preclude such resentment.

28. This kind of claim may look familiar to students of criminal jurisprudence, where there is a long-standing debate about whether the agent must be appropriately motivated to claim a legal justification for otherwise criminal conduct (Fletcher, 1978, pp. 559-65). On the approach Fletcher favors, a person who kills someone whom he correctly believes is threatening the life of a third party will only have a defense to murder if he was motivated by the protection of the third party, not, say, if he merely saw the attack as giving him a pretext for dispatching a hated rival. For Fletcher, this motivational requirement restricts the legal privilege of violating a prohibitory norm "to those who merit special treatment" by virtue of their motivation (1978, p. 565). Somewhat analogously, we might want to restrict the moral privilege of creating a person in an unavoidably harmful state to those who desire to create a child whom they believed could only exist in that state. A justification based on the good of the child's existence would then be available only to an agent who intended, with the requisite degree of particularity, to bring about its existence.

REFERENCES

Adams R.M. Must God create the best? Philosophical Review 1972; 81:317-32

Adams R.M. Existence, self-interest, and the problem of evil. The British Journal for the Philosophy of Science 1979; 29(4):53-65

Annis D. Abortion and the potentiality principle. Southern Journal of Philosophy 1984; 22:155-64

Baker L. What am I? Philosophy and Phenomenological Research 1999; LIX(1):151-9

Benatar D. Why it is better never to come into existence. American Philosophical Quarterly 1997; 34(3):345-55

Bennett, J. "Morality and Consequences. Lecture III: Intended as a Means." In The Tanner Lectures on Human Values. Vol II, S. McMurrin, ed. Salt Lake City: University of Utah Press, 1981.

Boorse, C. "A Rebuttal on Health." In What is Disease? (Biomedical Ethics Reviews), J. Humbar, R. Almeder, eds. Totowa, NJ: Humana Press, 1998.

Brennan, A., Conditions of Identity: A Study in Identity and Survival. Oxford: Clarendon Press, 1988.

Brock D.W. The non-identity problem and genetic harms—the case of wrongful handicaps. Bioethics 1995; 9(3/4):269-75

Elliot R. Identity and the ethics of gene therapy. Bioethics 1993; 7(1):27-40

Feinberg J. Wrongful life and the counterfactual element in harming. Social Philosophy & Policy 1982; 4(1):145-78

Fisher J.A. Why potentiality does not matter: a reply to Stone. Canadian Journal of Philosophy 1994; 24(2):261-80

Fletcher, G., Rethinking Criminal Law. Boston: Little, Brown and Co., 1978.

Hanser M. Harming future people. Philosophy & Public Affairs 1990; 19:47-70

Harman E. Creation ethics: the moral status of early fetuses and the ethics of abortion. Philosophy & Public Affairs 2000; 28(4):310-24

Heyd, D., Genethics: Moral Issues in the Creation of People. Berkeley: University of California Press, 1992.

Kavka G.S. The paradox of future individuals. Philosophy & Public Affairs 1982; 11(2):93-111

Kuhse, H., Singer, P. "The Issue of Moral Status." In *Embryo Experimentation*, P. Singer, H. Kuhse, K. Dawson, P. Kasimba, eds. Cambridge: Cambridge University Press, 1990.

McMahan J. The metaphysics of brain death. Bioethics 1995; 9(2):91-126

McMahan J. Cognitive disability, misfortune, and justice. Philosophy & Public Affairs 1996; 25(1):3-35

McMahan, J. "Wrongful Life: Paradoxes in the Morality of Causing People to Exist." In *Rational Commitment and Social Justice: Essays for Gregory Kavka,* J.L. Coleman, C.W. Morris, eds. New York: Cambridge University Press, 1998.

Noonan H. The necessity of origin. Mind 1983; 92:1-20

Oderberg D.S. Modal properties, moral status, and identity. Philosophy & Public Affairs 1997; 26(2):259-98

Olson E.T. Was I ever a fetus? Philosophy and Phenomenological Research 1997; 7(1):95-109

Olson E.T. Reply to Lynne Rudder Baker. Philosophy and Phenomenological Research 1999; 59(1):161-71

Parfit, D., *Reasons and Persons*. Oxford: Clarendon Press, 1984.

Persson, I. Genetic therapy, identity and the person-regarding reasons. Bioethics 1995; 9(1):16-31

Purdy, L.M. "Loving Future People." *Reproducing Persons: Issues In Feminist Bioethics*, Ithaca and London: Cornell University Press, 1996.

Roberts M.A. Present duties and future persons: when are existence-inducing acts wrong? Law and Philosophy 1995; 14:297-327

Ruddick W. "Ways to Limit Prenatal Testing." In *Prenatal Testing and Disability Rights*, E. Parens, A. Asch, eds. Washington, D.C.: Georgetown University Press, 2000.

Schaffner K. Genes, behavior, and developmental emergentism: one process, indivisible? Philosophy of Science 1998; 65(2):209-52

Shiffrin, S.V. Wrongful life, procreative responsibility, and the significance of harm. Legal Theory 1999; 5:117-48

Sober E. Evolution, population thinking, and essentialism. Philosophy of Science 1980; 47:350-83

Stone J. Why potentiality matters. Canadian Journal of Philosophy 1987; 17(4):815-30

Stone J. Why potentiality still matters. Canadian Journal of Philosophy 1994; 24(2):281-94

Tooley, M., *Abortion and Infanticide*. Oxford: Clarendon Press, 1983.

Wolf, C. "Person-Affecting Utilitarianism and Population Policy; or, Sissy Jupe's Theory of Social Choice." In *Contingent Future Persons,* N. Fotion, J.C. Heller, eds. Dordecht: Kluwer Academic Publishers, 1997.

Woodward J. The non-identity problem. Ethics 1986; 96:804-31

Wreen M. The power of potentiality. Theoria 1986; 52:16-40

Zohar N.J. Prospects for "genetic therapy"—can a person benefit from being altered? Bioethics 1991; 5(4):275-317.

CHAPTER 13

PAUL K. J. HAN

CONCEPTUAL AND MORAL PROBLEMS OF GENETIC AND NON-GENETIC PREVENTIVE INTERVENTIONS

1. INTRODUCTION

The burgeoning field of genetic medicine utilizes knowledge of the molecular basis of heredity to diagnose, predict the likelihood of, prevent, and treat disease. Because genetic medicine has the potential to push our diagnostic and therapeutic capabilities beyond present boundaries, it is appropriate to ask whether it also raises unique ethical problems. A prevailing view is that it does; many contributions in the popular press—for example, *Dangerous Diagnostics* (Nelkin and Tancredi, 1994), *Exploding the Gene Myth* (Hubbard and Wald, 1993)—suggest that genetic medicine traverses new territory, posing unique conceptual challenges and ethical risks that lie beyond the realm of ordinary, non-genetic medical interventions.

Expressions of this genetic exceptionalist view can also be found within the philosophy of medicine literature. Albert Jonsen *et al.*, for example, argue that genetic medicine poses a new form of disease risk having greater significance than conventional "epidemiological probability" (1996, pp. 622-3). Furthermore, they claim, genetic medicine will result in the creation of a new class of "unpatients"—persons neither diseased nor non-diseased—whose conceptual and moral status is ambiguous. In the same vein, Eric Juengst contends that some genetic interventions—what he calls "geno-preventive" interventions—entail a major shift in the goals and beneficiaries of medical interventions (1995, pp. 1595-6). Specifically, he sees geno-preventive interventions as being aimed at "preventing the intergenerational transmission of disease genes," and thus serving the interest of the public health, rather than the health of individual patients (Juengst, 1995, pp.

L.S. Parker and R.A. Ankeny (eds.), Mutating Concepts, Evolving Disciplines: Genetics, Medicine and Society, 265-286.
© 2002 *Kluwer Academic Publishers. Printed in the Netherlands.*

1595-6). He views this public health orientation as a problematic departure from the traditional "client-centered ethos" that legitimizes most other medical interventions.

The ethically cautious tenor of genetic exceptionalist arguments seems judicious given the rapidly evolving and novel nature of many genetic technologies. On closer examination, however, these arguments are difficult to maintain. For it is not obvious that distinctions commonly drawn between genetic and non-genetic preventive interventions are truly robust, or that genetic medicine's ethical problems are completely novel. For instance, while genetic technology might enhance the prognostic and diagnostic power of medicine, it is not clear how genetic risk is substantially different from traditional epidemiological probability. Likewise, the blurring of the conceptual distinctions between patients and non-patients, disease and health, can be seen as a consequence not of genetic technologies exclusively, but of all preventive interventions. Similarly, the shift in the goals and beneficiaries of intervention from individual risk reduction to the public health is a potentially problematic feature raised by non-genetic and genetic preventive interventions alike.

Prevailing expressions of the genetic exceptionalist view can be criticized as overstating the differences between genetic and non-genetic preventive interventions in two principal ways. First, they overlook the extent to which the difficult conceptual and moral problems of genetic preventive interventions are also raised by *non*-genetic preventive interventions. Second, they begin from debatable normative assumptions that themselves tend to ethically marginalize genetic as opposed to non-genetic interventions. These commitments include conceptual assumptions about genetic determinism and the nature of disease, and moral assumptions about the appropriate goals and scope of medical intervention and the good of medicalization. In the context of such assumptions genetic medicine becomes a uniquely problematic endeavor; however, the validity of these starting assumptions is open to question.

In this study, I develop these criticisms of the genetic exceptionalist view. I analyze some allegedly unique conceptual and moral characteristics of genetic preventive interventions: (1) the distinctive significance of genetic risk; (2) the blurring of the conceptual distinctions between patients and non-patients, disease and health; and (3) the orientation toward public as opposed to individual health. Considering these characteristics in turn, I argue that each is more properly construed not as unique, but as emblematic of all preventive interventions, non-genetic and genetic. I demonstrate how genetic exceptionalist arguments exaggerate their distinctiveness by disregarding the extent to which they are manifest by non-genetic preventive interventions. I then explore how genetic exceptionalist arguments depend upon particular conceptual and moral assumptions of a debatable nature. I advocate greater attention to the *similarities* as well as the differences between genetic and non-genetic preventive interventions, and to the normative assumptions that shape how genetic interventions are perceived.

2. GENETIC RISK, EPIDEMIOLOGIC RISK, AND GENETIC ESSENTIALISM

The claim that genetic medicine is ethically unique rests in part on the notion that genetic risk is conceptually unique. Jonsen *et al.*, for example, argue that genetic testing for disease susceptibility

> ...will *introduce* into the clinical transaction a *novel, almost unprecedented* form of prognosis....We still struggle to find the right words for the status conferred by the presence of these genes, calling it 'susceptibility,' 'predisposition,' 'propensity,' 'proclivity,' and referring to 'risk,' 'potential,' and 'likelihood.' Whatever the term, we are increasing the ability to inform a patient that he or she will suffer from a disease of greater or lesser severity, at some time in the future, with some probability, and in some cases almost with certainty. What is the ethical meaning of this sort of prognostic information? (Jonsen *et al.*, 1996, p. 622; emphasis mine)

They further assert that, while projections of risk in conventional, non-genetic medicine "often rest on epidemiological studies and are framed in statistical terms," the "tests of molecular medicine go further; beyond the educated guess or the epidemiological probability, these tests inform persons that they carry variant genes that confer what is often called 'an innate risk'" (Jonsen *et al.*, 1996, pp. 622-3). While they seem not to claim that genetic risk is entirely different from conventional epidemiologic risk—they acknowledge that the causal path from genes to disease is complex—they do regard genetic testing as conveying information of a substantially different kind.

The type of genetic exceptionalism expressed in these arguments conceives of genetic risk as distinct from epidemiological risk in two primary ways. In terms of its derivation, it is contrasted with the risk projections of non-genetic clinical medicine, which "rest on epidemiological studies" and represent a "statistical prediction derived from a population of patients" (Jonsen *et al.*, 1996, p. 622-3). For the derivation of non-genetic disease risks, the statistical method consists of correlating observed disease outcomes among individuals of a study population with the various baseline characteristics of these individuals; this process identifies 'risk factors,' a concept signifying characteristics associated with an increased probability of becoming diseased (Fletcher *et al.*, 1996). In this way, for example, cigarette smoking, hypertension, and hypercholesterolemia have been identified as prototypical non-genetic risk factors for coronary artery disease. The genetic exceptionalist view presumably sees genetic risk factors as being derived in some way other than conventional epidemiologic methods.

The second way in which genetic exceptionalist arguments distinguish between genetic and epidemiological risk is in terms of their meaning. The claim is that gene types detected by molecular testing confer an "innate" risk with a significance "beyond the educated guess or the epidemiological probability." By contrast, even the strongest of non-genetic coronary risk factors—e.g., extremely elevated serum LDL cholesterol—entail an indeterminate likelihood of future myocardial infarction; whether or not any particular individual will actually become diseased is at best an educated guess because of irreducible uncertainties and the limitations of aggregative reasoning (Asch and Hershey, 1995). The genetic exceptionalist view construes genetic risk as somehow transcending these limitations, as signifying more than an uncertain likelihood of future disease.

In both derivation and meaning, however, it is not clear that genetic and epidemiologic risk are substantially different. First of all, it is difficult to see how the disease risks associated with genetic mutations can be derived in any way other than the usual epidemiologic method of statistical correlation between putative risk factors and observed clinical outcomes in a study population. For genetic mutations no less than non-genetic characteristics, this method requires identifying individuals with the mutation, observing their clinical outcomes, and then comparing their outcomes to those of individuals without the mutation. While genetic mutations might be identified at the laboratory bench, their associated disease risks are ultimately quantified from epidemiological studies among populations of patients.

It is also difficult to see how genetic risk can *mean* anything substantially different from conventional risk in an epidemiological sense. It cannot merely be a quantitative difference that separates them, for there are no genetic risk factors that confer an absolute certainty—a 100 percent likelihood—of disease. Not even the adenomatous polyposis coli (*APC*) gene mutation on chromosome 5q21, which is commonly understood to be associated with an "inevitable" occurrence of colon cancer by age seventy in untreated mutation carriers (Giardiello, 1997; Peterson and Brensinger, 1996), can be interpreted as conferring this level of risk. The assertion of "inevitable" occurrence is an idealization of risk that assumes that mutation carriers will in fact live to age seventy rather than die of other causes. The magnitude of disease risk also accrues in time, rather than being immediately realized; an individual mutation carrier's statistical odds of developing colon cancer increase progressively as he or she lives longer, such that by age forty it is fifty percent, not approaching 100 percent until age seventy. The assertion of inevitability, furthermore, must be qualified by the fact that all statistics have an irreducible margin of error due to effects of chance and the inherent limitations of empirical observation. Thus "inevitable" can only mean something less than a 100 percent likelihood.

Moreover, whatever quantitative difference is alleged between the level of disease risks conferred by genetic and non-genetic risk factors is further diminished by the fact that there exist both ordinary non-genetic factors associated with very high disease risks—arguably on the same order of magnitude as genetic factors—and genotypic factors associated with relatively low disease risks. Many non-genetic cancer risk factors—e.g., exposure to chemical carcinogens, extreme elevations of serum cholesterol—are known to be associated with very high disease risks. At the same time, for most genotypes associated with neoplastic conditions—e.g., BRCA1 and BRCA2 mutations—the associated disease risks are far less than absolute (Couch *et al.*, 1997; Healy, 1997; Shattuck-Eidens *et al.*, 1997; Struewing *et al.*, 1997). The steps from genotype to phenotype are exceedingly complex; many steps in the pathogenetic process reduce gene penetrance and expressivity to such an extent that genetic constitution cannot confer a certainty of disease (Holtzman, 1996).

Might the crux of the genetic exceptionalist view simply be that these complex steps defy characterization at the present state of knowledge? If so, then the claim should only be that we need a better understanding of disease pathogenesis to obtain more accurate risk information. The genetic exceptionalist view, however, goes well beyond this weaker claim; it avers a popular perception that genetic risk is in fact distinct from epidemiologic risk (Nelkin and Tancredi, 1994). Jonsen *et al.*, for example, treat genetic risk as something unique; they conceive of this risk as "innate," even while acknowledging the subtleties of this claim (1996, p. 623).

The meaning of this innate-ness, then, becomes a critical issue in distinguishing between genetic and epidemiologic risk. In what sense can the disease risks conferred by genes be conceived as innate, by contrast to those conferred by non-genetic factors? The answer implicates a particular conceptual move, an adherence to a strong genetic determinism that assigns extraordinary causal significance and explanatory power to genes. As Nelkin and Lindee argue, this conceptual assumption leads to the equation of genetic risk factor with disease, correlation with causation, genotype with phenotype:

> In the quest to identify genetic predispositions, however, the statistically driven concept of correlation is often reduced to 'cause.' And possible future states, calculated by statistical methods, are often defined as equivalent to current status. (Nelkin and Lindee, 1995, p. 166)

Only in the context of such a strong determinism does it make sense to speak of a characteristic—genetic risk—that neither depends upon statistical derivation nor entails predictive indeterminacy. To characterize genetic risk as "innate" is to construe it as inexorable, immutable. Two questions must then be raised here: is strong determinism true, and is it a claim unique to *genetic* risk and thus genetic medicine? Definitively addressing the first question is beyond the scope of this chapter, although some doubts about the veracity of strong genetic determinism will be raised in addressing the second.

Strong determinism is not unique to genetic medicine and genetic risks. A similarly strong determinism can be discerned—and criticized—within the philosophy of non-genetic preventive medicine as well. Peter Skrabanek, for example, argues that preventive interventions as a whole are legitimated by the notion that risk factors represent more than characteristics statistically associated with future disease; instead, they become "reified into something real—part of the person's constitution" calling for intervention (1990, p. 162). Preventive medicine treats risk factors as *de facto* diseases in themselves; it endows risk factors with the same properties that Engelhardt views as defining disease in a functional sense:

> Choosing to call a set of phenomena a disease involves a commitment to medical intervention, the assignment of the sick role, and the enlistment in action of health professionals. (Engelhardt, 1974, p. 137)

The deterministic conceptual equation of risk factor and disease is illustrated by the examples of hypertension (Alderman, 1993, pp. 329-31) and hypercholesterolemia, risk conditions that involve a commitment to intervene which is indistinguishable from that directed to actual disease states. This is evidenced by the magnitude of healthcare resources devoted to the treatment of these risk factors, and the accepted and aggressive manner in which this treatment is sought after by patients and rendered by health professionals and industries. Their treatment also entails an assignment of the sick role, evidenced by the frequently-cited "labelling" and illness behavior that accompanies treatment of these risk factors (Alderman and Lamprot, 1990; Macdonald *et al.*,1984). Thus, the conceptual distance between risk factor and disease is greatly diminished—if not altogether eliminated—for many preventive interventions, genetic and non-genetic alike (Russell, 1994, p. 79).

Of course, this type of strong determinism can be criticized on various grounds. Strictly speaking from a conceptual standpoint, risk factors entail substantial predictive indeterminacy, and thus any simple equation of risk factor with disease state is unwarranted. Moral as well as conceptual objections can also be raised; some have argued that adherence to strong determinism has negative ethical consequences. With respect to genetic medicine, for example, Nelkin and Lindee see strong determinism as part of a larger philosophical worldview that they call "genetic essentialism," a reductive view of the self as a molecular entity, an equation of human beings with their genes (1995, p. 2). They argue that genetic essentialism has harmful social and political consequences: impoverishing human existence, encouraging fatalism, and discouraging social change by deflecting attention from social (as opposed to the individual and biological) conditions that need to be altered.

Just as claims relying on deterministic assumptions are not unique to genetic interventions, arguments against them are not restricted to genetic medicine. Skrabanek argues that deterministic thinking is one factor which serves, in effect, to exempt preventive medicine from traditional ethical constraints, and that it obscures the substantial uncertainties surrounding the effectiveness and safety of preventive interventions generally (1994a, pp. 31-6). Skrabanek sees "preventive interventionism," like genetic essentialism, as an ideology with negative social consequences (1986, pp. 133-4). It undermines personal liberty, because the preeminent value placed upon disease prevention legitimates public policies that subvert individuals' freedom to choose behaviors and lifestyles that pose health risks (Skrabanek, 1992, 1994a, 1994b; Zola, 1972, 1975). Deterministic beliefs in preventive medicine foster a moral viewpoint that blames victims for their disease, e.g., placing responsibility for lung cancer upon the person who smokes (Brett, 1989; Crawford, 1977; Skrabanek, 1994a; Svenson and Sandlund, 1990; Taylor, 1986; Veatch, 1980; Wikler, 1987). This "victim-blaming" excludes the import of social, economic, and political factors in the causation of disease, and diverts attention from the need to change these elements, e.g., supra-individual factors that encourage people to smoke, such as the political and economic influence of the tobacco industry and mass media. Rather than raising novel concerns, Nelkin and

Lindee's criticisms of genetic essentialism echo well-recognized moral problems of preventive medicine generally.

However the empirical, conceptual, and ethical issues surrounding deterministic thinking in preventive medicine are resolved, there will remain substantial difficulties in drawing clear distinctions between genetic and non-genetic disease risk. In both cases their derivation and meaning are strictly statistical and probabilistic. To attribute additional significance to genetic risk requires reliance upon a strongly deterministic view of disease risk that is not only open to moral, as well as conceptual, criticisms, but also similarly characterizes the conceptualization of non-genetic risk factors. Insofar as the purported conceptual distinctions between genetic and non-genetic risk are spurious, the genetic exceptionalist view remains difficult to uphold.

3. PATIENTS AND NON-PATIENTS: GENETIC MEDICINE, PREVENTION, AND THE CONCEPTS OF PATIENTHOOD AND DISEASE

A second main assumption embodied in the claim that genetic medicine is uniquely ethically problematic is the notion that genetic testing imparts an ambiguous conceptual and moral status to its subjects. Jonsen *et al.*, for example, argue that

> the ability to test for susceptibility of future disease has the potential to sweep into the world of medicine millions who experience no pain or discomfort or limitation. Some of these will be told that, based on genetic test results, they need not worry about a specific genetic risk...But many others will be put into the class of those who must wait and watch for a sign of disease, advised to organize their lives around colonoscopies and mammograms...Some others will take on the sick role and may develop psychosomatic symptoms of all sorts. Perhaps some may even live as invalids. But, whichever route is taken, those who are found to carry a genetic susceptiblity will constitute a new class of individuals for medicine: a class that might be designated as 'unpatients,' neither patients in the usual sense of being under treatment, nor non-patients, in the sense of being free of a medically relevant condition. (Jonsen *et al.*, 1996, p. 623)

Genetic testing for disease susceptibility supposedly creates a unique artifact: "unpatients" whose increased risk for disease places them somewhere between traditional conceptual categories of patient and non-patient, between the concepts of diseased and non-diseased.

Moral ambiguity supposedly follows from this conceptual ambiguity. Is it appropriate, on the one hand, for these inhabitants of a conceptual no-man's land to be treated as *bona fide* patients, knowing that such treatment has psychological, social, and economic costs—e.g., diminished sense of well-being, heightened vigilance and concern about disease, psychosomatic symptoms, and dependency upon medical professionals and interventions? Or on the other hand, should persons of ambiguous disease status be treated as non-patients; should the fact that these persons experience no pain, discomfort, or limitation free them from the reach of further medical intervention, and obligate health professionals simply to preserve their state of medically-innocent bliss? Jonsen *et al.* see genetic testing *per se* as placing physicians in a moral quandary over how to uphold their duty of beneficence. They believe physicians face a tension between dismissing these persons as members of the ranks of the "worried well," and at the same time being "only too willing to enroll those at risk in their lists of active patients, encouraging them to engage in incessant look-sees to measure the progress of that innate risk toward disease" (Jonsen *et al.*, 1996, p. 624).

This way of portraying genetic testing—as placing persons with genetic susceptibility to disease in a unique conceptual and moral class—is open to several criticisms. In the first place, one can argue that there are no clear, necessary, or ethically binding distinctions between patients and non-patients, disease and non-disease. In reality, persons forgoing or undergoing many medical interventions may not easily fit into one category or another. For example, should we regard as a non-diseased, non-patient the person who lacks objective evidence of cancer and yet fears disease and thus submits to cancer screening maneuvers? On the other hand, must we always label as a diseased patient the post-chemotherapy individual whose cancer remains in remission?

Moreover, no matter how starkly the lines are drawn, the distinctions between the concepts of patients and non-patients, disease and non-disease can be criticized as having limited ethical significance. One can argue that persons undertaking preventive measures need not base interventional decisions upon any *a priori* or fixed notions of patienthood or disease. In fact, some persons with symptomatically manifest disease nevertheless refuse to be treated like patients; they reject many facets of the sick role. Some even deny that they are ill and only receive medical treatment because it helps them to function better or live more fully. The "worried well," in contrast, embrace the sick role even as they fear the symptoms and diseases that would justify its adoption. The moral quandary over how any given person should be treated may in fact have little to do with the conceptual question of how the person should be regarded—as patient or non-patient, diseased or non-diseased. Rather than the question of which conceptual label should be applied to persons contemplating genetic and non-genetic medical interventions alike, there may be more salient issues—e.g., the magnitude of disease risk; the severity of the disease at issue; persons' values, perceptions, and preferences; what interventions are covered by their health insurance policies. The moral problem of what interventions should be enacted for or by any given person does not necessarily hinge upon an either/or choice between concepts.

In addition to these criticisms, the conceptual and moral problems that arise from positing stark dichotomies between patient and non-patient, disease and non-disease are by no means unique to genetic testing for disease susceptibility. Rather, they are emblematic problems of preventive medicine as a whole, insofar as all preventive interventions, by definition, are undertaken in persons who are either not actually diseased (primary preventive interventions) or are diseased but asymptomatic and hence "experience no pain or discomfort or limitation" (secondary preventive interventions) (Last and Wallace, 1992; Leavell and Clark, 1965). Whether it is tetanus immunizations, cholesterol-lowering therapy, screening mammography, or a host of other measures, preventive interventions treat non-diseased and/or asymptomatic persons who occupy an allegedly unclear conceptual and moral status somewhere between non-patients and patients. They are no longer medical virgins, but at the same time are also not fully initiated into the complete range of experience with disease and health care.

Thus, in the preventive medicine literature many analysts have noted conceptual and moral ambiguities paralleling those of genetic medicine's unpatients (Hulley *et al.*, 1980; Svenson and Sandlund, 1990). The well-worn term, "worried well," was itself coined as early as the 1960s by Dr. Sidney Garfield and other founding leaders of the Kaiser Permanente Health Plan, to describe the consequences of a prepaid healthcare system that implemented large-scale preventive interventions, such as periodic health examinations of healthy persons (Garfield, 1970). Speaking of preventive interventions generally, Allan Brett asserts that

> ...intervention based on risk factors differs qualitatively from the treatment of already manifest disease. It offers specific people therapeutic manipulations on the basis of statistical risks, not existing illness. An asymptomatic person is asked to consider accepting inconvenience, expense, and the risk of side effects to achieve benefits that are remote in time. (Brett, 1989, p. 679)

Robert Crawford (1980) argues that preventive medicine creates a new "potential-sick role" for persons. Skrabanek criticizes preventive medicine's "extension of 'health care' to the healthy" as a perversion of the goals of medicine (1994a, p. 31), while along the same vein, prolific critic Ivan Illich laments the fact that "people are turned into patients without being sick" (1976, p. 89).

It is thus mistaken to trace "the advent of the unpatients" as originating with genetic medicine. Unpatients are created by all preventive interventions involving non-diseased and asymptomatic persons. In all contexts, focusing upon a person's status as a patient or not obscures the question of which, if any, interventions are justified. One extreme response is an interventionistic approach, which treats persons as patients based on their risks for disease, and sees persons as diseased and deserving of medical intervention until proven otherwise. It views unsuspecting healthy persons as diseases waiting to happen. At the other extreme is a non-interventionistic approach, which treats persons as non-patients, viewing them as healthy until disease declares itself. The realistic approach of persons undertaking preventive interventions, however, may lie somewhere in between these extremes. In actual practice, a woman with the BRCA1 mutation is not treated the same as a woman with manifest breast cancer who is scheduled for surgery. A man with a cholesterol level of 300 cannot reasonably be given the same advice as someone with a level of 100; whether the advice the first person receives turns him into a patient depends on many factors, may not be decidable, and does not matter for his actual care.

Divergent and extreme positions regarding the definition of patienthood and the appropriateness of medical intervention not only create a false dichotomy, but also are grounded in very different conceptual assumptions, particularly regarding how disease is defined. In his book *The Strategy of Preventive Medicine*, Geoffrey Rose distinguishes between "categorical" and "continuum" conceptions of disease (1992, pp. 6-11). The categorical conception views disease as a discrete entity clearly separable from non-disease. Clinical medicine and research, which manifest a "demand for clear definitions" and assume that disease can be distinguished from normality, traditionally adopted the categorical view in which disease conditions have bright definitional boundaries—i.e., there is heart disease and non-disease, cancer and non-cancer (Rose, 1992, p. 8). These boundaries conventionally set the limits around which medical interventions are viewed as justified or not. Chemotherapy and bypass surgery are implemented only when the disease in question truly exists; only when intervention is justified do persons legitimately become patients. In this view, persons whose disease status is ambiguous present a moral dilemma.

In contrast to this dichotomous categorical view is the continuum conception of disease that Rose endorses. He argues that the categorical concept is "merely an operational convenience," that "the idea of a sharp distinction between health and disease is a medical artifact for which nature, if consulted, provides no support" (Rose, 1992, p. 6). Disease is nearly always a quantitatively defined, rather than a categorical or qualitatively distinct phenomenon. Pathological conditions such as cancer are conceived and diagnosed in quantitative terms—e.g., the number of neoplastic cells within a tissue sample, the extent of tissue invasion or lymph node involvement (Cutter, 1992; Rose, 1992). The natural history of most known diseases consists of a continuous progression of pathogenetic events that span a spectrum from minor preclinical changes to various gradations of symptomatic manifestations. Where the lines are drawn in defining disease within this continuum is a matter of social consensus.

Although more justified on conceptual grounds than the categorical view, the continuum concept of disease may be employed for either ethical or morally problematic ends. It might be used to lend pathological identity to non-diseased persons; since there are no bright lines between disease and non-disease, potentially everyone becomes a diseased patient worthy of medical intervention. The continuum view erects a slippery slope to patienthood, and a steep one at that. Clinical practice exemplifies how, for instance, even persons with average blood cholesterol might become *de facto* patients, subject to diagnostic and therapeutic measures (Downs *et al.*, 1998). Viewing the distinctions between disease and risk factor or between disease and non-disease as one of degree and of consensus expands the potential conceptual territory of the disease concept, thereby enlarging the domain of medical intervention. The territory gained could encompass risk factors whose causal connections to disease states might be distant; any risk factor could be viewed as a moment in the continuous history of disease, therefore warranting medical intervention. Not only the painful bony metastasis, the palpable breast lump, the mammographic abnormality, but even the most primordial molecular genetic events—the aberrant meiosis, the occult translocation or deletion, the earlier pathogenic steps that we cannot even imagine now—could become fair game for intervention, if the presence of disease, defined as the most minimal step along this continuum, by itself justified intervention.

One might argue, however, that such a view grants too much significance both to the way disease is defined and to the role of disease in grounding intervention. It makes defining a condition as a disease a necessary pre-condition for medical intervention. But these assumptions are precisely what is at issue; the crucial question is not whether disease should be defined in terms of category or continuum, but rather what is the proper scope of medical intervention in human life, and why. Whether or not persons should be treated as patients depends upon whether the scope of medical interventions should be narrowly restricted to the treatment of *disease*—however defined—or instead broadened to encompass other aims—e.g., the prevention of disease in asymptomatic or non-diseased persons. The broader view of the scope of medicine holds great currency, yet the contemporary discourse about unpatients and genetic medicine evinces the enduring legitimacy of the alternative narrow view.

The problem of how the unpatients created by preventive medicine should be treated can thus be conceived in terms of two different questions: on the one hand, whether disease should be defined as a continuum or a category, and on the other hand, whether or not the boundaries of medical intervention should be expanded to encompass preventive—in addition to curative—measures. The treatment of unpatients as *patients* is equally justified by either a continuum conception of disease or an expansive view of the scope of medicine. Both types of assumptions in fact blur the distinctions between disease, risk factor, and non-disease, between cure and prevention, between patients and non-patients. They each diminish whatever ethical import "the advent of the unpatients" might have. If disease is a continuous process with no temporal or spatial boundaries before or outside of which intervention is inappropriate, or else if medical interventions can legitimately be used for purposes other than the treatment of disease, then there is no necessary ethical problem implied by preventive interventions—genetic or otherwise. Indeed, from the perspective of either assumption it makes little sense to speak of preventive interventions as creating unpatients.

On the other hand, to even raise the issue of unpatients as a special ethical problem of genetic and non-genetic preventive medicine implies a sympathy with a very different viewpoint: that some persons can and should be declared off-limits to medical intervention. The very notion of unpatients and the argument that they should be treated as *non-patients* are justified by either a categorical conception of disease or a narrow view of the scope of medical intervention. The categorical view of disease maintains that there are clear temporal and spatial boundaries to disease, while the narrow view of the scope of medicine construes these boundaries as demarcating the moral limits of medical intervention. These are views with radical ethical import, for they challenge the *raison d'être* of preventive medicine—namely, the premise that the prevention of disease in non-diseased and asymptomatic persons is a legitimate goal of medicine.

Particular moral values inform the differing views of how persons of ambiguous disease status should be treated. Favoring acceptance of both the continuum disease concept and the broad view of medical intervention that support treating such persons as patients is an optimistic moral posture toward medical technologies, a view that emphasizes their potential benefits. Closely tied to this technological optimism is a favorable outlook toward 'medicalization'—a term signifying the expansion of medical diagnosis and therapy within human life (Barsky, 1988a; Conrad and Schneider, 1980, 1992; Crawford, 1980; Fox, 1977; Gallagher, 1996; Illich, 1975, 1976, 1982; Riessman, 1983; Zola, 1972). Genetic and non-genetic preventive interventions both engender medicalization by bringing more and more asymptomatic and non-diseased persons, who would otherwise be considered healthy, under the care of medical professionals and subjecting them to diagnostic and therapeutic measures. Preventive medicine also makes relevant and subject to medical attention an increasing number of aspects of everyday existence, not only symptoms and signs of illness, but also behaviors, lifestyles, and social, cultural, and even genetic characteristics that define personal identity. A favorable view of this medicalization sees its effects as potentially enriching and empowering human life (Gallagher, 1996; Riessman, 1983), or enabling persons to establish healing relationships and to satisfy fundamental human needs for caring and compassion (Marinker, 1975).

On the other hand, favoring acceptance of both the categorical disease concept and the narrow view of medical intervention is a skeptical ethical viewpoint that emphasizes the negative effects of medical technologies, eschewing their expansive application in human life. For many critics, this ethical position is founded upon a negative view of medicalization, which sees it as engendering great social and existential harms: dependency upon medical professionals; a privileging of health-related matters over other important concerns; overall diminished sense of well-being; and a diminished capacity of persons to lead autonomous lives and to deal with suffering and death in an authentic way (Barsky, 1988a, 1988b; Crawford, 1980; Illich, 1976; Skrabanek, 1988, 1994a). Critics of medicalization want to protect persons from becoming patients, to "unhook" patients from the medical system, as Illich puts it (1982, p. 470). Theorists from this perspective oppose preventive interventions because they promote medicalization.

Moral values regarding the good of medical intervention and medicalization thus determine how one conceives of disease and the proper scope of medicine, and influence the way in which the ethical problems of preventive interventions— genetic or not—are construed. Those who adopt a positive view of medicalization see little ethical problem with preventive interventions, for they begin with the assumption that all people are potential or real patients; non-patients are foreign notions. In contrast, those who view medicalization negatively cannot help but see preventive interventions as ethically problematic, for they begin by distinguishing between patients and non-patients, admitting the possibility that some persons are off-limits to medical intervention.

This influence can be discerned in the way that genetic exceptionalist arguments are framed. To even speak of "the advent of the unpatients" begs the question of whether genetic medicine is ethically problematic. This way of framing the ethical import of genetic medicine embodies assumptions of moral significance—either a highly categorical view of disease and of patienthood, or a narrow view of the domain of medical intervention—assumptions which already imply conclusions about the morality of intervening medically in people's lives. These assumptions are tied to still more fundamental moral presuppositions regarding the good of medical intervention and medicalization. It is difficult to imagine, by contrast, strong advocates of genetic testing even raising the notion of genetic medicine's creating unpatients, or regarding carriers of disease susceptibility mutations as less than deserving of medical intervention.

Which view of disease, patienthood, and medical domain is more valid is debatable; there are moral arguments in support of either view. But there is also an important epistemological criticism of the medicalization critique and the conceptual and moral viewpoint it manifests. Anthropologist Horacio Fabrega has noted that the critique of medicalization assumes that indeed "there is a distinct and more or less correct view of disease-illness and medical care," that not only what is disease and non-disease, patient and non-patient, but also what are medical and non-medical problems and institutions can be "clearly bounded, labeled unproblematically, and dealt with through delimitable institutions of control" (1980, p. 129). The medicalization critique thereby discounts the socially-negotiated, culturally-contingent, and value-driven manner in which disease and the realm of the medical are defined in the first place (Fox, 1977; Rosenberg, 1992). But if, in fact, diseases, patients, and medical problems are not given, but rather contingent, dynamic, and negotiated realities, then the issue raised by technologies that challenge traditional conceptions of disease, patienthood, and medical domain—e.g., genetic testing—is not whether such technologies are unjustified. Rather, the issue is whether our conceptions are expansive enough, and if not, whether there are compelling reasons to enlarge our conceptions to include these technologies as legitimate medical endeavors. Addressing this issue is a continually negotiated social process informed by moral and non-moral values.

The point of this analysis, however, is not to explore this process further or to defend any particular view regarding disease, patienthood, and the scope of medical intervention, but simply to identify the problems and underlying moral values of different views, and to show how the genetic exceptionalist position—that persons undertaking testing for genetic susceptibility to disease belong to a unique conceptual and moral class—itself represents a problematic value-laden position requiring further analysis.

4. INDIVIDUAL HEALTH, PUBLIC HEALTH, AND THE GOALS OF GENETIC AND NON-GENETIC PREVENTIVE INTERVENTIONS

A third main assumption embodied in the claim that genetic medicine presents special ethical problems is the notion that it is directed at unique goals and beneficiaries. Concerned specifically with germ-line gene therapy, Eric Juengst argues that this type of genetic intervention tests the notion of prevention as a goal of medicine. He claims that while advocates of germ-line gene therapy typically appeal to the concept of disease prevention to justify these interventions, in fact they are conflating two distinct senses of 'prevention' which have differing histories, philosophical assumptions, and moral authority. The first sense of prevention Juengst calls "phenotypic prevention" or "pheno-prevention," by which he means efforts to prevent the clinical manifestation of a genetic disease in patients at risk (1995, p. 1596). An example is newborn screening and dietary prophylaxis for phenylketonuria. Juengst contrasts this with what he calls "genotypic prevention" or "geno-prevention," by which he means efforts aimed at averting the transmission of particular genotypes to the next generation. An exemplary geno-preventive measure is selective pregnancy termination after intrauterine genetic diagnosis of Down syndrome.

In this argument, phenotypic prevention is ethically straightforward, simply recapitulating well-accepted goals, conceptual assumptions, and moral implications of non-genetic preventive medicine, which aims to further the health interests of individual patients by allowing them to avoid medical problems. Geno-preventive measures, in contrast, do not further the health interests of individual patients, but rather the "procreative interests of prospective parents, by allowing them to avoid the birth of individuals with foreseeable health problems" (Juengst, 1995, p. 1597). At the same time, geno-preventive interventions advance the social, economic, and public health interests of society by reducing the populational incidence of genetic disease. For Juengst, these goals diverge from those of traditional non-genetic, pheno-preventive measures; geno-prevention "broadens its practitioners' responsibilities beyond their presenting patients to the next generation's aggregate population" (1995, p. 1597).

According to this genetic exceptionalist argument, not only the goals, but also the conceptual assumptions and moral implications of geno-preventive measures diverge from those of traditional pheno-preventive ones. Genotypic prevention assumes a strong genetic determinism, manifested by a view that "the diseases it prevents are best understood at the level of the genotype, rather than through the pathophysiology of their expression" (Juengst, 1995, p. 1597). This strong determinism tends to be dangerously simplistic, placing persons at risk for unnecessary medical intervention. Moreover, genotypic prevention is highly contingent upon value judgments about the burden of disease for prospective parents and to society, and thus risks the "subordination of professional integrity to social ideology" (Juengst, 1995, p. 1600). It further subsumes individual interests beneath those of the collective, leading to a devaluation of human minorities and an infringement upon individual reproductive privacy. For all these reasons, Juengst argues, genotypic prevention is ethically dubious, and genetic interventions such as germ-line gene therapy should be rejected.

That the goals and beneficiaries of genetic interventions are unique represents another strong line of argument in support of the genetic exceptionalist view. However, this argument, like claims about the supposedly unique nature of genetic risk and of the subjects undergoing genetic testing, draws distinctions that are difficult to maintain. The argument for the uniqueness of the goals and beneficiaries of genetic medicine understates the extent to which traditional pheno-preventive interventions—both non-genetic and genetic—are directed toward the goal of future public health, and thus pose the same moral hazards as geno-preventive ones. It ignores well-described conceptual and ethical problems that make it difficult to construe preventive interventions as primarily advancing individual interests, and thereby obscures the debatable moral assumptions on which it depends.

First of all, one cannot neatly categorize geno-preventive measures as benefiting the aggregate, and pheno-preventive measures as benefiting individuals. The separation of goals is not pure; at least some geno-preventive interventions can in fact be construed as benefiting individuals in important ways. Juengst acknowledges that much prenatal genetic testing and counseling is "client-centered," advancing the "reproductive health"—i.e., the "ability to fulfill one's procreative ambitions"—of individual couples (1995, p. 1600). Genotypic prevention is thus not an exclusively public health-oriented endeavor, and the distinction between geno- and pheno-preventive interventions is less than absolute.

Even acknowledging this, however, one might maintain that the real point of the genetic exceptionalist argument is that geno-preventive measures benefit persons other than the ones receiving the intervention. But even this does not clearly distinguish geno-preventive from pheno-preventive interventions. For it is difficult to conceive of traditional pheno-preventive interventions as primarily advancing the health of the individuals receiving intervention. A phenomenon well-recognized in the preventive medicine literature is that preventive interventions cannot simply be construed as benefiting individual subjects, because these interventions are typically undertaken in populations with low baseline disease incidence and large variation in individual disease risks. For primary preventive interventions in particular—e.g., tetanus immunizations, cholesterol-lowering therapy, mammography—very large numbers of individuals need to undergo repeated interventions over prolonged time periods in order to effect improved outcomes for a very few. It is estimated, for instance, that saving just one life through the prevention of breast cancer with screening mammography in women less than age fifty requires 1000 women to receive annual mammograms for 12.5 years, amounting to 12,500 mammograms per life saved (Eddy, 1989, pp. 389-95). The 999 other individual women receiving annual mammograms gain nothing in terms of their final health outcomes. Epidemiologist Geoffrey Rose has famously described as the "prevention paradox" the "irony of preventive medicine that many people must take precautions in order to prevent illness in only the few," or conversely, the fact that "a preventive measure that brings large benefits to the community offers little to each participating individual" (1981, p. 1850, 1992, p. 12).

The ethical problems raised by the prevention paradox have been explored by various theorists. Rose himself reconciles the paradox by arguing that the principal justification for preventive interventions is the health of the public anyway, and that the paucity of benefit to individuals should simply be accepted (Rose, 1981, 1988, 1989, 1992). Other analysts have taken an alternative view, arguing that the lack of individual benefit makes many preventive interventions morally unjustified. Skrabanek quips that "the proverb, 'a stitch in time saves nine,' may be sound advice for mending socks but it makes little sense if a thousand people need one stitch (in its medical equivalent) to save one person from nine stitches" (1986, p. 188). Still other theorists describe the ethical difficulties of the prevention paradox in terms of justice issues, in light of a "transfer of benefit" that occurs between individual recipients of intervention—i.e., the fact that the particular population members who benefit from preventive intervention are not the same individuals who derive either no benefit or even iatrogenic harm (Blaney, 1985, pp. 19-24).

One can certainly debate the ethical implications of the prevention paradox, as well as the notion that final health outcomes are the only benefits of preventive interventions. A reduction in one's personal risk for future disease is, after all, another type of benefit secured by all individual recipients of preventive intervention, regardless of their ultimate health fates. Nevertheless, the point remains that it is difficult to conceive of pheno-preventive interventions as primarily benefiting individuals—whether as opposed to the aggregate or to other individuals. Such a view needs to be heavily qualified. It may be true that pheno-preventive measures *aim* at the benefit of individuals (and in particular, those persons in whom these measures are undertaken); however, for most of these individuals, these measures fail to *achieve* this benefit. In terms of the consequences of intervention, one cannot firmly distinguish the beneficiaries and goals of pheno-preventive and geno-preventive measures.

A more defensible statement of the genetic exceptionalist view would be that while neither pheno- nor geno-preventive measures primarily benefit the individuals undergoing intervention, pheno-preventive interventions have the virtue of at least aiming for this goal. One can argue that geno-prevention's primary *aim* of benefiting persons other than those undergoing intervention—i.e., parents and members of the wider community, instead of the unborn individuals potentially afflicted with disease—is itself morally problematic (Juengst, 1995, pp. 1595-8). But then the principal ethical problem is a conflict of interest not between individuals and society, but between different individuals, e.g., an intergenerational conflict between parents and their children. Confronting this problem is a matter of determining which individual's good should take moral precedence, not of weighing individual against public good.

An additional problem in distinguishing the morality of geno- and pheno-preventive measures in terms of their aims or intentions is that doing so risks granting insufficient moral weight to the consequences of intervention. One can legitimately ask whether a supposedly straightforward pheno-preventive measure like newborn phenylketonuria screening—which despite ostensibly aiming at individual health, will improve the outcomes of at most one out of every 10,000 to 25,000 babies screened (American Academy of Pediatrics, 1996)—is morally justified. Clearly, it does matter to what extent an intervention achieves its intended consequences; this consideration may have greater moral weight than the aims of the intervention.

Furthermore, not only the goals *vis a vis* individual and public health, but also the moral problems of geno-preventive and pheno-preventive measures are much more similar than the genetic exceptionalist view affirms. It is not clear that decisions to employ pheno-preventive measures are any less value-laden than those concerning geno-preventive measures, or that they carry any less risk of usurping professional integrity by social ideology. Indeed, exactly the same criticisms have been leveled against phenotypic preventive interventions. Skrabanek (1986, 1994a) has characterized preventive medicine's interventionist philosophy as a totalitarian ideology that grants primacy to the values of physical health and the welfare of the public over those of pleasure and individual freedom. Others fear the use of preventive interventions as instruments for the control of individuals by the state (Crawford, 1980; Zola, 1972, 1975). Even strong prevention advocates like Rose (1992) recognize that preventive interventions pose threats to personal liberty. Concerns that geno-prevention leads to a disvaluing of human minorities and a willingness to invade personal privacy on behalf of economic interests apply equally to pheno-preventive measures, insofar as they, too, subsume the interests of the individual beneath those of the collective.

But if the goals, beneficiaries, and moral problems of geno-preventive interventions are not, in fact, radically different from those of pheno-preventive interventions, then what drives efforts to maintain the distinction? One supporting factor is a legitimate view that geno-prevention yields an altogether different *type* of benefit from pheno-prevention. Juengst argues that geno-preventive interventions produce the benefit of "reproductive health," which "largely boils down to the ability to fulfill one's procreative ambitions" (1995, p. 1600). This is different from 'health' in the traditional (albeit also problematic) sense of freedom from disease, and this difference of benefit may have great moral import. The issue raises a larger question, which cannot be fully addressed here, of what are morally legitimate types of benefits yielded by medical interventions.

Another factor driving efforts to distinguish between pheno- and geno-prevention may be genetic essentialism—the reductive equation of human identity with genes—discussed previously (Nelkin and Lindee, 1995). Juengst notably asserts that "genes are not, like germs, external infectious agents that can be kept (or cleaned) out of a living person's body" (1995, p. 1597). Instead they are internal elements inseparable from the very identity of the persons who carry them. This is the basis of his concern over the negative discriminatory effects of geno-preventive interventions; he states that "genotypic prevention cannot help stigmatizing *genotypes, and (since they are inseparable) the people whom they mark,* as undesirable or pathological in themselves" (Juengst, 1995, p. 1599; emphasis mine). He also criticizes comparisons between pheno-preventive measures like immunization and germ-line genetic interventions meant to reduce the incidence of serious genetic defects in a population:

> But, again, this analogy is fatally flawed by the confusion over what germ-line case control would prevent. The 'specific protection' that germ-line case control programs would provide to the population does not guard against some external infectious agents. Germ-line case control programs would guard against a genotype and (by necessity) the class of people it characterizes. (Juengst, 1995, p. 1602)

Because he adopts such a strong genetic essentialist position—equating genotype with social class, gene with personal identity—Juengst is less apt to see germ-line interventions in any way other than as morally suspect.

This argument actually reproduces, rather than rejects, the strong genetic determinism that it criticizes as a stigmatizing force. In order to disallow social discrimination based on genetic constitution, it treats the particular genes and genotypes that happen to be associated *with disease* as essential constituents of personal identity and social class, respectively, a move that makes the prospect of genetic intervention morally problematic. It places genotypic disease prevention measures off-limits ethically, but at the cost of essentializing genes all the more, ironically reinforcing the assumption that genes—even disease-related ones—constitute the essence of human identity. The anti-deterministic argument simply perpetuates strong genetic determinism by essentializing genes that are associated with disease, in the same way that, as Susan Wolf (1995) argues, the antidiscrimination approach typically used to combat genetic discrimination "merely entrenches genetic bias" because it relies upon the same "thinking in genetic categories" that fuels discrimination in the first place.

It is possible, however, to take an alternate philosophical approach, to either reject genetic essentialism or reconceptualize the place of disease in human identity. One could view genes and genotypes as factors contributing to but nevertheless distinct from—i.e., external to—personal identity and social class. Such a view would imply that germ-line genetic interventions are not *ipso facto* morally suspect, for genes would then become just one more disease risk factor the modification of which poses no intrinsic threat to human identity. But one need not even reject genetic essentialism to question the ethical conclusions that follow from the genetic exceptionalist view. One might still adopt a strongly essentialist stance regarding genes in general, and yet ask why we cannot exclude from the realm of essential genetic factors those particular genes that cause disease; in other words, why do we need to regard *disease* as an essential, ethically-untouchable aspect of human identity? Can we not decide, for instance, that the genetically-determined condition of severe combined immune deficiency caused by adenosine deaminase deficiency is really an identity trait that its sufferers (or potential sufferers) are better off—physically, emotionally, existentially—doing without?

Of course, as Juengst avers, there are many more difficult moral problems raised here—e.g., who, if anyone, should make such evaluative decisions on behalf of others, and on what criteria—which may represent good reasons to reject germ-line therapies. Nevertheless, the salient issues are not whether germ-line or other genetic interventions are "geno-preventive" or "pheno-preventive." Rather, the morality of genetic interventions revolves around other questions: is reproductive health an appropriate goal of genetic interventions? What is the place of disease in human identity? The attempt to draw firm distinctions between genotypic and phenotypic prevention not only fails in important respects, but diverts attention from these more relevant questions.

5. CONCLUSION

Genetic medicine may raise many conceptual and ethical issues that are unique, and the purpose of this analysis has not been to suggest otherwise. What I have tried to show, rather, is that some of its attributes commonly understood to present special difficulties actually pose no problems that are not already raised by non-genetic preventive interventions. I have argued that the prognostic information yielded by genetic testing is not conceptually unique, that both the derivation and meaning of future disease risk are the same whether the associated factors are genetic or not. At the same time, the effect of genetic interventions upon the conceptual and moral status of its subjects is also not unique; I have argued that both genetic and non-genetic preventive interventions alike introduce similar ambiguities for persons receiving intervention. Finally, I have argued that the goals and beneficiaries of genetic and non-genetic preventive interventions do not necessarily differ, that both types of interventions aim just as much for the health of the public as for the health of individuals.

What accounts, then, for these various attempts to frame genetic medicine as conceptually unique? I have suggested that these depictions depend upon underlying commitments of an ethical nature. The attempt to portray genetic risk as distinct from epidemiologic risk manifests a strong philosophical determinism which some have called "genetic essentialism." To depict genetic medicine as creating "unpatients" expresses a particular view of patienthood, disease, and the scope of medicine that depends upon moral values regarding the good of medical intervention and medicalization. To argue that genetic medicine deviates from the traditional benefits and goals of preventive medicine depends upon particular views of the types of benefit for which medical interventions should aim, and of the moral implications of altering disease-related genes.

The import of these underlying conceptual and moral commitments is obscured when genetic medicine is simply portrayed as being distinct from preventive medicine. Strong philosophical determinism, a negative valuation of medicalization, a restrictive view of the goals of medicine, and a conception of disease-related genes as essential to human identity—all of these justifications informing various views of genetic medicine go unexamined when one fails to acknowledge the similarities, as well as the differences, between genetic and non-genetic preventive interventions. This conceptual bias not only overstates the true differences between genetic and non-genetic preventive interventions, but it also treats non-genetic interventions as morally unproblematic endeavors. It thus diverts attention from the difficult conceptual and ethical issues shared by all preventive interventions, genetic and non-genetic alike. A truly valid critique of genetic medicine would be a far more radical endeavor than the present efforts, for it would entail a critique of preventive medicine in general.

Notwithstanding these criticisms, one can also argue that the entire focus upon both the genetic and the preventive nature of genetic interventions is beside the point. The morality of interventions like genetic testing for disease susceptibility mutations or germ-line gene therapy may not depend on whether these interventions are "genetic" or "preventive" in nature, and thus it may not be worthwhile to debate what these and other concepts—e.g., 'patient,' 'disease'—signify. Rather, the morality of genetic interventions might depend more upon whether their uses—some of them truly novel—are justifiable. Neil Holtzman argues that the most important point raised by genetic testing for disease susceptibility is "not so much that a test is genetic as that it is used predictively" (1998, p. 128). Juengst acknowledges how "reproductive health" represents a distinctive aim of contemporary clinical genetics. The question is whether such novel uses of technology—i.e., for prediction (apart from disease prevention), enhancement of reproductive autonomy, and more—can be morally justified. Answering this question requires considering the morality of these goals in and of themselves, rather than assessing whether these goals approximate traditional aims of prevention and genetic medicine.

In any case, it seems that the focus of further work exploring the problems of genetic medicine could be fruitfully redirected. Greater efforts to examine the similarities rather than the differences between genetic and non-genetic preventive interventions might yield a better understanding of how to resolve the conceptual and moral dilemmas posed by both.

University of Pittsburgh
Pittsburgh, Pennsylvania, U.S.A.

ACKNOWLEDGMENTS

I thank Lisa Parker, John Mulvihill, and Robert Arnold for their comments on an earlier draft of this paper.

REFERENCES

Alderman M.H., Lamprot B. Labelling of hypertensives: a review of the data. Journal of Clinical Epidemiology 1990; 43:195-200
Alderman M.H. Blood pressure management: individualized treatment based on absolute risk and the potential for benefit. Annals of Internal Medicine 1993; 119:329-35
American Academy of Pediatrics Committee on Genetics. Newborn screening fact sheets. Pediatrics 1996; 98:473-501

Asch D.A., Hershey J.C. Why some health policies don't make sense at the bedside. Annals of Internal Medicine 1995; 122:846-50

Barsky, A.J., *Worried Sick: Our Troubled Quest for Wellness*. Boston: Little, Brown and Company, 1988a.

Barsky A.J. The paradox of health. New England Journal of Medicine 1988b; 318:45-9

Blaney, R. "Theoretical Basis of Disease Prevention." In *Ethical Issues in Preventive Medicine*, S. Doxiadis, ed. Dordrecht: Martinus Nijhoff, 1985.

Brett A.S. Treating hypercholesterolemia. How should practicing physicians interpret the published data for patients? New England Journal of Medicine 1989; 321:676-80

Conrad P., Schneider J.W. Looking at levels of medicalization: a comment on Strong's critique of the thesis of medical imperialism. Social Science and Medicine 1980; 14A:75-9

Conrad, P., Schneider, J.W., *Deviance and Medicalization: From Badness to Sickness*. Philadelphia: Temple University Press, 1992.

Couch F.J. *et al.* BRCA1 mutations in women attending clinics that evaluate the risk of breast cancer. New England Journal of Medicine 1997; 336:1409-15

Crawford R. You are dangerous to your health: ideology and politics of victim blaming. International Journal of Health Services 1977; 87:663-80

Crawford R. Healthism and the medicalization of everyday life. International Journal of Health Services 1980; 10:365-88

Cutter, M.A.G. "Value Presuppositions of Diagnosis: A Case Study in Diagnosing Cervical Cancer." In *The Ethics of Diagnosis*, J.L. Peset, D. Garcia, eds. Dordrecht: Kluwer, 1992.

Downs J.R. *et al.* Primary prevention of acute coronary events with lovastatin in men and women with average cholesterol levels: results of AFCAPS/TexCAPS. Air Force/Texas Coronary Atherosclerosis Prevention Study. Journal of the American Medical Association 1998; 279:1615-22

Eddy D.M. Screening for breast cancer. Annals of Internal Medicine 1989; 111:389-99

Engelhardt, H.T., "The Concepts of Health and Disease." In *Evaluation and Explanation in the Biomedical Sciences*, H.T. Engelhardt, S.F. Spicker, eds. Dordrecht: Reidel Publishing Co., 1974.

Fabrega H. The idea of medicalization: an anthropological perspective. Perspectives in Biology and Medicine 1980; 24(1):129-42

Fletcher, R.H., Fletcher, S.W., Wagern, E.H., *Clinical Epidemiology: The Essentials*. Baltimore: Williams & Wilkins, 1996.

Fox R.C. The medicalization and demedicalization of American society. Daedalus 1977; 106:9-22

Gallagher, E.B. "The Medical Dignity of the Individual: A Cultural Exploration." In *Society, Health, and Disease: Transcultural Perspectives*, J. Subedi, E.B. Gallagher, eds. Upper Saddle River, NJ: Prentice Hall, 1996.

Garfield S.R. Multiphasic health testing and medical care as a right. New England Journal of Medicine 1970; 283:1087-9

Giardiello F.M. Genetic testing in hereditary colorectal cancer. Journal of the American Medical Association 1997; 278:1278-81

Healy B. BRCA genes—bookmaking, fortunetelling, and medical care. New England Journal of Medicine 1997; 336:1448-9

Holtzman N. Are we ready to screen for inherited susceptibility to cancer? Oncology 1996; 10:57-64

Holtzman N.A. Bringing genetic tests into the clinic. Hospital Practice 1998; Jan. 15:107-28

Hubbard, R., Wald, E., *Exploding the Gene Myth: How Genetic Information Is Produced and Manipulated by Scientists, Physicians, Employers, Insurance Companies, Educators, and Law Enforcers*. Boston: Beacon Press, 1993.

Hulley S.B. *et al.* Epidemiology as a guide to clinical decisions. The association between triglyceride and coronary heart disease. New England Journal of Medicine 1980; 302:1383-9

Illich I. Clinical damage, medical monopoly, the expropriation of health: three dimensions of iatrogenic tort. Journal of Medical Ethics 1975; 1:78-80

Illich, I., *Medical Nemesis: The Expropriation of Health*. New York: Random House, 1976.

Illich I. Medicalization and primary care. Journal of the Royal College of General Practitioners 1982; 32:463-70

Jonsen A.R. *et al.* The advent of the "unpatients." Nature Medicine 1996; 2:622-4

Juengst E.T. "Prevention" and the goals of genetic medicine. Human Gene Therapy 1995; 6:1595-1605

Last, L.M., Wallace, R.B., eds., Maxcy-Rosenau-Last Public Health and Preventive Medicine. Norwalk, CT: Appleton & Lange, 1992.

Leavell, H.R., Clark, E.G., Preventive Medicine for the Doctor in His Community. New York: McGraw-Hill, 1965.

Macdonald L.A. *et al.* Labelling in hypertension: a review of the behavioural and psychological consequences. Journal of Chronic Disease 1984; 37:933-42

Marinker M. Why make people patients? Journal of Medical Ethics 1975; 1:81-4

Nelkin, D., Lindee, M.S., *The DNA Mystique: The Gene as a Cultural Icon*. New York: W.H. Freeman and Co., 1995.

Nelkin, D., Tancredi, L.T., *Dangerous Diagnostics: The Social Power of Biological Information: With a New Preface*. Chicago: University of Chicago Press, 1994.

Peterson G.M., Brensinger J.D. Genetic testing and counseling in familial adenomatous polyposis. Oncology 1996; 10:89-98

Riessman C.K. Women and medicalization. Social Policy 1983; 14:3-18

Rose G. Strategy of prevention: lessons from cardiovascular disease. British Medical Journal (Clinical Research Edition) 1981; 282:1847-51

Rose G. Mass prevention: population-directed approach. Acta Cardiologica Supplementum 1988; 29:131-9

Rose G. High-risk and population strategies of prevention: ethical considerations. Annals of Medicine 1989; 21:409-13

Rose, G., *The Strategy of Preventive Medicine*. Oxford: Oxford University Press, 1992.

Rosenberg, C. "Framing Disease: Studies in Cultural History." In *Framing Disease: Illness, Society, and History*, C. Rosenberg, J. Golden, eds. New Brunswick, NJ: Rutgers University Press, 1992.

Russell, L.B., *Educated Guesses: Making Policy About Medical Screening Tests*. Berkeley: University of California Press, 1994.

Shattuck-Eidens D. *et al.* BRCA1 sequence analysis in women at high risk for susceptibility mutations. Risk factor analysis and implications for genetic testing. New England Journal of Medicine 1997; 278(15):1242-50

Skrabanek P. Preventive medicine and morality. Lancet 1986; 1:143-4

Skrabanek P. The physician's responsibility to the patient. Lancet 1988; 1:1155-7

Skrabanek P. Why is preventive medicine exempted from ethical constraints? Journal of Medical Ethics 1990; 16:187-90

Skrabanek P. Health policing erodes civil liberties. Practitioner 1992; 236:988-90

Skrabanek, P., *The Death of Humane Medicine and the Rise of Coercive Healthism*. Suffolk: Social Affairs Unit, 1994a.

Skrabanek P. The coercive altruism of "saving lives" on roads. Lancet 1994b; 343:1176

Struewing J.P. *et al.* The risk of cancer associated with specific mutations of BRCA1 and BRCA2 among Ashkenazi Jews. New England Journal of Medicine 1997; 336:1401-8

Svenson T., Sandlund M. Ethics and preventive medicine. Scandinavian Journal of Social Medicine 1990; 18:275-80

Taylor, R.C. "The Politics of Prevention." In *The Sociology of Health and Illness: Critical Perspectives*. 2nd ed., P. Conrad, R. Kern, eds. New York: St. Martin's Press, 1986.

Veatch R.M. Voluntary risks to health. The ethical issues. Journal of the American Medical Association 1980; 243:50-5

Wikler D. Who should be blamed for being sick? Health Education Quarterly 1987; 14:11-25

Wolf S.M. Beyond "genetic discrimination": toward the broader harm of geneticism. Journal of Law, Medicine, and Ethics 1995; 23:345-53

Zola I.K. Medicine as an institution of social control. Sociological Reviews 1972; 20:487-504

Zola I.K. In the name of health and illness: on some socio-political consequences of medical influence. Social Science and Medicine 1975; 9:83-7

CHAPTER 14

JOHN H. ROBINSON AND ROBERTA M. BERRY

UNRAVELING THE CODES: THE DIALECTIC BETWEEN KNOWLEDGE OF THE MORAL PERSON AND KNOWLEDGE OF THE GENETIC PERSON IN CRIMINAL LAW

1. INTRODUCTION

The criminal code is both enduring and fragile. The reasons it is bound to endure, in some form, are obvious. We cannot hope to conduct our lives in community without some minimal assurance of safety and public order. We are, unavoidably, social beings, and our lives in community require that we protect ourselves against unwarranted force, unjustified deception, and all the other ways in which harm can be done to our basic interests. Whatever else we might say about the criminal code, we surely would agree that it was invented and sustained to serve this core purpose. It is no wonder that it endures.

But the criminal code, at least as we now know it, is fragile as well. The reason for this is less obvious, but can be brought to mind if we examine it more closely. Consider how it is that we make effective our shared desires to preserve the public peace. First, of course, we draw up a list of prohibited acts. While there is controversy at the margins, there is consensus at the core that we should prohibit unjustified incursions upon the life, liberty, or property of others. But how should we go about accomplishing our goal of preserving the public peace?

Consider for a moment what our answer might be if the only sources of disruption to the public peace were ferocious animals that, from time to time, invaded our communities and caused us harm. We wouldn't linger too long in devising and choosing our means for preserving the public peace. Some might argue

L.S. Parker and R.A. Ankeny (eds.), Mutating Concepts, Evolving Disciplines: Genetics, Medicine and Society, 287-317.
© *2002 Kluwer Academic Publishers. Printed in the Netherlands.*

that we should destroy the animals if they were found in close proximity to human communities, given our knowledge of their natures and history. Others might argue that we should instead round them up and release them in an isolated place so that they would not threaten us. Still others might suggest that we should render them harmless by some other means, such as by permanent confinement. The choice among these options might be controversial. We might disagree about how we should weigh our interest in efficiency in achieving our goal of public peace against our interest in preserving the lives, and the ways of living, of these animals who, after all, merely acted in accordance with their nature. Despite these potential controversies, consensus on an appropriate response to the threat these animals posed would be relatively easy to reach, and, just as significantly, the institutional form that the response took would be relatively uncomplicated.

When the targets of our prohibitions are our fellow humans, however, the relevant considerations are vastly different. We are far more concerned about whether it would be right to kill those who have killed others, or to remove them to some isolated place with no prospect of rejoining the human community, or otherwise to disable them from doing harm, regardless of the efficiency of these means to our end. We are, at the same time, far readier to blame them, on the grounds that they need not have done what they did, should not have done what they did, and have wronged their victims and all of us by what they did. But we also want to hold out hope, however dim, of reconciliation with them, provided that they will acknowledge what they have done, seek forgiveness, and undertake to mend their ways.

Reflection of this sort quickly persuades us that the criminal code was not invented and is not sustained solely to secure the public peace. The criminal code is a creature of a vast and complex extended political community that we have come to know as the state. States coexist in more or less healthy tension with the significantly more amorphous social groups into which their citizens are arrayed, and both political and social communities are constituted, in part, by shared beliefs about our lives as persons and as members of communities. The beliefs that sustain our political community's criminal code have been developed and refined through generations of challenges mounted on behalf of those for whom their consequences are most significant: those who face judgment and punishment as criminals due to their alleged defiance of the criminal laws.

In this crucible of judgment and punishment, our political community, by its lawmakers, judges, and those who execute its laws, has arrived at a working truce, if not a comprehensive agreement, about these beliefs. We generally agree, first, that all of us are persons who are entitled to respect as such and that almost all of us are possessed of moral agency and the capacity to choose, even as philosophical debates about free will and determinism continue to rage.[1] Accordingly, none of us deserves to be victimized by others of us, and all of us, in the absence of mental defect, incapacity, or duress, are capable of choosing to honor or defy the norms expressed in the criminal code. Secondly, there is widespread agreement that we are ordinarily obligated to honor these norms, their obligatory force deriving, in part, from our shared goal of securing the public peace and, in part, from our shared goal of securing the ties between our political community and the shared beliefs that sustain it. Thirdly, we also widely agree that we are subject to judgment and punishment for defying these norms, and that the punishment should aim at some mix of the following: incapacitation, deterrence, denunciation, and a reaffirmation of the norms that are expressed in our criminal code and that capture our shared beliefs about our lives as persons and as members of the political community. We achieve this last objective by punishing according to desert and by seeking both to rehabilitate those whom we punish and to facilitate their successful return to the community.

These beliefs help explain our complex concerns when faced with a threat posed to the community by one of our own. We do not simply seek the most efficient means to secure the public peace. We are reluctant to kill, isolate, or confine individuals who, whatever their crime, remain persons entitled to respect and who continue to belong to our community. Yet, we must reaffirm the norms that they have in some sense chosen to defy, thereby wronging both their victims and the community to which they belong. Keenly aware of how rarely we will succeed, we are even allowed to hope that judgment and punishment will go beyond mere incapacitation and will bring home to wrongdoers the wrongfulness of their conduct, leading perhaps to their reconciliation with the community whose norms they have violated.

These beliefs also lead us to the reason for the fragility of the criminal code. While some sort of mechanism for preserving the public peace surely will endure for as long as human beings remain social animals who live in close proximity with one another, the criminal code, as it has been constructed over the centuries, will unravel quickly if we yank out one or more of the strands of belief that sustain it. If, for example, we conclude that it is not true that human beings are possessed of moral agency and can choose whether to defy or abide by moral norms, then it is hard to know what it would mean to say that we are "obligated" to honor community norms. It would also be hard for us to justify the claim that punishment can be deserved, or that it could educate some wrongdoers, opening for them the door to reintegration into the community. It would even be hard to show that punishment is defensible as a vindicator of the norms the general observance of which makes community of any sort possible.

In this chapter, we anticipate what may well constitute the strongest challenges yet to the criminal code as we now conceive it. These challenges will issue, at least initially, from two separate quarters: from "below"—the realm of behavioral genetics as allied with neurophysiology; and from "above"—the realm of behavioral genetics as allied with sociobiology. The immediate stimulus for these challenges will arise from the unraveling of the genetic code by the Human Genome Project and related research programs, and the continuing development of the research program of sociobiology.

We begin our discussion of these challenges with an analysis in section 2 of the relationship between criminal codes and the process by which some offenders are selected out for punishment under them. In section 3, we turn to a discussion of the challenge from below, from claims associated with neurophysiology. In section 4, we discuss the challenge posed from above, from the claims associated with sociobiology. And in section 5, we offer concluding thoughts about the continuing dialectic between knowledge of the moral person and knowledge of the genetic person in criminal law.

2. THE CRIMINAL CODE, THE PRACTICE OF PUNISHMENT, AND MORAL PERSONHOOD

A criminal code is so many black marks on white paper until it is embodied in the social and political practice known as punishment, and that practice requires an elaborate set of institutions for its maximal functioning.[2] Reflection on the practice of punishment reveals it to be a remarkably *assertive* practice, and it is to the contents and implications of those assertions that we will call attention here. To say of a practice that it is assertive is, we concede, quite odd: persons make assertions, practices do not, and we could always translate peculiar claims about the assertiveness of punishment into more acceptable claims about how things must stand in the world if the practice of punishment is to be intelligible. It remains true, however, that if we attend to the assertive significance of punishment (and of the crime to which it is a response), we can learn an enormous amount about crime, punishment, and the sorts of creatures that can be capable of the former and deserving of the latter.

From the point of view of the practice of punishment—admittedly, another strange locution—criminals in committing crimes deliver a message to their victims and to the relevant community. In committing a crime, paradigmatic criminals assert, so to speak, either their exemption from the constraints of morality, or their victims' exclusion from its coverage, or some mix of the two. In punishing criminals, the state denies these assertions. It asserts the subjection of each of us to the constraints of morality and the inclusion of each within its protection (Robinson, 1996, pp. 4-5, 1991, p. 261). Where the wrongfulness at issue is wholly execrable, as in the Hutu slaughter of the Tutsis in the spring of 1994 or in the Bosnian Serbs' slaughter of the Bosnian Muslims later that year, the necessity of appropriate punishment—and even of those institutions that are necessary to the imposition of just and defensible punishment—becomes obvious. But when the wrongfulness at issue is less revolting, or when we are not even sure that the conduct in question is really wrongful, we routinely detect less enthusiasm for punishment itself and less support for those institutions that punishment requires. The failure of those institutions to impose punishment where punishment is thought to be appropriate is as profoundly demoralizing as is the imposition of punishment upon those who are perceived as not deserving it.

The assertiveness of punishment is not exhausted by its denial of the intolerable assertions implicit in the conduct of the paradigmatic criminal. It implies, if it does not state, the correctness of the understanding of the person that we have just ascribed to the criminal code itself. Punishment depicts us—or the practice of punishment depends for its intelligibility on a depiction of us—as simultaneously susceptible to the allure of moral evil and capable of appreciating its ultimate repulsiveness, even as capable of understanding how it is that we sometimes succumb to that allure and sometimes resist it. Punishment introduces us, therefore, to rationalizations and to their repudiation; it depicts us as experiencing, describing, and explaining human conduct as motivated by reasons—sometimes noble, sometimes base, usually mixed—and it depicts those reasons as ultimately accessible to both agents and observers, even if at times we are inclined to agree with the old English judge who said, "The devil himself knoweth not the thought of man."[3] The person whom punishment depicts is the same sort of person whom we encounter in Scripture, in drama, in novels, across the dining room table, and in social and political life generally.

Because this sort of person is ultimately free to choose between good and evil—free to recognize others as making moral claims on him or her and free also to rationalize unwarranted self-preference—he or she can be judged and, sometimes, punished. Where the appropriate conditions are met, punishment legitimizes the authority of those institutions that impose it and redresses the harm that the person being punished has inflicted upon the moral order that those institutions ought to vindicate. Just what those conditions are has proven difficult to specify, but a reasonable consensus exists as to their basics: the norms the violation of which risks punishment must be specified with reasonable clarity before anyone is required to answer for his or her conduct; the person being punished must ordinarily have been morally culpable in acting as he or she did at the time of the conduct in question;[4] the process used to determine suitability for punishment must be reasonably productive of accurate results and must treat the accused humanely throughout; the punishment must to some extent fit the crime as regards its severity; and the form the punishment takes must not affront the humanity of the offender. Each of these conditions for appropriate punishment tends to save punishment from itself, purging it of mindless vindictiveness and permitting it to vindicate the moral claims that the one being punished has sought to deny, while at the same time it reasserts the personhood of both the offender and the victim.

The whole idea of "saving" punishment from vindictiveness (or anything else) requires a few words of explanation. As a practical matter, punishment is often mistakenly regarded as a morally indefensible practice, as if it consisted in doing moral evil to those who are thought to have done moral evil in a vain attempt to undo the evil that they are thought to have done, thereby duplicating the evil to which it responds. Brief reflection on crime and punishment reveals why this mistake is so likely to be made. Crime, after all, almost always consists in the wrongful deprivation of another's life, liberty, or property, and punishment almost always involves a deprivation of the same. Viewed materially, therefore, crime and punishment look very much alike—hence, the attractiveness of the caricature. To discredit the caricature, punishment must be ritualized and solemnized by the institutions that impose it. It must also honor those constraints that, when they are honored, reveal punishment for what it really is or can be—namely, the polar opposite of evil, a response to evil that does not so much negate an already committed evil deed as reassert the claims of morality that the evildoer has denied in his or her evildoing.

Chief among the constraints that the practitioners of punishment must honor if they are to save it from caricature and self-destruction is the culpability constraint. This constraint ordinarily requires that the practice of punishment deprive no one of life, liberty, or property unless that person has been wrongfully involved in an effort—successful or not—to deprive another of his or her life, liberty, or property. Beyond determining on the basis of imperfect evidence that the accused (and not somebody else) did what he or she is accused of doing, no forensic task is more difficult than determining whether or not the accused acted with the mental state required both by the relevant criminal code and by the demands of the culpability constraint itself. It is, however, precisely this determination that must be made—and must be seen to be made—if punishment is to be "saved." Failure to make this determination would make punishment intolerably *retaliatory*, when morality requires it to be instead *retributive* in how it responds to wrongdoing; failure to make this determination would also make the *harm* done to the victim the focal point of the process leading to punishment, when morality requires the focal point to be the *wrong* done to the victim.

We advert to two contrasts here: one between retaliatory punishment and retributive punishment; the other between harm-focused proceedings and wrong-focused proceedings. We see these two contrasts as intimately related conceptually, and perhaps historically as well. To understand the contrasts and to perceive the nexus between them, imagine a time when the state left punishment wholly in the hands of victims and their families, imposing no limits on the severity of the punishment imposed. Such a regime would be utterly intolerable, leading to endless vendettas and mocking the very idea of punishment as a vindicator of community norms. Now imagine a later time when the state still left punishment largely in the hands of victims and their families, but when it gave itself the role of policing the excesses of this self-help regime: when a person lost an eye or a tooth to the causal acts of another, that person would be entitled to deprive the other of an eye or a tooth, but would be forbidden to impose a more severe sanction, such as beheading. We can easily imagine how much more protective of the public peace this regime would have been over one in which self-help was not so constrained. A moment's reflection would, nevertheless, reveal the great potential for serious injustice in such a regime. It is, to be sure, a terrible thing to lose an eye to the causal acts of another, but there is a morally relevant difference between losing an eye to an intentional or reckless act and losing an eye to an innocent one. A purely retaliatory system is, however, blind to this difference. What is needed, then, is a shift in focus from the harms that trigger and cap retaliatory punishment to the wrongs by which retributive punishment is measured.

Nothing could be more destabilizing to self-help punishment regimes than this shift in focus. I am a very good judge of the harms that I have suffered, but I am a fairly poor judge of the wrongs that I have suffered. My judgment, therefore, can be trusted in a harm-focused self-help regime but cannot be trusted in a wrong-focused regime. Even if I were capable of superhuman objectivity with regard to the mental state that informed the acts that caused irreparable harm to me, no one—not the person whom I punished and not our fellow citizens who learned of the punishment—would have much confidence in my capacity for objectivity. The shift in focus from harms to wrongs entails, therefore, a massive institutional shift from self-help to third-party "administration of justice," that is, a shift to courts (and perhaps to juries) as the ultimate decision makers with regard to punishment. As far as my fellow citizens are concerned, a neutral and conscientious third party is sure to be a more *credible* judge of whether or not I have been wronged at all, and, if so, of the extent to which I have been wronged. A third party may even be a *better* judge of those matters than I am, as I, in my reflective moments, may be willing to admit.

We punish, then, for wrongful conduct rather than harmful conduct, as retribution rather than retaliation, in an effort to vindicate community norms rather than to square our own accounts. It is the wrongfulness of conduct that justifies punishment and makes what would otherwise be a paradigm case of the forbidden— namely, the deprivation of another's basic interests—morally permissible and, in some cases, even morally mandatory. It is also the wrongfulness of conduct that makes it possible for that deprivation to function as a vindication of the moral order that was disturbed by the wrongful acts. It almost goes without saying, furthermore, that in reserving punishment for those judged to have acted wrongly, a political community commits itself in its most solemn and public moments to an account of its citizens in which wrongfulness is possible and intelligible. It commits itself, that is to say, to the ordinary individual's having at least as much freedom to choose good over evil as justifies the political community in selecting out for punishment some, at least, of those who have failed to do so when the criminal code has required that of them. It commits itself to those presuppositional beliefs in our personhood that sustain the criminal code.

The judges who more or less directly oversee the trials that select out for punishment those thought to deserve it have long realized how crucial judgments of wrongfulness are to the proper functioning of the practice of punishment and thereby to saving it from the caricature of itself to which it is permanently at risk. Working in league with their academic allies, state and federal judges in the American system have used their common law power to protect from punishment those defendants who are not really culpable with regard to the harms that they have caused or threatened. Members of the legislative and executive branches, on the other hand, have for almost as long a time realized how seriously the public peace may be endangered by excessive preoccupation with the extent to which an offender's conduct was truly wrongful. Recent generations, therefore, have witnessed a sustained tussle between judges intent on refining the standards of wrongfulness requisite to justified punishment and legislatures and executives intent on proscribing conduct that endangers the public peace.

From Queen Victoria's fury at the psychotic who killed her Prime Minister's secretary but then succeeded on an insanity plea at trial, to the congressional outrage that greeted the not-guilty-by-reason-of-insanity verdict that John Hinckley's lawyers won for him, courts have been in conflict with the legislative and executive branches of government as to how strong the culpability constraint on punishment should be.[5] A generation ago, courts massaged relatively opaque criminal codes to make the culpability constraint particularly demanding when mental disease or defect was at issue, and their academic allies generated a model penal code that exhibited heightened sensitivity to the ways in which mental disease can nullify culpability. Since then legislatures have been hard at work reversing that trend, making it relatively easy to convict, and then to punish, a person who has harmed others in ways that the criminal code prohibits even if that person's capacity for choice between a specious evil and a real good was seriously compromised. When these legislative initiatives have been challenged in the courts, the courts have usually acquiesced in the legislatures' balancing the risks to the practice of punishment that are inherent in punishing individuals whose wrongfulness with respect to the conduct in question is itself seriously in doubt against the need to protect the public.[6]

Recently courts have been rather busy adding to the list of situations in which otherwise criminal conduct is either justified or excused, as the creation and expansion of the battered woman syndrome illustrates.[7] It is impossible to say with any certainty today just how the legislatures of the several states will respond to these judicial innovations. What is fairly certain, though, is the structure that the tussle between court and legislature is likely to have. For both, the important point will be that most of us most of the time are capable of genuine choice between a specious evil and a real good. What will separate court from legislature, if anything does, will be how far removed from the norm a person or a person's situation must be in order to qualify for an exemption from punishment. In those debates, no strategy will be more likely to doom a claim to exemption than one that suggests that the normal case is nonexistent and that all harm-causing or harm-threatening conduct can be exhaustively analyzed in nonmoral terms. Whatever orthodoxy obtains among particular expert subcultures, the working assumptions of the American courtroom and legislative chamber will not countenance the abandonment of the claim to moral agency for the unimpaired defendant (Robinson, 1983, pp. 310-1).

To this point, the functional truce that we have sketched has resisted all efforts to work fundamental revisions to its terms. Novel claims about certain features of our moral lives, whether about the cognitive processes of the insane or the syndromic behavior of the battered woman, successively merge into the mainstream, incorporated by a dialectical process of assertion, deliberation, and accommodation. We revise or understand in a fresh way what it is that we mean by justification or excuse; our concept of personhood is informed and enriched—but not supplanted—by the content of these novel claims about our nature and experience as moral persons.

From time to time in human experience, however, a new orthodoxy overtakes and replaces an old one—the new regime underwritten by novel beliefs that prove both irresistibly powerful and fundamentally inconsistent with the beliefs that sustained the old. A serious postmodern challenge of this sort seems unlikely to emerge from the subcultures of the social sciences or the humanities. The social sciences continue to suffer embarrassment in their efforts to yield reliable insights and guidance with respect to the issues that concern the political community, and the humanities labor to persuade the broader culture of the relevance of their deliverances to these issues. Rather, the challenge is likely to come from the most successful subculture of the modern era, natural science. It is to the challenges we might anticipate from this subculture that we now turn.

3. THE CHALLENGE FROM BELOW

If a challenge to an established orthodoxy is to succeed, then, it must present claims that are both irreconcilable with those of the prevailing orthodoxy and powerful enough to displace them. We consider, first, whether the claims of the new genetic science allied with neurophysiological science —what we call the challenge from below—might satisfy these conditions.

The challenge from below derives from claims that genetic material may exert a significant influence upon neurophysiological structures and processes which, in turn, significantly influence behavioral traits, including predispositions to engage in violent conduct (Walters and Palmer, 1997, pp. 118-9; Stoff and Cairns, 1996, pp. 3-63). Research that supports this view has proceeded on two fronts: studies of genetically related individuals for the purpose of arriving at heritability estimates with respect to certain behavioral traits, and biochemical studies seeking to identify potential biochemical mechanisms linking genetic material to certain behavioral traits (Walters and Palmer, 1997, pp. 118-9; Stoff and Cairns, 1996, pp. 3-63).

In the 1970s and 1980s, research focused upon two sets of genetically related individuals: (1) monozygotic (identical) twins and (2) adoptees and their biological parents. These studies sought to isolate genetic factors from environmental factors to arrive at heritability estimates of liability to criminal conduct[8]—'heritability' referring here to the proportion of variation in a specified trait that is statistically correlated with genetic variation (Carey, 1996, p. 3; Berkowitz, 1996, pp. 46-7; Lancet, 1995, pp. 466-7; Cloninger and Gottesman, 1987, p. 92; Mednick, Gabrielli, Jr., and Hutchings, 1987, p. 74).

Biochemical studies that have followed upon these studies of genetically related individuals have attempted to trace the causal links that would account for a genetic influence upon violent conduct. The research includes DNA linkage studies (Brunner *et al.*, 1993) and studies correlating concentrations of certain neurotransmitter metabolites in cerebrospinal fluid with violent conduct or criminal acts (Berman and Coccaro, 1998; Coccaro and Kavoussi, 1996; Virkkunen and Linnoila, 1996; Stoff and Vitiello, 1996). Some of these studies have hypothesized a link between gene structure and neurotransmitter levels, and between the neurophysiological effects of neurotransmitter levels and violent conduct.

The methodologies and conclusions of some of these heritability and biochemical studies have been challenged and remain scientifically controversial.[9] The controversy has spilled over into the social and political realm as well. In 1992, Frederick Goodwin, Director of the Alcohol, Drug Abuse, and Mental Health Administration during the administration of President George H.W. Bush, spoke before a meeting of the Mental Health Advisory Council. He noted the results of a 1992 survey showing that eighty percent of youthful offenses were committed by seven percent of the population and then analogized the behavior of young urban males to the behavior of male rhesus monkeys in the wild, suggesting that the loss of social structure in urban areas encouraged reversion to a more "natural" kind of conduct. Goodwin made no reference to the race of these urban males, and, in recounting these events later, Goodwin explained that the 1992 survey figures had shown no correlation between violence and race when socioeconomic factors were discounted.[10] But some members of the audience understood Goodwin to be talking about violent conduct by African-American males in particular. A firestorm ensued. Louis Sullivan, President Bush's Secretary of Health and Human Services, reprimanded Goodwin, and Goodwin issued a public apology for his "insensitivity" and the "inappropriateness" of his comments.[11]

A few months after the Goodwin incident, the program for a 1992 conference funded by the National Institutes of Health (NIH) was released. The program, entitled "Genetic Factors in Crime: Findings, Uses & Implications," noted that the search for genetic markers for criminal conduct was underway, motivated in part by "the apparent failure of environmental approaches to crime—deterrence, diversion, and rehabilitation" (Williams, 1994, p. 98). The program also stated that genetic research "holds out the prospect of identifying individuals who may be predisposed to certain kinds of criminal conduct, of isolating environmental features which trigger those predispositions, and of treating some predispositions with drugs and unintrusive therapies" (Williams, 1994, p. 98).

Release of the program reopened and aggravated the controversy surrounding Goodwin's remarks. Amid accusations of government-sponsored eugenic efforts directed at African-Americans, the NIH withdrew its funding.[12] The withdrawal of funding, in turn, generated vigorous protests from conference organizers, who adamantly insisted that they intended to highlight the dangers of genetic research in this area, not to promote eugenic or racist approaches to criminology. Others protested as well, concerned about the cancellation of funding for controversial subject matter areas.[13] NIH funding was restored after appeal, and a somewhat revised version of the conference went forward three years later amid continuing controversy (Williams, 1994, p. 100; Roush, 1995, p. 1808). Six of the conference participants issued a formal protest statement to fellow participants, reading, in part, "Scientists as well as historians and sociologists must not allow themselves to be used to provide academic respectability for racist pseudoscience." Other conference participants expressed concerns about hasty conclusions that might be drawn from inconclusive studies and the potential for eugenic policies in the form of genetic therapies devised and administered in a racially discriminatory fashion (Roush,

1995, p. 1809). And outside protestors invaded the conference hall, chanting "Maryland conference, you can't hide—we know you're pushing genocide" (Roush, 1995, p. 1808). The worry for some and the hope for others is that the criminal genetics research program will usher in a new era in which our current model of criminal culpability and punishment will be supplemented or perhaps supplanted by a preventive/therapeutic model of crime control. On the most worried account, the research program is fundamentally racist and eugenic in its motivations. Its pseudoscientific claims play to the public's sense of helplessness and fear in the face of violent crime and threaten to divert public concern and dollars away from beneficial social programs and toward a future eugenic program of labeling, stigmatizing, and then coercively intervening to "treat" the genetically "at risk," who will turn out to be, in disproportionate numbers, members of minority groups.[14] On the most hopeful account, the research program will enable future generations to screen their newborns for genetic predisposition to violence and to "correct" for this predisposition by counseling, conventional drug therapy, or somatic-cell genetic therapy.[15] Some day, the need for "correction" might be virtually eliminated if scientific advance enables us to engage in a program of germ-line genetic therapy to eliminate genetic predisposition to violence.[16]

But there is good reason to think that neither of these accounts accurately forecasts the future. Consider, first, that the strongest claims that this research program might yield could not include a claim of genetic determinism. Multiple environmental factors—both within the human body and in the external world—undeniably play a complex, interrelated, causal role in the development of human behavioral traits. And, given the complexity of these factors, no researchers anticipate that our new genetic knowledge will enable us to predict whether an individual possessed of a particular genotype will, in fact, commit an act of violence in the future.[17]

Perhaps, though, a reductionist criminal genetics research program of the not-too-distant future will penetrate to the neurophysiological structures and events underlying the choice to engage in violent conduct and will redescribe this act of choice in purely naturalistic terms. Perhaps purely naturalistic references—to brain structures and brain states, to elements of the surrounding environment with which the brain interacts, to the laws of chemistry and physics—will supplant our current-day references to a thinking agent's intentional states (beliefs, desires, thoughts) and their purported causal and explanatory role in the moral agent's act of choice.[18]

Reductionist programs have underwritten enormous advances in human powers of prediction and control in other domains. For example, we once were limited to describing the work of the steam engine in terms of the observable consequences of the proximity of burning coal to water, the pressure of the steam, and the work accomplished. Nineteenth-century advances in chemistry and physics enabled us to substitute reductionist references to molecular motion, to energy, to the laws of thermodynamics, and to Boyle's law, and this new reductionist knowledge supported the invention of machines far more powerful than the steam engine. Perhaps, then, knowledge generated by a reductionist criminal genetics research program would enable us to invent crime-control mechanisms far more powerful than our current practice of punishment.

But, for several reasons, there appears to be no good prospect of a reduction of the acts of the moral agent to purely naturalistic terms. While there surely is a naturalistic story to be told in connection with our intentional states, the relationship between the two appears to be one of supervenience of the intentional upon the naturalistic rather than of the definability of the intentional in terms of the naturalistic; the same intentional states conceivably could supervene upon an infinite number of sets of naturalistic elements involving slightly different brain structures and states and slightly altered environmental conditions.[19] Furthermore, any effort to accomplish a reduction would have to cope with the utterly puzzling question of how it is that the matter of the brain could represent the propositional content of an agent's beliefs, desires, and thoughts,[20] such as: "Jones believed that Smith had a lot of money," "Jones believed that he was entitled to take Smith's money," "Jones believed that it was true that he was entitled to take Smith's money," "Jones wanted to kill Smith so that Jones could take his money and become rich." The problems multiply with the observation of the *intensionality* of intentional states—unlike statements about the natural world, it is not the case that we can freely substitute apparently coreferential or coextensive terms for the terms of our intentional statements and preserve their truth value.[21] If, for example, Jones did not know that Smith was his son, we could not substitute, 'his son' for 'Smith' and preserve the truth-value of the statement "Jones wanted to kill *his son* so that Jones could take his money and become rich."[22]

An even more ambitious future version of the criminal genetics research program—a program of eliminativism—might be motivated, in part, by these troublesome problems with a reductionist account. On the eliminativist account, the reductionist is unable to reduce our intentional states to naturalistic terms because "intentional states" are no more real than "vital spirits"; believing them to be real does not make them so. It is the physical stuff of the human organism and the behaviors of the human organism that are real.[23] A future eliminativist criminal genetics research program might point to our new genetic knowledge as the blueprint connecting the physical stuff of the human organism to its violent behaviors. If such a program were to succeed, its consequences for our criminal justice system might be as radical as the consequences for our child-rearing practices would have been if the Skinnerian eliminativist research program had succeeded.[24] But eliminativists of all stripes face two daunting obstacles: persuading us that our beliefs and desires do not bear a causal and explanatory relationship to our conduct, and providing a plausible account of the real causes of our conduct. For a time, adherents of the Skinnerian research program were persuaded that they could surmount these obstacles. Skinnerian behaviorism ultimately foundered, however, and alternative eliminativist accounts have yet to attract many adherents (Rosenberg, 1994, pp. 146-7, 1995, p. 218).

If the criminal genetics research program will not yield claims of genetic determinism, if a reductionist version of the program would face challenges whose resolution is not currently imaginable, and if an eliminativist version of the program appears even less plausible, what might we realistically expect from a viable criminal genetics research program? Two new sorts of information might be produced: (1) a set of reliable statistical correlations between certain genotypes and high incidences of violent behaviors, and (2) a good biochemical causal account of why certain genotypes contribute to predispositions to violence.

New information of either or both sorts surely would test our current conception of the criminal code, but it is far from obvious that it would push us to revise its terms in the ways envisioned either by the worried or the hopeful. We know right now that those who are intoxicated are statistically much more likely to commit violent acts than those who are sober,[25] and researchers are attempting to piece together a biochemical account of genetic predispositions that contribute to substance abuse and to engaging in violence when intoxicated (Barros and Miczek, 1996; Virkkunen and Linnoila, 1996). We already recognize that the choices for some moral agents—to resist drinking, to refrain from violence when drunk, to obtain help for a drinking problem—may be especially hard. But, to this point, we continue to find those who commit crimes while intoxicated guilty and deserving of punishment.[26] For that matter, we know that a wide range of environmental experiences that are entirely beyond the control of the individual—being abused as a child, growing up in a neighborhood in which violence is a commonplace—are strongly correlated with violent conduct and we suspect that there are good causal explanations for why this is so. We nonetheless punish these individuals when, as adults, they choose to adopt the ways of their parents and peers.[27] We are reluctant to conclude that influence constitutes excuse, that hard choices are determined choices, or that victimization conveys absolution.

Perhaps the challenge would be posed in a somewhat different way, however. If the research program yielded reliable statistical information correlating genotypes with incidences of violent behavior, then we would know in advance that individuals constituting a certain percentage of a certain genotype would commit acts of violence in the coming year. We would know this even though we knew nothing of the particular moral struggles of any of these individuals, just as insurance companies now know in advance that individuals constituting a certain percentage of those of certain ages will die in the coming year even though they know nothing of the particular struggles with mortality of any of these individuals. Perhaps we would come to think of ourselves as subject to irresistible social regularities and to believe that the real story of criminal conduct is told at the social level and in the language of social laws rather than at the individual level and in the language of resistance or surrender to the allure of evil.[28]

We now recognize, however, even if some in the not-too-distant past have imagined otherwise, that statistical correlations are not the tracings of social laws that somehow govern our conduct.[29] We generally understand these statistical correlations as either reducing to or supervenient upon the aggregate underlying choices and conduct of the thinking agents of intentional psychology, the same agents as are depicted in the criminal code's conception of the moral agent.[30] Furthermore, while the practical benefits of statistical correlations are obvious in the case of insurance, they are not so obvious in the case of the criminal justice system. Insurers need to know how much to charge individuals of different ages for life insurance policies in light of the risk undertaken in insuring these individuals. For business-of-insurance purposes, it does not matter which individuals die, when they die, or how they die. For purposes of the basic "business" of the criminal justice system—preserving the public peace—however, it matters very much which individuals commit violent acts, when, and against whom. In the absence of this particular knowledge, we cannot prevent individuals from committing criminal acts or protect their potential victims, or, if a crime is committed, we cannot know whom we should punish.

Perhaps, though, it is precisely in virtue of their practical appeal that these statistical correlations might tempt us to abandon our commitment to the current criminal code. While we might continue to believe that the genetically predisposed who committed violent acts were blameworthy in some sense, we might also conclude that there was little return in dwelling upon the moralistic fine points. Instead, we might opt for humane, early interventions in the lives of those genetically predisposed to violence—by drug therapy, somatic-cell genetic therapy, counseling, or, some day, by germ-line genetic therapy—to reduce or eliminate their predispositions before they engaged in violent conduct. In this way, we might hope to benefit both these individuals and the rest of society, which would be spared the disruption of the public peace that otherwise predictably would ensue.

There would be serious problems with a regime of this sort, however. Consider the fact that we already know of a strong statistical correlation between the male genotype and criminal conduct. One day, we may be able to develop a good biochemical account of why this is so.[31] There might well be public health/therapeutic interventions that could be administered to the male population that would be effective in reducing the incidence of violent crime. But many males likely would decline to undergo these interventions, if given the choice, because of the time, effort, or expense involved or because they would prefer not to have their personalities altered even if this might benefit them or others. Perhaps we would not allow them this choice; after all, in times of epidemic, we force inoculation upon the recalcitrant for their own good as well as the good of all of us.[32] There likely would be fierce opposition to a mandatory program, however. And even if policymakers— about half of whom likely would be members of the target group—were tempted to enact such a program, they might hesitate before another set of concerns: the unintended consequences of the program, both foreseen and unforeseen. Would individual males derive less enjoyment from life? Would creativity diminish, would birthrates fall, would athletic competition fall into decline, would we no longer have a viable military, would our gross national product decline?

Of course, it is unlikely that a successful criminal genetics program would recommend to us a target population consisting of about half of the total population.

To the extent the research program succeeded in identifying a relatively narrow target group, the foregoing political concerns would be diminished, although ethical concerns about forced personality alteration would persist. While resistance to personality alteration might well be very intense within the target group, as a practical matter, the resistance of a small group would hardly register on the political scales unless others chose to take up their cause. Unintended consequences within the target group might be troublesome for them, but their troubles would hardly amount to a drop in the bucket of the troubles of the wider world. A new set of political concerns would be introduced, however. While identifying the target group would be a relatively easy task if the group consisted of all phenotypic males, it would be quite challenging if instead it consisted of individuals identifiable only by genotype. We would need to develop a vast system of universal, mandatory, genetic testing—and of certifying the accuracy of test results, especially in light of the incentives for tampering by the target group.

We might limit the concerns associated with this public health/therapeutic approach by adopting instead an approach that emphasized environmental interventions rather than therapeutic interventions. We might, for example, require "genotyping" of customers at bars or gun shops and prohibit proprietors from rendering service to those with "listed" genotypes.[33] But this system of constraint through blacklisting by genotype would raise other troublesome concerns. Everyone would be forced either to undergo genetic testing or to forgo certain social activities. The genetically targeted would be forced either to accept their status as members of a genetic underclass barred from or constrained in their participation in certain social activities, or to undergo "corrective" intervention to take themselves off the blacklist. Predictably, some would pursue a higher risk but more rewarding alternative: circumvention of enforcement mechanisms through bribery or sabotage.

For that matter, we could right now identify a target group whose members were "at risk" of future violent conduct by invoking some statistically powerful combination of "risk" factors: previous record of criminal violence, age, gender, race, socioeconomic status, area of geographic residence. We could pursue a program of universal, mandatory "at risk" profiling and of intervention by drug or "talk" therapies that were aimed at blunting or bringing under control violent impulses. Or we could engage in on-site mandatory profiling to exclude individuals with these "at risk" profiles from admission to bars or gun shops. The repugnancy of programs of this sort is due, at least in part, to our continuing commitment to the beliefs that sustain the criminal code. The advent of additional genetic "risk" factors is unlikely to persuade us that we should abandon these beliefs or that hoped-for efficiency gains in preserving the public peace should override our commitment to them.

We anticipate that our new genetic knowledge will open the door to a deeper understanding of genetically influenced mental disorders. This understanding might enable us to develop more effective treatments one day[34] and, in the meantime, might persuade us to excuse or reduce the criminal liability of those who suffer from these disorders.[35] In this way, our new knowledge may well exert a significant influence upon our criminal code, but an influence that contributes to its dialectical growth rather than to its deconstruction.

4. THE CHALLENGE FROM ABOVE

Frederick Goodwin's 1992 remarks before the Mental Health Advisory Council suggested his sympathy with another controversial research program: the research program in sociobiology. At one point, Goodwin noted that one-half of male monkeys in the wild die by violence, and asserted, "that is the natural way of it for males, to knock each other off—and, in fact, there are some interesting evolutionary implications of that because the same hyperaggressive monkeys that kill each other are also hypersexual, so they copulate more" (Williams, 1994, p. 97). Here, Goodwin was suggesting, we could see the law of natural selection at work, favoring survival, or, more precisely, survival long enough to reproduce, of those monkeys predisposed both to hyperaggressive and hypersexual behavior. And when Goodwin compared the violent ways of inner city youth to the conduct of monkeys in the wild, he was suggesting that the law of natural selection had favored the survival of certain hyperaggressive humans as well, a fact ordinarily hidden from view by the constraints imposed by our social structures. This is a distinctively sociobiological view of human nature.

On the mainline view in sociobiology, the process of natural selection operates at the level of the individual organism.[36] Each individual is a vessel, a temporary carrier of a collection of genes. If an individual's genes support functional features that contribute to its procreative success, then that individual is likely to pass along some of its genes to one or more members of the next generation. These functional features may include structural features, such as the polar bear's white fur or the giraffe's long neck, or they may include behavioral features, such as the intricate dance of the honeybee or the complex behaviors that constitute the social lives of ants or, more controversially, human behaviors (Wilson, 1975, 1978, 1998; Crippen, 1994; Fisher, 1992). Sociobiologists recognize that functional features, including behaviors, are not determined solely by genetic material; genes and environment must interact. Moreover, different genetic material may support the same functional features. So the story of natural selection, on the sociobiological account, is complex. Nevertheless, the process generally ensures that the genes of successful procreators are retained in the gene pool and the genes of the not-so-successful are eliminated from the gene pool so that the gene pool retains an increasing proportion of genes that support procreation-favoring functional features (Wilson, 1998, 1978, 1975; Crippen, 1994; Fisher, 1992).

Edward O. Wilson's pioneering work, *Sociobiology: The New Synthesis* (1975), aimed at showing the operation of natural selection on behavioral features in animals and explored, in its final chapter, the thesis that natural selection might operate on the behaviors of humans as well. *On Human Nature* (1978) developed in detail this selectionist thesis with respect to humans. In its wake, the sociobiological research program has directed much of its effort toward developing persuasive accounts of selection on a broad array of human behavioral traits. The sociobiological program has developed accounts of why it might be that we see widespread evidence of altruism, a trait predisposing us to cooperative and self-sacrificing behavior toward others—behavior that would not appear to convey a survival benefit upon us.[37] The program has also developed accounts of why it might be that we see widespread evidence of aggressiveness. For example, there is evidence of a high, cross-cultural incidence of rape of women of child-bearing age and of rape among other animals, including other primates. One hypothesis suggests that, at one time in our evolutionary history, genes predisposing males to engage in forcible copulation persisted in the gene pool because this behavior enhanced the likelihood of reproductive success for these males (Ellis, 1990, pp. 63-5).

Wilson has published another ambitious and pioneering work, *Consilience: The Unity of Knowledge* (1998), in which he attempts to carry the account of evolutionary selection on behavioral traits several steps further. Wilson sketches a selection process involving a complex, genetically mediated interaction of genes and culture: our genes determine "epigenetic rules," and these epigenetic rules, in turn, determine our behavioral traits by directing the cognitive processes by which we learn from our cultural environment. Wilson writes, "Culture is created by the communal mind, and each mind in turn is the product of the genetically structured human brain. Genes and culture are therefore inseverably linked." (Wilson, 1998, p. 127; italics omitted) However, he points out, the link is flexible (though it is not yet known to what degree) as well as 'tortuous': "Genes prescribe epigenetic rules, which are the neural pathways and regularities in cognitive development by which the individual mind assembles itself." A mind grows by selectively absorbing certain facets of the surrounding culture, guided by the brain's inherited epigenetic rules. As a result, when genes and culture coevolve, culture is "reconstructed each generation collectively in the minds of individuals." Culture can grow indefinitely when writing and art supplement oral tradition, but because the epigenetic rules are "genetic and ineradicable", their influence remains constant. (p. 127; italics omitted)

Natural selection, Wilson explains, selects for those individuals whose genetic inheritance—in the form of epigenetic rules—yields relatively successful behaviors, that is, behaviors that enable them to survive and reproduce. He argues that individuals who inherit certain epigenetic rules are more successful in both survival and reproduction than others who either lack such rules or in whom the rules have a weaker expression. By implication, he argues, these epigenetic rules spread over generations through the population, "...along with the genes that prescribe the rules. As a consequence the human species has evolved genetically by natural selection in behavior, just as it has in the anatomy and physiology of the brain." (Wilson, 1998, p. 127; italics omitted)

There is another layer of complexity in the gene-culture interaction, on Wilson's account. Although we remain subject to this process of natural selection, we are also unique among animal species in the distance to which we can extend our "genetic leash" through the development, selection, and transmission of cultural norms. We can, to the limits of our genetic leash, invent our own cultural norms, and these cultural norms will survive and be transmitted to subsequent generations according to how adaptive they are to changing environmental conditions. Wilson suggests that this 'genetic leash' and the role of culture can be understood in the following way: some cultural norms, like certain genetic traits, are reproduced more successfully and survive better than do other norms. This leads to a 'cultural evolution' which outpaces genetic evolution. As he argues,

> [t]he quicker the pace of cultural evolution, the looser the connection between genes and culture, although the connection is never completely broken. Culture allows a rapid adjustment to changes in the environment through finely tuned adaptations invented and transmitted without correspondingly precise genetic prescription. (Wilson, 1998, p. 128; italics omitted)

He points out that this is a fundamental difference between humans and other animals.

All of that which we call 'ethics,' Wilson explains, is comprehended by this sociobiological narrative.[38] On Wilson's account, our genes predispose us to certain "moral sentiments" about the rightness and wrongness of behaviors. Behaviors that are consistently favored by our culture are captured in our moral codes.[39] Those individuals who conform to their culture's moral codes are more likely to survive and reproduce,[40] and certain moral codes support cultural flourishing better than others.[41] When we fully understand the genetic origins of our "moral sentiments" and the selectionist advantage that our capacity for commitment to certain moral tenets conveyed upon our ancestors, then we will be in a position to engage in the cultural fine-tuning of these tenets that best assures our cultural flourishing.[42] But our new knowledge should not lead us to jettison our old rituals, Wilson advises. These will continue to possess significance for us because we evolved to believe in their power. Wilson writes, in the voice of an imagined defender of his own point of view:

> '...It would be a sorry day if we abandoned our venerated sacral traditions. It would be a tragic misreading of history to expunge *under God* from the American Pledge of Allegiance. Whether atheists or true believers, let oaths be taken with hand on the Bible, and may we continue to hear *So help me God*. Call upon priests and ministers and rabbis to bless civil ceremony with prayer, and by all means let us bow our heads in communal respect. Recognize that when introits and invocations prickle the skin we are in the presence of poetry, and the soul of the tribe, something that will outlive the particularities of sectarian belief, and perhaps belief in God itself.' (Wilson, 1998, pp. 247-8)

Many of the claims of sociobiology are vigorously disputed and the prospects for its long-term success are difficult to predict.[43] Also, although Wilson's most recent version of the narrative sets forth the most ambitious and comprehensive statement of the sociobiological project to date, other sociobiological researchers may carry the program in a different direction, rejecting some or many of Wilson's commitments. We will assume for purposes of our discussion that the program at some point succeeds by the following two measures, neither of which depends upon the theoretical or empirical details of the sociobiological claims involved, but both of which generally track Wilson's version of the narrative: (1) sociobiologists provide evolutionary explanations for the persistence of genes predisposing individuals to a wide range of violent conduct, from rape to child abuse to murder, and these explanations are widely believed to be correct; and (2) sociobiologists develop an account of morality in which all of our moral motivations, beliefs, norms, and judgments are subsumed by the sociobiological narrative, and this account is widely believed to be correct.

If we come to believe the first claim, that genes predispose certain individuals to behave violently and that the persistence of these genes reflects the operation of the law of natural selection, this would appear to pose a powerful challenge to our current conception of the criminal code. We might come to think that, while the law of natural selection may grind relentlessly, indifferent to the consequences for us, we need not remain indifferent to the fate of those of us evolutionarily "designed" to be predisposed to violence. Rather than judge and punish these individuals, we might decide that we should intervene before they fall into violence by a public health/therapeutic or a public health/environmental response of the sort discussed in section 3 of this chapter. Alternatively, or in addition to these responses, we might decide that we should respond to the commission of a crime in a compassionate rather than a judgmental way, by confining the offender in comfortable surroundings or by monitoring the offender to forestall repeat offenses. By these responses we might both preserve the public peace and display compassion toward those fated by the laws of nature to their violent predispositions.

A successful program in sociobiology, however, would not and could not claim that some of us were evolutionarily determined to commit acts of violence. The influence of culture and environment on our behavior is obvious and widely acknowledged by sociobiologists as by others. And, while we cannot escape the operation of the law of natural selection, neither can we escape the operation of a number of other causal factors that likely influence us to engage in violence: the ill-treatment we suffered at the hands of our parents, or the mean streets we endured as adolescents. As we have argued, our beliefs in these influences have not dislodged our beliefs in moral agency and the capacity to choose. So, if it is true that particular genes influence the choice to engage in violent conduct and that this influence is owing, in some sense, to the law of natural selection, then when we know this, we will have a deeper understanding of how it is that a genetic influence to violence was visited upon some of those who face judgment and punishment for acts of wrongdoing. But we will not have a new reason for abandoning the beliefs that sustain the criminal code.

As we have observed, these beliefs have proved quite resilient in the face of novel challenges. We should note, however, the dependence of these beliefs, for their persuasive power and their very intelligibility, upon widely shared background assumptions about normativity in the world: these beliefs refer to our personhood and our capacity to choose between right and wrong conduct, to our obligation to abide by moral norms expressed in the criminal code, and to the vindication of these moral norms by the practice of punishment. The sociobiological account of morality is fundamentally inconsistent with these background assumptions. To see why this is so, imagine a future in which the sociobiological account of morality were widely believed.

In this imagined future, punishment could no longer assert our subjection to the constraints of morality nor depict us as agents who were susceptible to the allure of evil yet capable of appreciating its repulsiveness, as agents who were capable of understanding how we could succumb and how we could resist, or as agents whose conduct was motivated by reasons. Punishment would make no sense applied to human animals whose behavioral traits were fashioned by the interaction of genes and culture in a universe ruled by the law of natural selection, a universe in which our ideas about good and evil were no more than delusions that had proved adaptive during human pre-history.

Perhaps the lingering appeal of longstanding rituals of punishment to minds genetically predisposed to find them meaningful might persuade policymakers that they should engage in only modest tinkering with our current criminal code. Even if more radical reform efforts might, in some sense, be desirable, policymakers might fear that reform efforts would rouse the citizenry to contemplate the emptiness behind our rituals with potentially counterproductive consequences for public safety. But there would be difficulties in preserving even a sacred shell for any length of time.[44] For one, if we could no longer understand the lawbreaker as culpable wrongdoer, as deserving of punishment for having chosen evil over good in defiance of the moral order, punishment would quickly collapse into its caricature. In light of these difficulties, policymakers might choose, instead, to revamp our approach to crime control in line with our post-modern self-awareness, opting perhaps for the public health/therapeutic or public health/environmental models that we have described in section 3

As these reflections suggest, if a sociobiological account of morality came to be widely accepted, our current conception of the criminal code would quickly unravel. But how likely is it that such a sociobiological account might succeed? We consider next why the account might appeal to ethicists, scientists, and the wider public. Non-cognitivists in ethics hold that our moral claims should not be viewed as assertions about the world that could be literally true or false. Instead, they say, we should understand that these claims express our feelings about objects or states of affairs in the world, or that these claims express our prescriptions for action in the world.[45] The sociobiological account might add new weight to these non-cognitivist accounts of moral language by linking them to an evolutionary explanation of why it is that we seem to engage in cognitive discourse when we use moral language; why it is, that is, that we seem to seek evidence for our beliefs and to believe that moral claims can be true or false or objectively justified. The sociobiologists would explain that we have been evolutionarily designed to believe, mistakenly, that our moral claims have cognitive content; this mistaken belief and the actions that flowed from it, they would explain, enhanced our ancestors' reproductive success in our evolutionary past. [46]

Alternatively, a sociobiological account might hold that, although our moral claims do indeed have cognitive content, all of them are false,[47] including the implicit claim that our moral judgments refer to something that is objectively prescriptive.[48] A sociobiological account, again, could provide a good explanation for why it is that we believe in the truth of our moral claims—including the truth of the claim that there is in fact normativity in the world—but are, as it turns out, hugely mistaken. Again, we believe because it enhanced our ancestors' chances of survival to believe

Finally, a sociobiological account might hold that our moral claims do indeed have cognitive content and that some of them are true, but that their truth or falsity is entirely subjective, a matter of what is good for me,[49] or is entirely intersubjective, a matter of what happens to be established by the customs of my culture. If morality consists only in principles drawn from genetically directed behaviors favored in the evolutionary process, then judging the truth or falsity of moral claims by reference to my own genetically directed preferences or to the genetically directed practices of my culture becomes an at least plausible version of cognitivism

The sociobiological narrative also might appeal to scientists, in particular to those scientists hoping to achieve a conceptual and methodological unification of science and other domains. Biologists, psychologists, and ethicists could join with physicists in quest of a unified theory of everything, including all of our experience of morality. And, while Wilson claims that we will one day reduce all of the realms of biology and psychology to the realm of physics,[50] an attractive feature of sociobiological theory is that we might grant ourselves permission to bracket the problems presented by efforts at a simple reduction of the phenomena of our biological and mental lives to the physical.[51] Instead, we might be satisfied with a sociobiological account that provides plausible explanatory theories for why we believe and behave as we do. The sociobiological account also might have broad appeal because it appears to make sense of threatening and otherwise apparently inexplicable behavior. The sociobiologist explains the criminal's choice in the language of human genetics, neurophysiology, and evolutionary science, rather than in the "unscientific" language of the moral choice to do evil. The sociobiologist offers to make sense of violent human conduct in the same way that meteorologists now make sense of the violent movements of hurricanes, and the sociobiological account is fraught with the same expectation of scientific progress toward ever-more complete understanding of the phenomena and ever-more effective techniques for minimizing their unpleasant consequences. In this way, a sociobiological account of morality would draw upon the capital earned by science, the most successful of modern enterprises.

In addition, certain features of the sociobiological account of morality would render its claims peculiarly resistant to refutation.[52] Sociobiological method consists of gathering the data of human experience, fashioning sociobiological hypotheses as to why we might expect to find these data if the law of natural selection is at work, and pursuing confirming evidence in heritability studies and biochemical causal accounts of genetic influence. Because its method assumes that all of morality is to be treated as data demanding sociobiological explanation, and because its explanatory hypotheses must discount the force of our moral reasoning and intuition by treating them as delusional if they are to be subsumed by the sociobiological account, that account of morality is impervious to challenges that appeal to moral reasoning or intuition.

Finally, a sociobiological account of morality would share with many post-modern doctrines a deep appeal to our vanity. It would confirm that members of the current generation are, at last, smart enough to see through the delusions that ensnared their forebears.

As we have already suggested, we find the appeal of the sociobiological explanation of the moral life in non-moral terms to be fundamentally meretricious. We do not for a moment deny that sociobiological research and reflection can shed considerable light on the mystery of iniquity that has troubled humankind since the dawn of history. Neither do we doubt that, as a result of this research and reflection, therapeutic interventions may take the place of punishment in particular cases with great gains for the treated individuals and for society as a whole. We do, however, have grave misgivings about the reductionist thrust of the sociobiological project. We see the idea of personhood that informs our criminal code and our practice of punishment as a major cultural achievement that should not be rejected unless evidence and argument make it rationally untenable. As the title of our chapter suggests, we favor a dialectical resolution of the conflict between the moral and the genetic person rather then the defeat of one by the other. By that we mean that we want to see both images of the person—the moral and the genetic—kept in productive tension, one with the other. From that tension we will surely learn, perhaps by deepening our appreciation for the struggles involved in moral choice and perhaps by increasing our compassion for the person who stands before the community accused of wrongdoing. But we should resist the temptation to quick surrender of the moral knowledge that infuses our beliefs and practices in connection with the criminal law, moral knowledge won over generations of reflection and expressive of beliefs fundamental to our sense of ourselves.

5. CONCLUSION

Under our current criminal code, legislators seek to safeguard the community and vindicate its moral norms by prospectively establishing standards for punishment, and judges and juries mete out punishment to the guilty, whose own stories, typically, include victimization by parents or spouses, addiction or other severe health problems, educational or vocational failure, or some combination of these. Soon, genetic predispositions to violence might be included in their stories. In their reflective moments, those who write the laws and those who sit in judgment on individual wrongdoers, even though persuaded that the guilty deserve punishment, must wonder if they themselves could have done better under similar circumstances. Those who legislate, those who judge, and those who ultimately execute punitive judgments continue the long labor of constructing our criminal code by a process that takes in novel theories and novel stories of victims and victimizers, and produces ever-richer re-affirmations of our beliefs in moral agency and choice, the obligation to abide by community norms, and punishment for wrongdoing.

It might appear that the greatest threat to the criminal code will come from direct challenges to our beliefs that we "choose" whether or not to honor community norms and, hence, that we deserve to be punished for failing to properly execute our obligations. But, for the reasons set forth above, we think that challenges of this sort, launched either from below or above or both, are unlikely to succeed.

Far more plausible, in our estimation, is that a sociobiological account of morality, if its appeal proved strong enough, might contribute to the dissolution of our background beliefs in a world in which we can objectively justify by reference to moral norms the distinctions we draw between right and wrong conduct. The consequences for our criminal code would be profound. While Wilson assumes, and hopes, that our "rituals" would continue pretty much as they have for these many generations, that assumption, we believe, is mistaken. It fails to take into account the extent to which our current beliefs and our practices in furtherance of these beliefs draw upon this background normativity, and it fails to take into account how difficult it is for us to judge and punish our own. Our code does not endure in virtue of the ease or simplicity of its application. Rather, it endures in virtue of the strength of its constitutive beliefs and their embeddedness in our deepest moral convictions.

We believe that the sociobiological account of morality is mistaken. But we also urge that the appropriate response to its challenge is not to argue that its proponents are motivated by anything other than the search for truth or that the program should be shut down because of the dangers its findings might pose. We should welcome this challenge, as well as other challenges from below and above, as likely to yield valuable contributions to the dialectic by which our criminal code continues to grow. The appropriate response to the particular challenge posed by a sociobiological account of morality is to pursue with renewed vigor the justifications for our judgments of wrongdoing.

Notre Dame Law School
South Bend, Indiana, U.S.A.
and
Georgia Institute of Technology
Atlanta, Georgia, U.S.A.

NOTES

1. For recent discussions of genetic research and the philosophical questions surrounding the concept of free will, see Brock and Buchanan (1999) and Parker (1999). For the best work on the role that personhood should play in analyses of punishment, see Morris (1976).
2. When we speak of the practice of punishment here, we refer both to the imposition of a sanction upon an individual because he or she is thought to have engaged in prohibited wrongdoing *and* to the process by which that determination is reached. Later, we will have occasion to distinguish between convicting a person of a crime and punishing him or her, but here we are calling attention to certain features of both those phenomena taken together. We wish to thank Lisa Parker for calling our attention to the need for this, and other, clarificatory notes.

3.   This quote comes from Sir Thomas Brian, a judge during the reign of Edward IV, who ruled from 1461 till 1470, then from 1471-1483. It can be found in the Year Book for the seventeenth year of Edward's reign, on p. 2 of the cases for Easter Term of that year (Y.B. Pasch. 17 Edw. IV, f. 2a, p. 2 (1477)).

4.   We say "ordinarily" here because we want to bracket all those difficult questions that relate to the punishment of those who engage in conscientious violation of a particular law and to other cases (e.g., strict liability offenses) that deviate from the paradigm that we are articulating here.

5.   The psychotic who killed the Prime Minister's secretary was Daniel M'Naghten. The rules to which his acquittal (by reason of insanity) gave rise were promulgated in 1843 and can be found in Volume 8 of the English Reports starting at page 718 (M'Naghten's Case, 10 Cl. & F. 200, 8 Eng. Rep. 718 (1843)). The statutory changes in federal law that resulted from John Hinkley's acquittal (by reason of insanity) can be found in Title 18 of the United States Code, in Section 17 (18 U.S.C.A. Sec. 17 (West Supp. 1999)).

6.   For one account of the transformation in the insanity defense that various legislatures have effected in recent decades, see Steadman (1993).

7.   The battered woman syndrome entered legal parlance in 1979 with the publication of Lenore Walker's *The Battered Woman*. Since that time both courts and legislatures have had occasion to address its propriety as a defense to criminal charges in a variety of contexts. For one account of this defense and of the problems incident to its recognition, see Downs (1996).

8.   E.g., Cloninger and Gottesman, reporting on the results of a twin study, find that "heritability of liability to property offenses is 78%, and heritability of liability to crimes against persons is 50%" (1987, p. 100).

9.   Regarding heritability studies, Berkowitz argues that even when a trait has a high heritability within each of two groups of genotypes, this does not tell us the reason for any differences between the two groups: "In short, heritability cannot be measured accurately in human studies and, even if it could be, it would not indicate the relative importance of genes and environments" (1996, p. 46).

     With respect to twin studies, Berkowitz claims that researchers may fail to take into account the possibility that the environment of monozygotic (identical) twins is more alike than that of other siblings. Also, while researchers assume that monozygotic twins have twice as many genes in common as dyzygotic (fraternal) twins, it may be the case that monozygotic twins have more than twice as many functional genes in common; heritability estimates may be inflated for this reason as well (Berlpwotz, 1996, pp. 46-7)

     Berkowitz notes, finally, that these studies do not tell the causal story of how genes contribute causally to the expression of a trait (1996, p. 47). And any such causal account would be very complex. It would begin with the complex role of a gene in the production of a protein, which may involve a wide variety of contributory influences, including other bits of DNA, RNA, other proteins, hormones, substances from the diet, the history of cell division, and the location of the cell in the body.

     Regarding biochemical studies, inconsistent results have been noted (Berman and Coccaro, 1998, p. 308). Berkowitz observes that the interaction among networks of neurons, neurotransmitters, neurotransmitter receptors, and genes is not well understood, limiting our ability to draw causal conclusions (1996, pp. 48-9). Alper notes that there are difficulties in determining whether abnormalities in neurotransmitter levels cause abnormal behaviors, or abnormal behaviors cause the observed alterations in metabolite levels, or a third factor is the cause of both (1995, p. 272).

     For other expressions of reservations regarding current scientific understanding of the neurophysiology of violence, see Lancet (1995) and Pallone and Hennessy (1998).

10. Goodwin, in a 1994 interview with journalist Juan Williams, explained:

     [The survey results of youthful offenders showing that 80 percent of the offenses were committed by 7 percent of the population were] an incredible concentration. So the first thing we looked for was correlation—an association between this 7 percent and some other factor. And I want to make clear there was no correlation between violence and race at all, when you took socioeconomic status out of it—in fact, black middle-class kids, we'd previously found, were less likely to abuse drugs than white middle-class kids and were more socially responsible. There *was*, however, a strong association of violence with low socioeconomic class. Nevertheless, there were a lot more people of low socioeconomic status who were *not* violent than who were violent, so class was not deterministic. There had to be something else influencing violent behavior. (Williams, 1994, p. 97; emphases in original rendering of interview remarks)

11. Williams describes these events as they were later recounted by Louis Sullivan, Secretary of Health and Human Services under President Bush from 1989 to 1993, and Frederick Goodwin, at that time Director of the Alcohol, Drug Abuse, and Mental Health Administration (1994, pp. 95, 97). Sullivan organized the "Violence Initiative" during his tenure as Secretary in an effort to address the public health implications of criminal violence, especially for African-Americans who are both victimized by crime and convicted for committing crimes in numbers disproportionate to their representation in the general population. The Violence Initiative focused upon social factors thought to be significant in the incidence of crime, including unemployment, poverty, and drugs. A very small amount of the funding for the Violence Initiative, 0.5%, was devoted to explorations of biological factors in crime, including links between aggression and low levels of the neurotransmitter serotonin.

12. The NIH is an agency within the Department of Health and Human Services and, hence, was under the ultimate direction of Sullivan, although Sullivan explained later that he knew nothing of NIH's original decision to fund the conference (Williams, 1994, p. 98).

13. The editors of a volume of scientific studies, published after the controversy surrounding the canceled 1992 conference, wrote in their information:

> This volume was conceived at a time when biological research on aggression and violence was drawn into controversy because of sociopolitical questions about its study. A group of dedicated scientists recognized the importance of freedom of inquiry, deemed it vital to address the state of the art of biological studies in the field, and generated this volume. The challenge for biological research in aggression and violence will be the courage to pursue responsible research so that opinions and political decisions can be informed by reliable and valid facts. (Stoff and Cairns, 1996, p. ix; citations omitted)

14. See Powledge (1996), Roush (1995), Mann (1994), and Williams (1994) reporting the views of opponents of research into the genetic causes of violence.

There has been a strong historical interest in applying eugenic theory and practice to the problem of crime. Charles Davenport, a leader in the early twentieth century eugenics movement in America, emphasized the connection between undesirable genetic constitutions and crime: "We are breeding too many people with feeble inhibitions and without proper social instincts...Satisfactory progress will be made only when we understand how those with congenital criminalistic make-up are bred and try to prevent such breeding" (Davenport, 1928, p. 313). Also at the beginning of the twentieth century, psychologist Henry Herbert Goddard developed a highly influential theory of hereditary criminality as a function of hereditary "feeblemindedness," and called for a eugenic response (Rafter, 1997, pp. 134-48). Scholarly consideration of possible biological influences on criminality tailed off dramatically with the collapse of the international eugenics movement after the Holocaust. Scholarly attention gradually returned to these questions in the post-War period, but associations with the eugenic past continue to generate controversy (Walters and Palmer, 1997, p. 118).

Berman and Coccaro note the current salience of the issue of genetic influence upon violent behavior given the impact of violent crime on society and given that about half of the state prison population consists of violent offenders (federal prisoners are largely drug offenders) (1998, p. 304).

15. Walters and Palmer suggest that we will discover multiple genes that interact in complex ways to produce predispositions to violent conduct (1997, p. 125). They acknowledge that some sorts of aggressive behavior may be morally laudable, but suggest, with qualifications and caveats, that somatic-cell or germ-line genetic therapy aimed at revising human behavior might be acceptable or desirable in two cases. The first case would involve "clearly anti-social or sociopathic behavior," in which case the goal would be to "move behavior that is outside the range of species-typical human functioning to some point within the normal range." Therapy in this case would be "at least *akin* to therapy for disease." The second case would involve behavior of the kind "that characterizes most human beings, at least in our dispositions and often in our overt actions or failures to act." They explain that, in this case, "the goal of the intervention would be to move species-typical functioning itself toward a more conscientious and agreeable norm."

For a discussion of the potential ethical implications of a new era of psychiatric genetics, see Farmer and Owen (1996), Coffey (1993), and Lelling (1993).

16. Walters and Palmer forecast the arrival of this day (1997, p. 125).

17. As Parens writes:

> Even if there are strong correlations between single-gene defects and certain dispositions to some complex behaviors, such correlations will never provide anything approximating a full account of those behaviors. To begin with, a fuller account will require thinking in terms of nonlinear and dynamic interactions among many genes, hormones, nutrients, and other biological factors in the 'internal' environment that is the body. And as genetics always will be only one important part of biology, biology always will be only one important part of any richer account of human behavior. Such an account will have to consider not only interactions *among* social and biological factors in the 'external' environment, but will have to consider the complex interactions *between* the internal and external environments....In light of such complexity, it should be clear that even if there were a correlation between, say, a single-gene defect and a predisposition to impulsivity, no amount of behavioral genetics could predict whether such a predisposition will gain expression in a bar room—or a board room—fight. (Parens, 1996, pp. 13-4; emphases in original)

18. For a recent philosophical account of the psychology of the thinking intentional agent, see Pettit (1996, pp. 54-108).

19. As Pettit explains, the supervenient relationship means that if we hold constant the particular set of naturalistic elements that pertains to a particular set of intentional states, the intentional states will not change (1996, pp. 25-8).

20. See Rosenberg for a discussion of this problem and the virtues and shortcomings of notable efforts to address it (1994, pp. 141-56).

21. See Rosenberg for a discussion of this problem and its significance for reductionist projects (1994, pp. 156-68).

22. Rosenberg offers the following example:

> The intensionality of psychological states consists at least in this: take any true attribution of the belief that object *b* has property *F*, and substitute into the attribution terms coreferential with *b* and/or coextensive with *F*. Sometimes the result will be a false attribution of belief. For example, 'Oedipus believed that Jocasta was the queen of Thebes.' But if we substitute for 'Jocasta' the coreferring term 'Oedipus' mother', we shall produce the presumably false statement that 'Oedipus believed his mother was queen of Thebes.' And if we substitute for 'queen of Thebes' the coextensive predicate 'Oedipus' bride,' we will produce the equally false statement that 'Oedipus believed that his mother was his bride.' By contrast, such substitutions do not enable us to produce a false statement from a true one in biological, chemical, or physical theory. (Rosenberg, 1994, pp. 156-7)

23. See Churchland, an "eliminative materialist" who concludes that we will not succeed in reducing our current psychological notions of intentional states to neurophysiological terms and that we will one day jettison our intentional-state talk in favor of an entirely neurophysiological account of our cognition and behavior (1988, pp. 43-9). See also the discussions of eliminativism in Pettit (1996, pp. 45-6) and Rosenberg (1994, pp. 141-2).

24. For discussions of the eliminative materialist project and its potential implications for the criminal code, see Reider (1998) and Lelling (1993).

25. Citing previous studies, Barros and Miczek write, "Epidemiological data show that more than half of the persons known to have committed violent behaviors have consumed ethanol" (1996, p. 237; citations omitted).

26. Montana v. Egelhoff, 518 U.S. 37 (1996) (the Fourteenth Amendment to the U.S. Constitution does not require states to allow juries in criminal cases to hear evidence about a defendant's allegedly intoxicated condition at the time of the crime in question before the jury decides on the defendant's guilt or innocence); Powell v. Texas, 392 U.S. 514 (1968) (chronic alcoholism does not constitute a legal excuse from culpability for the crime of public intoxication). For recent work on this issue, see Watson (1999).

27. Parens writes:

> The idea of an absolutely free and disembodied will is not a necessary condition for having...a concept of responsibility. Indeed, our society has for some time now accepted that some actions are not simply freely chosen insofar as economic and social

forces constrain individual actions, while simultaneously insisting that citizens must be held responsible for their actions. Today we need to grapple with the respect in which, like economic and social factors, some genetic factors to some extent constrain how we act. (Parens, 1996, p. 16)

28. See discussions and critiques in Pettit (1996, pp. 11-164) and Boudon (1982, pp. 153-227) of various versions of social determinism.

29. See discussion of Emile Durkheim's views in Pettit (1996, pp. 129-32).

30. For a recent discussion of the relationship between social regularities and our intentional psychology, see Pettit (1996, pp. 143-55).

31. Powledge recounts the explanation of a participant in the 1995 conference on crime and genetics as to the complex story of one hormone—testosterone—and its influence on violence in male animals:

[Margaret McCarthy, of the University of Maryland School of Medicine] reviewed several studies of the role of testosterone in aggressive behavior among experimental animals. Testosterone, it turns out, unlike cocaine or heroin, does not act directly on the brain to trigger behavior. Like other steroid hormones, testosterone instead regulates gene expression, acting on many different sites in a cell's DNA. 'We have very little clue as to what these sites are, but they are multiple in the brain,' she said. 'That turns on the gene products, and it is these gene products that then alter the behavior.'

Everyone who reads newspapers, McCarthy pointed out, thinks that there is a direct relationship between testosterone levels and aggression. The real story is considerably more complicated. When two male experimental animals with similar genes and similar testosterone levels fight, the winner's testosterone rises and the loser's falls, resulting in different levels of gene expression. The fight also stimulates production of other steroids, the glucocorticoids, the so-called stress hormones, which turn on another set of genes, and initiate another set of gene products.

The two animals, whose genetic endowment is similar, are now in quite different states. If the aggression does not recur, a stable social hierarchy will be established and the testosterone levels of the two animals will return to approximately their original similarity. If this social hierarchy is disrupted, however—by repeatedly introducing strange males, or by limiting resources, for example—testosterone will end up having a large effect on gene expression, as will the glucocorticoids.

'Given enough of these encounters, you can exert more-or-less permanent effects on gene expression in these animals,' McCarthy said. The result: animals that are genetically similar respond to the same stimulus quite differently.

'We keep talking about genes, and genetic variability, but genes are not static,' McCarthy noted. 'It doesn't matter a whit if you have a gene if it doesn't get turned on. It has to be regulated. What genes we inherit are only relevant in terms of their expression.' (Powledge, 1996, p. 8)

32. See Jacobson v. Commonwealth of Massachusetts, 197 U.S. 11 (1905).

33. We thank Lisa Parker for prompting us to contemplate this alternative scenario.

34. Farmer and Owen predict significant changes in clinical psychiatric practice in the wake of our newly acquired genetic knowledge:

The ability to correct the effects of faulty genes could lead to prevention or cure. The knowledge base will be radically altered and psychiatry may become more allied to mainstream medicine, since the pathogenic mechanisms underlying psychiatric disorders like those of other common diseases such as coronary artery disease and hypertension will be understood at the molecular level. Enhanced knowledge of the genetics of abnormal behavioural traits will almost certainly provide effective means for altering such behaviour. (Farmer and Owen, 1996, p. 138)

35. Berman and Coccaro explore the idea that criminal responsibility might be reduced "if it can be shown that a particular neurobiologic defect reliably interferes with one's ability to appreciate the wrongfulness of an act or to self-regulate behavior" (1998, p. 305). They note that few studies directly address the connection between neurotransmitter functioning and crime, but "a rich literature exists to support the notion that neurotransmitter functioning is linked to a history of aggressive behavior in adults." The authors acknowledge the inferential leap involved in extrapolating from studies of "aggressive tendencies in non-criminal samples to criminal acts of violence."

36. This is the mainline view in evolutionary biology as well (Rosenberg, 1995, p. 167). But others hold different views. See, for example, Sober and Wilson (1998) envisioning selection at the level of genes, individuals, and groups.

37. Sociobiologists theorize, generally, that altruism directed toward biological kin improves "inclusive fitness" through "kin selection": our genes are preserved by the procreative success of the genetic kin we assist by our altruistic behavior. Reciprocal altruism (displayed among non-kin) brings survival benefits to those who both accept altruistic aid and reciprocate; those who do not reciprocate are eventually identified and thereafter excluded from the circle of mutual assistance (see Rosenberg, 1995, pp. 168-71; Crippen, 1994; Fisher, 1992; Wilson, 1978, pp. 149-67; Sober and Wilson, 1998).

38. Other sociobiologists offer similar accounts of the development, nature, and role of moral norms. Sociobiologists Lumsden and Gushurst describe the development of moral codes:

> In ethical situations, the individual will often feel strongly that one or another course of action is the 'right' thing to do. However, there is generally also an accompanying rationalization or...logical justification of this choice. This is of particular importance in a social context, where the individual must explain to others the reasons behind his or her choice of action. Social approval is important to individual adaptation in human societies and it is essential to the development of moral knowledge....
>
> The ethical center in human social life can therefore be expected to create selection pressures favoring certain forms of the moral passions. The associated genetic basis will in turn influence the sorts of ethical knowledge assimilated during enculturation. A culturally fit ethical system is one that can be reliably transmitted and upheld. If a system existed in which values or actions could not be verbally defended, that set of beliefs would be open to questioning and alteration. Coevolutionarily successful moral passions will therefore tend to have self-evident propositions as culturgenic counterparts. These propositions are readily attached to lore of rationalization through which value is interpreted and persuasively justified. Given the expansive tendencies of human reason, the consistency and covering power of these propositions under logical derivation and (possibly) generalization will also test their cultural fitness in competition with alternative ethics. (Lumsden and Gushurst, 1985, pp. 18-9; citations omitted)

39. Wilson explains, "Ethics...is conduct favored consistently enough throughout a society to be expressed as a code of principles. It is driven by hereditary predispositions in mental development...causing broad convergence across cultures, while reaching precise form in each culture according to historical circumstance" (1998, pp. 246-7).

40. Wilson explains: "For more than a thousand generations [moral codes] have increased the survival and reproductive success of those who conformed to tribal faiths. There was more than enough time for epigenetic rules...to evolve that generate moral and religious sentiments. Indoctrinability became an instinct" (1998, pp. 246-7).

41. Wilson writes that "the [moral] codes, whether judged by outsiders as good or evil, play an important role in determining which cultures flourish, and which decline" (1998, p. 240).

42. Wilson writes:

> From a convergence of these several approaches [to understanding], the true origin and meaning of ethical behavior may come into focus. If so, a more certain measure can then be taken of the strengths and flexibility of the epigenetic rules composing the various moral sentiments. From that knowledge, it should be possible to adapt the ancient moral sentiments more wisely to the swiftly changing conditions of modern life into which, willy-nilly and largely in ignorance, we have plunged ourselves. (Wilson, 1998, p. 256)

43 See, e.g., critiques in Rosenberg. (1995, pp. 167-86) and Kitcher (1985).

44. Alasdair MacIntyre writes of the demise of certain taboos in Polynesian culture:

> In the journal of his third voyage, Captain Cook records the first discovery by English speakers of the Polynesian word *taboo*....When they enquired why men and women were prohibited from eating together, they were told that the practice was *taboo*. But when they enquired further what *taboo* meant, they could get little further information....What this *suggests* is...that the native informants themselves did not

really understand the word they were using, and this suggestion is reinforced by the ease with which Kamehameha II abolished the taboos in Hawaii forty years later in 1819 and the lack of social consequence when he did.

[Steiner and Douglas suggest] that taboo rules often and perhaps characteristically have a history which falls into two stages. In the first stage they are embedded in a context which confers intelligibility upon them....Deprive the taboo rules of their original context and they at once are apt to appear as a set of arbitrary prohibitions, as indeed they characteristically do appear when the initial context is lost, when those background beliefs in the light of which the taboo rules had originally been understood have not only been abandoned but forgotten.

In such a situation the rules have been deprived of any status that can secure their authority and, if they do not acquire some new status quickly, both their interpretation and their justification become debatable. When the resources of a culture are too meagre to carry through the task of reinterpretation, the task of justification becomes impossible. Hence perhaps the relatively easy, although to some contemporary observers astonishing victory of Kamehameha II over the taboos....(MacIntyre, 1984, pp. 111-2)

45. See discussion of non-cognitivism in Sayre-McCord (1988, p. 8).

46. Ruse explains the position of the sociobiologist and links this position to Hume's account:

Objective ethics, in the sense of something written on tablets of stone (or engraven on God's heart) external to us, has to go. The only reasonable thing that we, as sociobiologists, can say is that morality is something biology makes us believe in, so that we will further our evolutionary ends. Hume argued: "The hypothesis which we embrace is plain. It maintains that morality is determined by sentiment. It defines virtue to be whatever mental action or quality gives to a spectator the pleasing sentiment of approbation: and vice the contrary." This is precisely the position of the sociobiologist. Where he/she has the edge on Hume is that he/she can show precisely why we have moral sentiments at all. Such sentiments further our biological advantage.

Note that neither Hume nor the sociobiologist is plunged into moral relativism. Because we are all members of the same species, with a common evolutionary heritage, we have shared moral standards. And these can serve as the basis for rational argument, at least as well as can any supposedly objective ethical absolutes. Nor is there any question that we simply make up morality to suit us, as we go along. We are what we are, namely possessors of a psychology which feels there is an objective morality external to us and which we must obey. There may not be such objectivity, but evolution has us thinking otherwise. (Ruse, 1985, p. 237; citation omitted)

47. See discussion of cognitivism in Sayre-McCord (1988, pp. 9-13).

48. See discussion of J.L. Mackie's error theory invoking this point in Sayre-McCord (1988, pp. 11-3).

49. Variants of subjectivism might refer to what is good for another or for an idealized other. See discussion of subjectivism in Sayre-McCord (1988, pp. 16-8).

50. Wilson writes of this ultimate reduction: "The central idea of the consilience world view is that all tangible phenomena, from the birth of stars to the workings of social institutions, are based on material processes that are ultimately reducible, however long and tortuous the sequences, to the laws of physics" (1998, pp. 264-5).

51. See Rosenberg (1994) for an account of why reductionist approaches in biology cannot hope to succeed.

52. Some argue that the entire sociobiological narrative is "unscientific" in that its claims are tautological or not "falsifiable." For a discussion and refutation of these claims, see Ruse (1985, pp. 111-9).

REFERENCES

Alper J. Biological influences on criminal behaviour: how good is the evidence? British Medical Journal 1995; 310:272-73

Barros, H.M.T., Miczek, K.A. "Neurobiological and Behavioral Characteristics of Alcohol-Heightened Aggression." In *Aggression and Violence: Genetic, Neurobiological, and Biosocial Perspectives*, D.M. Stoff, R.B. Cairns, eds. Mahway, NJ: Lawrence Erlbaum Associates, Inc., 1996.

Berkowitz A. Our genes, ourselves? Roles of genes in determining human characteristics. BioScience 1996; 46:42-51

Berman M.E., Coccaro E.F. Neurobiologic correlates of violence. Behavioral Sciences and the Law 1998; 16:303-18

Boudon, R., *The Unintended Consequences of Social Action*. New York: St. Martin's Press, 1982.

Brock, D.W., Buchanan, A.E. "The Genetics of Behavior and Concepts of Free Will and Determinism." In *Genetics and Criminality: The Potential Misuse of Scientific Information in Court*, J.R. Botkin, W.M. McMahon, L.P. Francis, eds. Washington, D.C.: American Psychological Association, 1999.

Brunner H.G. *et al.* Abnormal behavior associated with a point mutation in the structural gene for monoamine oxidase A. Science 1993; 62:578-80

Carey, G. "Family and Genetic Epidemiology of Aggressive and Antisocial Behavior." In *Aggression and Violence: Genetic, Neurobiological, and Biosocial Perspectives*, D.M. Stoff, R.B. Cairns, eds. Mahway, NJ: Lawrence Erlbaum Associates, Inc., 1996.

Churchland, P.M., *Matter and Consciousness: A Contemporary Introduction to the Philosophy of Mind*, rev. ed. Cambridge: MIT Press, 1988.

Cloninger, C.R., Gottesman, I.I. "Genetic and Environmental Factors in Antisocial Behavioral Disorders." In *The Causes of Crime: New Biological Approaches*, S.A. Mednick, T.E. Moffitt, S.A. Stack, eds. New York: Cambridge University Press, 1987.

Coccaro, E.F., Kavoussi, R.J. "Neurotransmitter Correlates of Impulsive Aggression." In *Aggression and Violence: Genetic, Neurobiological, and Biosocial Perspectives*, D.M. Stoff, R.B. Cairns, eds. Mahway, NJ: Lawrence Erlbaum Associates, Inc., 1996.

Coffey M.P. The genetic defense: excuse or explanation? William & Mary Law Review 1993; 35:353-99

Crippen T. Toward a neo-Darwinian sociology: its nomological principles and some illustrative applications. Sociological Perspectives 1994; 37:309-36

Davenport C.B. Crime, heredity, and environment. Journal of Heredity 1928; 19:307-13

Downs, D.A., *More Than Victims: Battered Women, The Syndrome Society, and the Law*. Chicago: University of Chicago Press, 1996.

Ellis, L. "The Evolution of Violent Criminal Behavior and its Nonlegal Equivalent." In *Crime in Biological, Social, and Moral Contexts*, L. Ellis, H. Hoffman, eds. New York: Praeger Publishers, 1990.

Farmer A., Owen M.J. Genomics: the next psychiatric revolution? British Journal of Psychiatry 1996; 169:135-8

Fisher A. Sociobiology: science or ideology? Society 1992; 29:67-80

Jacobson *v.* Commonwealth of Massachusetts, 197 U.S. 11 (1905).

Kitcher, P., *Vaulting Ambition: Sociobiology and the Quest for Human Nature*. Cambridge: MIT Press, 1985.

Lancet (editorial). Is it "all in the genes?" Lancet 1995; 345:466-7

Lelling A.E. Comment: eliminative materialism, neuroscience and the criminal law. University of Pennsylvania Law Review 1993; 141:1471-1564

Lumsden, C.J., Gushurst, A.C. "Ethical Epistemology: Coevolution and the Cultural Superstructure." In *Sociobiology and Epistemology*, J.H. Fetzer, ed. Dordrecht: D. Reidel Publishing Co., 1985.

MacIntyre, A., *After Virtue*, 2nd ed. Notre Dame, IN: University of Notre Dame Press, 1984.

M'Naghten's Case, 10 Cl. & F. 200. Eng. Rep. 1843; 8:718.

Mann C. War of words continues in violence research. Science 1994; 263:1375

Mednick, S.A., Gabrielli, W.F., Jr., Hutchings, B. "Genetic Factors in the Etiology of Criminal Behavior." In *The Causes of Crime: New Biological Approaches*, S.A. Mednick, T.E. Moffitt, S.A. Stack, eds. New York: Cambridge University Press, 1987.

Montana *v.* Egelhoff, 518 U.S. 37 (1996).

Morris, H., *On Guilt and Innocence*. Berkeley: University of California Press, 1976.

Pallone N.J., Hennessy J.M. Brain dysfunction and criminal violence. Society 1998; 35:21-7

Parens E. Taking behavioral genetics seriously. The Hastings Center Report 1996; 26:13-8

Parker, L.S. "Genetics, Social Responsibility, and Social Practices." In *Genetics and Criminality: The Potential Misuse of Scientific Information in Court*, J.R. Botkin, W.M. McMahon, L.P. Francis, eds. Washington, D.C.: American Psychological Association, 1999.

Pettit, P., *The Common Mind: An Essay on Psychology, Society, and Politics*. New York: Oxford, 1996.

Powell *v*. Texas, 392 U.S. 514 (1968).

Powledge T. Genetics and the control of crime. BioScience 1996; 46:7-10

Rafter, N.H., *Creating Born Criminals*. Urbana, IL: University of Illinois Press, 1997.

Reider L. Comment: toward a new test for the insanity defense: incorporating the discoveries of neuroscience into moral and legal theories. UCLA Law Review 1998; 46:289-342

Robinson, J.H. "Madness and the criminal law." Book Review of N. Morris: *Madness and the Criminal Law*, Chicago: University of Chicago Press, 1983. Notre Dame Law Review 1983; 59:297-311

Robinson J.H. Crime, culpability, and excuses. Notre Dame Journal of Law, Ethics & Public Policy 1996; 10:1-10

Rosenberg, A., *Instrumental Biology, or, The Disunity of Science*. Chicago: University of Chicago Press, 1994.

Rosenberg, A., *Philosophy of Social Science*, 2nd ed. Boulder, CO: Westview Press, 1995.

Roush W. Conflict marks crime conference. Science 1995; 269:1808-9

Ruse, M., *Sociobiology, Sense or Nonsense?*, 2nd ed. Dordrecht: D. Reidel Publishing Company, 1985.

Sayre-McCord, G. "Introduction: The Many Moral Realisms." In *Essays on Moral Realism*, G. Sayre-McCord, ed. Ithaca: Cornell University Press, 1988.

Sober, E., Wilson, D.S., *Unto Others: The Evolution and Psychology of Unselfish Behavior*. Cambridge: Harvard University Press, 1998.

Steadman, H.J. *et al.*, *Before and After Hinckley: Evaluating Insanity Defense Reform*. New York: Guilford Press, 1993.

Stoff, D.M., Cairns, R.B. "Introduction." In *Aggression and Violence: Genetic, Neurobiological, and Biosocial Perspectives*, D.M. Stoff, R.B. Cairns, eds. Mahwah, NJ: Lawrence Erlbaum Associates, Inc., 1996.

Stoff, D.M., Cairns, R.B., eds., *Aggression and Violence: Genetic, Neurobiological, and Biosocial Perspectives*. New Jersey: Lawrence Erlbaum Associates, Inc., 1996.

Stoff, D.M., Vitiello, B. "Role of Serotonin in Aggression of Children and Adolescents: Biochemical and Pharmacological Studies." In *Aggression and Violence: Genetic, Neurobiological, and Biosocial Perspectives*, D.M. Stoff, R.B. Cairns, eds. Mahway, NJ: Lawrence Erlbaum Associates, Inc., 1996.

Virkkunen, M., Linnoila, M. "Serotonin and Glucose Metabolism in Impulsively Violent Alcoholic Offenders." In *Aggression and Violence: Genetic, Neurobiological, and Biosocial Perspectives*, D.M. Stoff, R.B. Cairns, eds. Mahway, NJ: Lawrence Erlbaum Associates, Inc., 1996.

Walker, L.E., *The Battered Woman*, 1st ed. New York: Harper & Row, 1979.

Walters, L., Palmer, J.G., *The Ethics of Human Gene Therapy*. New York: Oxford University Press, 1997.

Watson G. Excusing addiction. Law and Philosophy 1999; 18:589-619

Williams J. Violence, genes, and prejudice. Discover 1994; 15:92-102

Wilson, E.O., *Sociobiology: The New Synthesis*. Cambridge: Harvard University Press, 1975.

Wilson, E.O., *On Human Nature*. Cambridge: Harvard University Press, 1978.

Wilson, E.O., *Consilience: The Unity of Knowledge*. New York: Alfred A. Knopf, Inc., 1998.

Y.B. Pasch. 17 Edw. IV, f. 2a, p. 2 (1477).

18 U.S.C.A. Sec. 17 (West Supp. 1999).

NOTES ON CONTRIBUTORS

Douglas Allchin is an historian and philosopher at the University of Minnesota, Minneapolis, Minnesota, U.S.A.

Garland E. Allen is an historian of biology in the Department of Biology, Washington University, St. Louis, Missouri, U.S.A.

Rachel A. Ankeny is Lecturer and Director, Unit for History and Philosophy of Science, University of Sydney, Australia.

Roberta M. Berry, a legal scholar and philosopher, is Associate Professor of Public Policy and Director of the Law, Science and Technology Program, Georgia Institute of Technology, Atlanta, Georgia, U.S.A.

Licia Carlson is Assistant Professor of Philosophy, Seattle University, Seattle, Washington, U.S.A.

Fred Gifford is Professor of Philosophy, Michigan State University, East Lansing, Michigan, U.S.A.

Joseph L. Graves, Jr. is Professor of Evolutionary Biology, Arizona State University West, Phoenix, Arizona, U.S.A.

Paul K. J. Han, a clinician-educator, is Assistant Professor of Medicine, Section of Palliative Care and Medical Ethics, University of Pittsburgh, Pittsburgh, Pennsylvania, U.S.A.

Manfred D. Laubichler, a biologist and historian, is Assistant Professor of Biology and Affiliated Assistant Professor of Philosophy in the Department of Biology and the Biology and Society Program at Arizona State University, Tempe, Arizona, U.S.A.

Helen Longino is a philosopher in the departments of philosophy and women's studies at the University of Minnesota, Minneapolis, Minnesota, U.S.A.

Lisa S. Parker, a philosopher-bioethicist, is Associate Professor of Human Genetics and Director of Graduate Education of the Center for Bioethics and Health Law, University of Pittsburgh, Pittsburgh, Pennsylvania, U.S.A.

Diane B. Paul is Professor of Political Science and Director or the Program in Science, Technology, and Values, University of Massachusetts, Boston, Massachusetts, U.S.A.

John H. Robinson is Associate Dean for Academic Affairs and Associate Professor of Law at the Notre Dame Law School, South Bend, Indiana, U.S.A.

Sahotra Sarkar is a philosopher in the department of philosophy and the Director of the Program in the History and the Philosophy of Science at the University of Texas, Austin, Texas, U.S.A.

Anita Silvers, Professor of Philosophy, teaches bioethics at San Francisco State University, San Francisco, California, U.S.A.

David Wasserman, who writes on ethical and policy issues in biotechnology and disability, is a Research Scholar at the University of Maryland, College Park, Maryland, U.S.A.

INDEX

Philosophy and Medicine

Philosophy and Medicine

21. G.J. Agich and C.E. Begley (eds.): *The Price of Health.* 1986
ISBN 90-277-2285-4
22. E.E. Shelp (ed.): *Sexuality and Medicine.* Vol. I: Conceptual Roots. 1987
ISBN 90-277-2290-0; Pb 90-277-2386-9
23. E.E. Shelp (ed.): *Sexuality and Medicine.* Vol. II: Ethical Viewpoints in Transition.
1987 ISBN 1-55608-013-1; Pb 1-55608-016-6
24. R.C. McMillan, H. Tristram Engelhardt, Jr., and S.F. Spicker (eds.): *Euthanasia and the Newborn.* Conflicts Regarding Saving Lives. 1987
ISBN 90-277-2299-4; Pb 1-55608-039-5
25. S.F. Spicker, S.R. Íngman and I.R. Lawson (eds.): *Ethical Dimensions of Geriatric Care.* Value Conflicts for the 21th Century. 1987 ISBN 1-55608-027-1
26. L. Nordenfelt: *On the Nature of Health.* An Action-Theoretic Approach. 2nd, rev. ed. 1995 SBN 0-7923-3369-1; Pb 0-7923-3470-1
27. S.F. Spicker, W.B. Bondeson and H. Tristram Engelhardt, Jr. (eds.): *The Contraceptive Ethos.* Reproductive Rights and Responsibilities. 1987
ISBN 1-55608-035-2
28. S.F. Spicker, I. Alon, A. de Vries and H. Tristram Engelhardt, Jr. (eds.): *The Use of Human Beings in Research.* With Special Reference to Clinical Trials. 1988
ISBN 1-55608-043-3
29. N.M.P. King, L.R. Churchill and A.W. Cross (eds.): *The Physician as Captain of the Ship.* A Critical Reappraisal. 1988 ISBN 1-55608-044-1
30. H.-M. Sass and R.U. Massey (eds.): *Health Care Systems.* Moral Conflicts in European and American Public Policy. 1988 ISBN 1-55608-045-X
31. R.M. Zaner (ed.): *Death: Beyond Whole-Brain Criteria.* 1988
ISBN 1-55608-053-0
32. B.A. Brody (ed.): *Moral Theory and Moral Judgments in Medical Ethics.* 1988
ISBN 1-55608-060-3
33. L.M. Kopelman and J.C. Moskop (eds.): *Children and Health Care.* Moral and Social Issues. 1989 ISBN 1-55608-078-6
34. E.D. Pellegrino, J.P. Langan and J. Collins Harvey (eds.): *Catholic Perspectives on Medical Morals.* Foundational Issues. 1989 ISBN 1-55608-083-2
35. B.A. Brody (ed.): *Suicide and Euthanasia.* Historical and Contemporary Themes.
1989 ISBN 0-7923-0106-4
36. H.A.M.J. ten Have, G.K. Kimsma and S.F. Spicker (eds.): *The Growth of Medical Knowledge.* 1990 ISBN 0-7923-0736-4
37. I. Löwy (ed.): *The Polish School of Philosophy of Medicine.* From Tytus Chałubiński (1820–1889) to Ludwik Fleck (1896–1961). 1990
ISBN 0-7923-0958-8
38. T.J. Bole III and W.B. Bondeson: *Rights to Health Care.* 1991
ISBN 0-7923-1137-X

Philosophy and Medicine

Philosophy and Medicine

Philosophy and Medicine

74. H.T. Engelhardt, Jr. and L.M. Rasmussen (eds.): *Bioethics and Moral Content: National Traditions of Health Care Morality*. Papers dedicated in tribute to Kazumasa Hoshino. 2002   ISBN 1-4020-6828-2
75. L.S. Parker and R.A. Ankeny (eds.): *Mutating Concepts, Evolving Disciplines: Genetics, Medicine, and Society*. 2002   ISBN 1-4020-1040-0
76. W.B. Bondeson and J.W. Jones (eds.): *The Ethics of Managed Care: Professional Integrity and Patient Rights*. 2002   ISBN 1-4020-1045-1

KLUWER ACADEMIC PUBLISHERS – DORDRECHT / BOSTON / LONDON